High-Power Microwave Measurement

Wenxiang Wang · Lingna Yue · Yanyu Wei

High-Power Microwave Measurement

国防工业出版社
National Defense Industry Press

Springer

Wenxiang Wang
University of Electronic Science
and Technology of China
Chengdu, Sichuan, China

Lingna Yue
University of Electronic Science
and Technology of China
Chengdu, Sichuan, China

Yanyu Wei
University of Electronic Science
and Technology of China
Chengdu, Sichuan, China

ISBN 978-981-96-7762-7 ISBN 978-981-96-7763-4 (eBook)
https://doi.org/10.1007/978-981-96-7763-4

Jointly published with National Defense Industry Press
The print edition is not for sale in China (Mainland). Customers from China (Mainland) please order the print book from: National Defense Industry Press.
ISBN of the Co-Publisher's edition: 978-7-118-10474-5

Translation from the Chinese Simplified language edition: "High-Power Microwave Measurement" by Wenxiang Wang et al., © National Defense Industry Press 2022. Published by National Defense Industry Press. All Rights Reserved.

This Springer imprint is published by the registered company Springer Nature Singapore Pte Ltd.
The registered company address is: 152 Beach Road, #21-01/04 Gateway East, Singapore 189721, Singapore

Preface

High-power microwave technology has attracted much attention due to its prominent role in military and civilian electronic technology and has become a research hotspot in major industrial countries.

In the development process of high-power microwave technology, it is bound to involve its measurement issues. Only through measurement we can understand the output of high-power microwave devices, the transmission of high-power microwaves, the transformation of high-power microwaves, the characteristics of high-power microwave components, and determine whether it meets the requirements, put forward improvement measures to meet the needs of high-power microwave application. It can be seen that the measurement of high-power microwave will be related to the determination and improvement of the performance of each link of the system. Therefore, high-power microwave researchers have adopted various measurement methods in their respective work to measure high-power microwaves, and a large number of papers have been published. However, on the other hand, these documents are currently mainly scattered in various academic journals and academic conference proceedings, and scholars from various countries have proposed various measurement schemes, so that there is still no comprehensive book at home and abroad of works dealing with high-power microwave measurement.

The parameters that high-power microwave needs to measure are similar to conventional microwave measurements. However, due to the characteristics of extremely high-power microwaves (hundreds of megawatts to more than ten gigawatts), extremely short pulses (usually tens of nanoseconds), and working in high-order mode of the high-power microwave, their specific measurement methods are different from conventional microwave measurement. This is mainly manifested in the measurement methods of power and frequency and in the instruments and equipment commonly used for their measurement. These instruments and equipment are converters that convert microwave power into measurable physical quantities and are frequency measurement devices that can compare with the frequency of microwave signals. Generally, they are difficult to respond to nanosecond level pulse, also difficult to withstand the power levels ranging from hundreds of megawatts to gigawatts. Therefore, the conventional measurement methods of power and frequency are not suitable for high-power microwave measurements, and new measurement methods must be found. The high-order mode output or the multi-mode output of high-power microwave devices, as well as the overmoded transmission and conversion of high-power microwaves, makes it inevitable for high-power microwave workers to pay attention to the measurement and analysis of mode, in order to understand the composition of the output mode of the high-power microwave device, the mode composition and the quality of the mode converter in the high-power microwave transmission system, etc. The measurement of other parameters of high-power microwaves, such as pulse width, pulse waveform, and repetition frequency, is the same as that of ordinary microwave measurement and can be directly measured by oscilloscope display, and the gain (amplification) and bandwidth of high-power microwave amplifiers can be determined according to the power measurement results. Therefore, the measurement of high-power microwaves currently mainly refers to the measurement of power, frequency and mode, which are different from that of conventional microwave measurements. As far as the power measurement is concerned, conventional gyrotrons are generally classified as the category of high-power microwave, although its output is continuous microwave or continuous microwave pulse (pulse width for microseconds, milliseconds), the output power level is much lower, but from the integrity of the content of this book, considering the conventional gyrotrons measurement requirements, the measurement of continuous wave power and continuous pulse power is also properly introduced in this book.

While engaged in high-power microwave research, the authors deeply felt the importance of high-power microwave measurement; therefore, we began to gradually compile some relevant information on high-power microwave measurement and began to conduct some research in the field of high-power microwave measurement. Moreover, the authors have introduced the measurement of power, mode and frequency of high-power microwave in various training courses for high-power microwave professionals, graduate summer classes and some academic lectures. In this process, the authors have gradually enriched and improved the content of the lectures on the measurement of power, mode and frequency of high-power microwave. At the same time, when the authors were engaged in research work at home and abroad, they also paid high attention to the measurement of high-power microwave.

Based on many original academic reports of the authors, this book comprehensively summarizes the various methods of high-power microwave power, frequency and mode measurement that have been proposed by research units in various countries, as well as the authors' own research results in this field. The book not only introduces the theoretical basis and principles of each measurement scheme, but also specifically gives the measurement system, measurement method and scope of application.

The readers of this book are mainly scholars, researchers, engineering technicians engaged in high-power microwave technology, high-power microwave sources and high-power microwave application research, as well as university teachers and graduate students engaged in high-power microwave teaching and research. As the development of high-power microwave has attracted the attention of more and more scholars, its application fields are becoming more and more extensive. Therefore, anyone engaged in high-power microwave technology and its application, microwave measurement and other related work personnel are also suitable for reading this book.

The preparation of this book refers to a large number of scholars' published literature from various countries and some postgraduate theses, and some of the illustrations in this book are also quoted from references. The authors of this book should first express their most sincere gratitude to the authors of these papers. Without their contribution to the field of high-power microwave measurement, this book would not have been finished. In addition, some scholars and engineering technicians provided some information or compilation opinions to the authors, so the authors also express our heartfelt thanks to them. The authors would like to express special thanks to Prof. Duan Zhaoyun, Prof. Yu Guofen, Dr. Dong Jibo, Dr. Yin Pengcheng, and Teacher Yang Dengwei for their assistance during the writing of this book.

Chengdu, China
June 2024

Wenxiang Wang
Lingna Yue
Yanyu Wei

Contents

Chapter 1
Overview

1.1 High-Power Microwaves

1.1.1 Definition of High-Power Microwaves

According to the definition proposed by J. Benford and J. A. Swegle in the book ***High Power Microwave*** [1], high-power microwaves (HPM) refer to coherent electromagnetic radiation with the frequency range of 1–300 GHz and a peak power exceeding 100 MW. However, in practice, this definition is not strictly followed. Some traditional microwave tubes with peak powers exceeding 100 MW, such as the klystron (although in most cases, the power of the klystron is less than 100 MW), are not classified as a high-power microwave device. On the other hand, some other microwave devices with peak powers generally less than 100 MW, such as the ordinary gyrotrons, are always considered to be high-power microwave devices. In general, in many references and books on high power microwave, it also contains the content of ordinary gyrotrons. J. Benford et al. themselves also pointed out that the above definition is not very strict [2].

For the vast majority of high-power microwave sources, they use pulsed power sources (accelerators) to obtain high voltages ranging from several hundred kilovolts to megavolts, and achieve large currents ranging from hundreds of amperes to tens of kiloamperes through explosive emission cathodes. Therefore, it is entirely possible for the output peak power to reach tens of megawatts to gigawatts. Due to the very high voltage, these devices use an electron beam that already exhibits significant relativistic effects, so it seems more appropriate to name them as relativistic devices or relativistic electron beam devices, including relativistic gyrotrons. However, for ordinary gyrotrons, they use a thermionic cathode electron gun, with operating currents usually only ranging from a few amperes to tens of amperes, and voltages ranging from a few kilovolts to tens of kilovolts. Therefore, except for a few

W. Wang et al., *High-Power Microwave Measurement*,
https://doi.org/10.1007/978-981-96-7763-4_1

special gyrotrons, in general, the peak power does not exceed 100 MW. However, because their operating principles also involve considering relativistic effects (weak relativistic effects), people generally still classify them as high-power microwave devices [3, 4].

1.1.2 Characteristics of High-Power Microwaves and Their Measurement

This book will mainly discuss the measurement of single or low-repetition-rate pulsed high-power microwaves generated by relativistic devices, with appropriate consideration of the measurement of continuous wave or continuous pulse from ordinary gyrotrons. For high-power microwaves output by relativistic devices, they have the following main characteristics [5]:

(a) High peak power, generally, it is on the order of 100 MW or above, or even more than 10 GW;
(b) The pulse duration is short, mostly only a few tens of nanoseconds;
(c) The repetition frequency is low, with most high-power microwave devices operating as single-pulse, and even for devices with repetitive pulse operation, the repetition frequency is only in the range of tens to hundreds of hertz;
(d) High-power microwave devices output high order modes or even multimode output, as well as the need for high-power transmission, thus high-power microwave transmission systems must use overmoded waveguides, leading to the presence of multimode transmission in the system;
(e) The above characteristics determine that the average power of high-power microwaves is very low, but the pulse power is very high. For energy efficiency defined by the ratio of microwave pulse energy to electron beam pulse energy, due to the presence of the unique pulse shortening phenomenon in relativistic devices, resulting in a much narrower microwave pulse width compared to high-voltage pulses (or currents), and thus the energy efficiency is also low.

It is these characteristics of high-power microwaves that make their measurement different from conventional microwave measurements.

Firstly, the parameters that need to be measured for high-power microwaves are mainly its power, frequency, and mode. As for the measurement of pulse width, pulse waveform, repetition frequency, and other parameters, it is the same as conventional microwave measurements and can be directly measured using an oscilloscope display. The gain and operating bandwidth of high-power microwave amplifiers can be determined based on power measurement results. Power and frequency are the most important parameters of any microwave device, so the importance of measuring them is self-evident. However, due to the characteristics of high-power microwaves, their measurement methods for power and frequency are significantly different from conventional microwave measurements methods. As for mode measurement, this is unique to high-power microwave devices, as this characteristic is determined by the fact that their output modes are typically high-order modes and the use of overmoded waveguide transmission. Therefore, the discussion of high-power microwave measurement in this book will be limited to power, frequency, and mode measurements. Obviously, conventional microwave measurements involve many more measurement parameters such as gain characteristics, phase characteristics, noise characteristics, etc., which are often not yet developed or difficult to carry out in high-power microwave measurements.

Secondly, the measurement methods are different. Because of the extremely short pulse duration of high-power microwaves, the power and frequency measurement methods and measurement equipment generally applicable to conventional microwave measurement are difficult to respond to it. So other power and frequency measurement methods suitable for the characteristics of high-power microwaves must be used. And the mode measurement method is not typically involved in the general microwave measurements.

Finally, since the measurement methods for high-power microwave measurements are different from conventional microwave measurements, the components and equipment used for measurement may also be different, often requiring some specialized measurement devices.

1.2 Introduction to the Power Measurements of High-Power Microwave

1.2.1 Microwave Power Measurement

1. **Concept of microwave power measurement**

Microwave power is one of the most important parameters that characterize the characteristics of microwave signals, so the measurement of microwave power has received significant attention, leading to the development of many measurement methods. Unlike the measurement of typical electrical power, where power can be directly calculated by measuring the voltage and current in low-frequency or DC power measurements, microwave power cannot be directly measured in the same way due to the lack of a strictly defined single voltage and current. The solution is to indirectly measure microwave power by utilizing various effects produced by the interaction of microwaves with other media or devices. That is, by the effects generated when microwaves interact with certain functional materials or devices through the microwave's electric field, magnetic field, power, or energy, the microwave is converted into measurable physical quantities of electrical, thermal, force, or acoustic etc. After detecting these measurable physical quantities, the microwave power can be calculated. Therefore, it can be said that the measurement of microwave power is measured through the microwave effect.

The methods for measuring microwave power vary depending on the characteristics of the microwave signal. The factors that affect power measurement mainly include the frequency range of the microwave signal, the dynamic range of power, power levels, signal spectrum, signal modulation methods, particularly pulse width, pulse repetition frequency, etc. Therefore, it is necessary to be familiar with the basic principles of various microwave power measurement methods and have a deep understanding of the modulation methods and frequency spectra of the signals being measured in order to select the appropriate power measurement method to obtain accurate measurement results [6].

Similar to general microwave power measurement, the measurement of high-power microwaves is also achieved through microwave effects. However, the measurement of high-power microwaves places higher demands on the components generating microwave effects, requiring them to respond very rapidly to microwave energy or signals, with a very short response time, in order to accurately measure the power level of nanosecond-level high-power microwaves and reduce measurement errors caused by response delays. Therefore, any power measurement method with a sufficiently fast response time to microwave pulses can be used for the measurement of high-power microwaves. As for the issue of extremely high power in high-power microwaves, various measures to reduce power can be implemented to meet the requirements of measurement methods, which is not difficult to achieve in the field of microwave technology.

2. **The unit of power**

It's well known that the unit of power is the watt, abbreviated as W. When there is a need to represent larger magnitudes of power, it is common to use kilowatts (kW, 10^3 W), megawatts (MW, 10^6 W), and gigawatts (GW, 10^9 W) as units. For smaller magnitudes of power, milliwatts (mW, 10^{-3} W), microwatts (μW, 10^{-6} W), and nanowatts (nW, 10^{-9} W) are used as units.

It is also common for people to express the magnitude of a power by using the logarithm of the ratio of that power to a reference power unit. The formula is given as follows:

$$A = 10\lg\frac{P}{P_0} \quad (\text{dB}) \tag{1.1}$$

where, P is the power to be measured, and P_0 is the reference power unit used for comparison, is often taken 1 W (1W) or 1 milliwatt (1 mW) as reference power. In this case, A is correspondingly represented in dBW or dBm, known as decibel watts or decibel milliwatts. dBW or dBm are also commonly used as units for power, for example, 0 dBm represents $P = 1$ mW, 13 dBW represents $P = 20$ W, − 20 dBm represents $P = 0.01$ mW (or 10 μW), and so on.

1.2.2 *Definition of Microwave Power*

Due to the distinction between continuous and pulsed operation modes of microwave tubes, there are corresponding different definitions of power as well.

1. **Power in continuous wave state**

   (1) **Power of continuous wave**

If the microwave tube operates in a continuous wave state, then at a certain operating voltage, current, and input power, its output power is the continuous wave power (Fig. 1.1), where "CW" in the figure represents continuous wave.

(2) **Average power**

If the microwave tube outputs a continuous wave that is periodically modulated by a signal, then we should use the average power to describe the output power of the microwave tube. Its definition is the average value $\overline{P}$ of the microwave (carrier wave) power over several periods of the modulated signal (Fig. 1.2)

$$\overline{P} = \frac{1}{nT}\int_{0}^{nT} P(t)\mathrm{d}t \tag{1.2}$$

where n is the number of periods taken to calculate the average power, T is the period of the modulated signal, and $P(t)$ is the function of microwave power that varies with time.

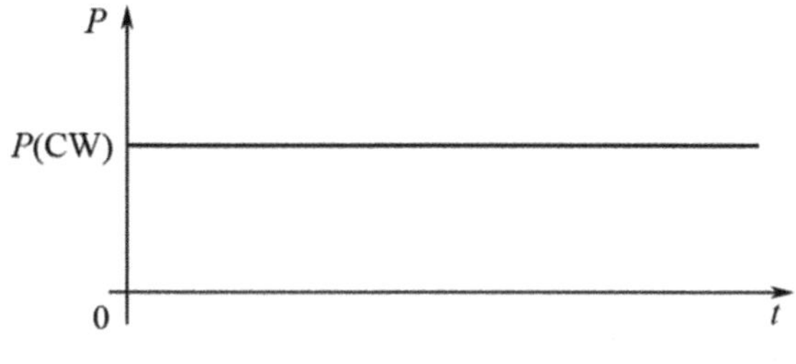

Fig. 1.1 The output power in continuous wave state

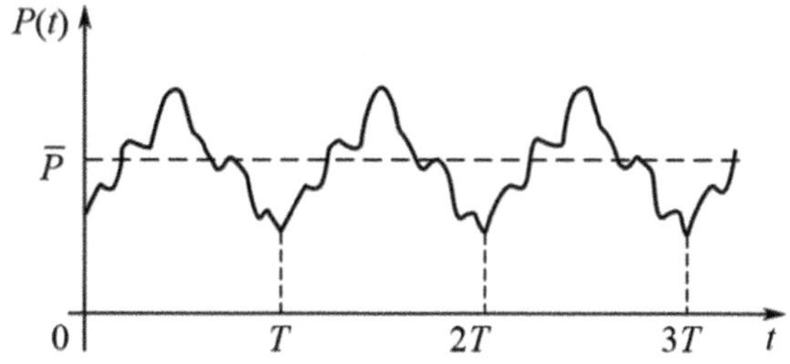

Fig. 1.2 Average output power of modulated wave

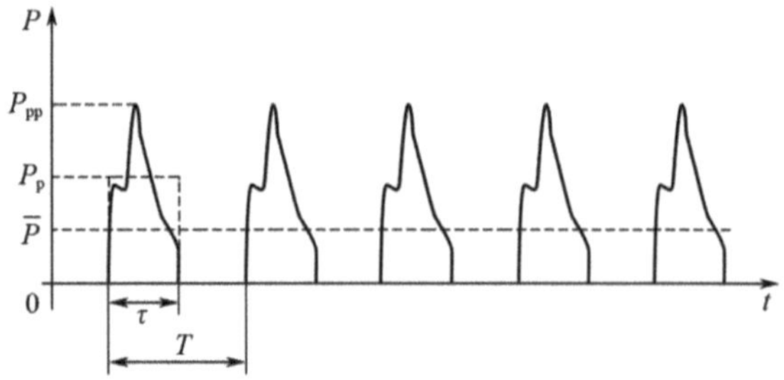

Fig. 1.3 Power in pulse operation state

2. **Power in pulse state**

When a microwave tube is operating in a pulsed state, there are several different defined powers (Fig. 1.3).

(1) **Peak power**

Pulse peak power refers to the microwave power at the peak point of the pulse modulation signal. Since the fundamental definition of power involves work done over a unit of time, power must be associated with a certain time interval to be meaningful. However, strictly speaking, a "point" at the peak does not have a time interval. Therefore, strictly speaking, the power at the pulse peak point is meaningless. So, peak power should actually be the average power over an extremely short time interval at the pulse peak. It can be determined by the peak voltage on the load resistance:

$$P_{pp} = \frac{\hat{V}^2}{R} \tag{1.3}$$

where, $\hat{V}$ is the peak voltage of the modulated pulse on the constant load resistance R, often indicated by an oscilloscope.

The Chinese national standard SJ 20769-1999 provides the definition of pulse peak power P_{pp} as the average power within one carrier cycle corresponding to the position of peak top of the RF pulse modulation signal, i.e. [7],

$$P_{pp} = \frac{1}{T_0}\int_0^{T_0} P(t)\mathrm{d}t \tag{1.4}$$

where, T_0 is the period of the pulse carrier, i.e., the microwave signal period.

Clearly, the peak power defined by Eq. (1.4) no longer has the contradiction of lacking a time interval and inconsistent with the definition of power itself. It takes a carrier cycle as the time interval for averaging, thus Eq. (1.4) strictly adheres to the definition of power as work done over a unit of time. However, in practical measurements, due to the microwave period T_0 being much smaller than the duration of the modulated pulse τ, the value of $P(t)$ over one T_0 interval is generally impossible to determine accurately. Therefore, Eq. (1.4) is only a theoretical definition and has no practical significance in measurements. Consequently, the specific measurement method recommended in the Chinese national standard SJ 20769-1999 for pulse peak power is not based on the definition in Eq. (1.4).

(2) **Pulse power**

Pulse power is the average microwave power in one duration τ of a modulated pulse.

$$P_p = \frac{1}{\tau}\int_0^{\tau} P(t)\mathrm{d}t \tag{1.5}$$

Obviously, if the modulated pulse is an ideal rectangular pulse, then $P(t)$ is a constant, $P(t) = P_p$, and its peak power P_{pp} is equal to the pulse power P_p. For conventional pulse microwave tubes, they typically output pulses that are close to an ideal rectangular pulse, so only the pulse power needs to be considered, and the peak power no longer needs to be considered. However, for relativistic devices, since their pulse modulation voltage is generated by an accelerator, it is difficult to obtain an ideal rectangular pulse, so the peak power and the energy contained in a pulse ($W = P_p\tau$) are often used to simultaneously represent its output capability.

For an ideal rectangular modulated pulse, $P(t)$ becomes a constant, and in this case, the pulse power P_p is equal to the pulse peak power P_{pp}

$$P_p = P_{pp} \tag{1.6}$$

(3) **Average power**

In general, except for single-pulse relativistic microwave tubes, the pulse modulation of the microwave tube's output power is repetitive. The concept of average power can also be introduced for repetitive pulse-modulated microwave.

$$\overline{P} = P_p\frac{\tau}{T} = \frac{P_p}{Q} \tag{1.7}$$

where, T is the pulse repetition period, $T = 1/F$, F is the pulse repetition frequency. We define

$$Q = \frac{T}{\tau} = \frac{1}{F\tau},\quad Q^{-1} = \frac{\tau}{T} \tag{1.8}$$

where, Q^{-1} is called the pulse duty cycle.

For an ideal rectangular pulse, it is obvious that there is a simple relationship between average power and pulse power:

$$\overline{P} = P_pQ^{-1} \quad \text{or } P_p = \overline{P}Q \tag{1.9}$$

In power measurements of relativistic device, the parameter of duty cycle is rarely used due to the characteristics of single-pulse or low-repetition-frequency pulses.

1.2.3 Fundamental Composition of Microwave Power Measurement System

The measurement system (power meter) for microwave power measurement generally includes power convertor, measuring devices, and indicators. They are combined to form a power meter, which can be further divided into three measurement methods: absorption-type measurement (termination measurement), pass-type measurement (online measurement), and radiation-type measurement.

1. **Basic components of measurement system**

The microwave power measurement system generally consists of convertors, measuring devices, and indicators.

(1) **Convertor**

In the microwave power measurement system, the microwave signal should first be converted into a measurable physical quantity. The convertor is the device that achieves this purpose, and its function is to convert microwave energy, or power, or field strength into a measurable physical quantity. This part is often referred to as a power probe, and it roughly includes the following types:

(a) Thermal convertor: such as dry heat load, water load, thermoelectric resistances, thermocouples, resistive sensors, etc.
(b) Electronic convertor: mainly including crystal detection diodes and vacuum diodes.
(c) Electromagnetic effect convertor: such as microwave pressure convertor, Hall-effect convertor, Kerr- effect convertor, magnetoresistiv-effect convertor, etc.

For high-power microwaves, due to their pulse duration of only a few tens of nanoseconds, the presence of thermal inertia means that general thermal convertors cannot react to it in a timely manner, and therefore cannot accurately reflect the magnitude of the pulse power. Therefore, the power measurement of high-power microwaves mostly uses electronic convertor or electromagnetic effect convertor. Only a few special thermal convertors can be applied, such as resistive sensor. Of course, for continuous pulses (and high repetition frequency pulses) or continuous microwaves, thermal convertors are generally still used for power measurement. This type of measurement is actually completed after the thermal convertor reaches thermal equilibrium. It's just that the time for this equilibrium process is much shorter than the operational time of the personnel conducting the measurements, so the measurement personnel will not noticeably feel it, and in this case, using a thermal convertor is more convenient and reliable.

(2) **Measurement device**

The physical quantity output from the converter is often unable or difficult to be directly measured or directly indicated. The function of the measuring device is to amplify, compare, or further transform the signal output from the converter into a more convenient signal for measurement and indication.

Typically, such measuring devices may include thermocouple, balanced bridges, amplifiers, calibrators, and other secondary converters.

(3) **Indicator**

The function of the indicator is to indicate the physical quantity output from the measurement device, and after transformation and calibration, it can also directly indicate microwave power. At this point, the indicator is often directly referred to as a power indicator.

Indicators generally come in analog and digital types. Specifically, in high-power microwave measurements, oscilloscopes are commonly used.

2. **Classification of measurement systems**

Microwave power measurement systems can be classified into three categories according to the measurement method [8–10].

(1) **Absorption-type measurement (termination-type measurement)**

The block diagram of this measurement system is shown in Fig. 1.4. In this measurement system, all microwave power will enter the converter and be absorbed by the converter and transformed into other physical quantities, which are then indicated by the indicator for the power level. The power measurement system (power meter) itself becomes the terminal of the microwave system, and the system no longer outputs microwaves to the load, hence it is also called terminal type measurement. Obviously, the terminal of this type of measurement system cannot be connected to the original working load, which means that the system must be operated separately and dedicatedly for power measurement.

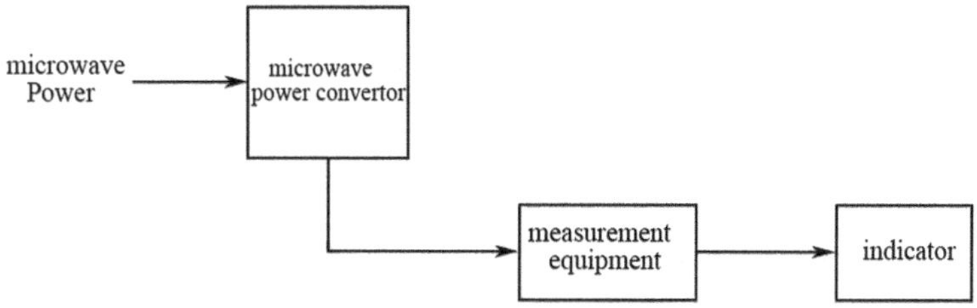

Fig. 1.4 Block diagram of absorption-type power measurement system

To expand the power range of absorption-type measurement, an attenuator with a pre-calibrated attenuation coefficient can also be added before the converter to reduce the microwave power entering the converter. After measuring the attenuated microwave power, the attenuation coefficient of the attenuator can be used to calculate the original input microwave power. This method is frequently used in absorption-type measurement, but the accuracy of power measurement is limited by the calibration accuracy of the attenuator's attenuation coefficient, and the range of power that can be extended is also limited by the allowable power capacity of the attenuator.

(2) **Pass-type measurement (Online measurement)**

In the measurement system of pass-type measurement, only a small part of the microwave power is coupled out through the coupling device for measurement. Since the coupled power is very small, it can be directly measured with a small power meter, and if necessary, a precision attenuator can be used to further reduce the power to fit within the measurement range of the small power meter (Fig. 1.5). Clearly, in this case, because the coupled power only accounts for a very small part of the total power transmitted by the system, the measurement system can still be connected to the terminal load and operate normally. Therefore, this type of measurement is an online measurement, where measurement and normal system operation can be carried out simultaneously. In addition, the total power in the system can be obtained according to the measured coupling power by using the coupling coefficient of the coupling device and the attenuation amount of the attenuator, that is, allowing for large power measurement using a small-power meter. However, it is necessary to calibrate the coupling coefficient of the coupling device and the attenuation coefficient of the attenuator strictly during measurement, and calibration error is often a major source of error in this type of power measurement.

The coupling device in the pass-type measurement method may be a probe, directional coupler, power divider, etc.

Due to the high-power level of high-power microwaves, only a very small part of the power is coupled out from the high-power transmission system for measurement through the coupling device. If the coupled power is still too large and exceeds the range of the converter, measurement device, or indicator, an attenuator needs to

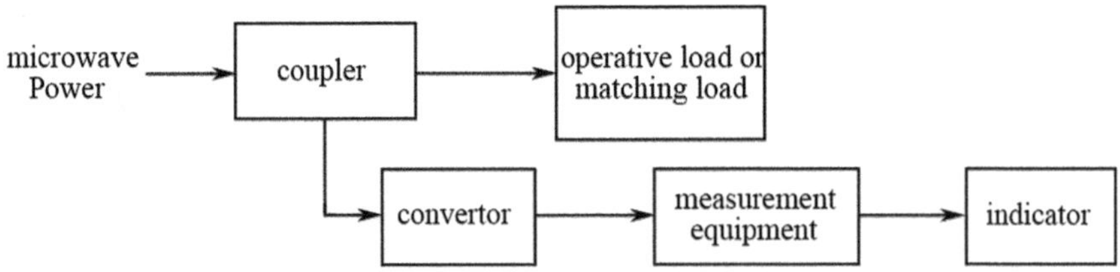

Fig. 1.5 Block diagram of pass-type power measurement system

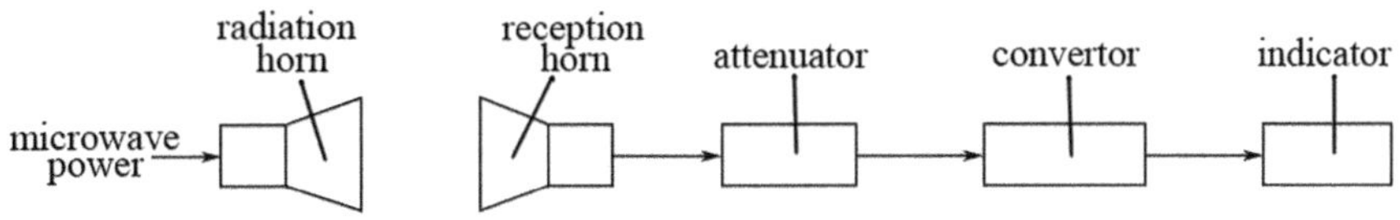

Fig. 1.6 Block diagram of radiative-type power measurement system

be added at the appropriate position of the system. It can be seen that high-power microwave power measurement generally only uses pass-type measurement, which can also be called coupling-type measurement.

(3) **Radiative-type measurement**

For high-power microwaves, even with coupled measurement, the coupled power is still too high due to the limitations of the coupling coefficient of the coupling device. When using high-power attenuators, there are limitations on their power capacity and the difficulty of calibrating their attenuation level at high power levels. Therefore, the most commonly used method for measuring the power of high-power microwaves is the radiative-type measurement method, as shown in the schematic diagram in Fig. 1.6.

For radiative-type measurement, high-power microwaves are radiated into space through a radiating horn. A receiving horn is placed at a certain distance from the radiating horn to receive the radiated power. The received power is much smaller than the power radiated into space, and this allows for convenient measurement using conventional attenuators, converters, measurement devices, and indicators.

The main difficulty in radiative-type measurement comes from the calibration of the coupling coefficient (or attenuation) between the received power of the receiving horn and the radiated power of the radiating horn. Due to the large coupling coefficient (attenuation), the calibration error can be relatively large. Similar to the pass-type measurement, it may be necessary to further reduce the power using attenuators to ensure that the converter is not damaged, which can also lead to even larger errors in the calibration of the total attenuation.

1.2.4 Classification of Microwave Power Meters

In a microwave power measurement system, the converter, measurement device, and indicator are often assembled together to form an independent instrument called a power meter. If the converter has already converted the microwave power into an electrical signal, the converter part can be separated from the measurement device and indicator parts and connected by a cable. In this case, the converter part is called a power probe, while the remaining two parts combined together form the power meter indicator. Obviously, this type of power meter is more convenient and flexible to use. Generally, small power meters used for measuring small power microwaves adopt this configuration. In high-power microwave measurements, after various forms of coupling and attenuation, the high-power microwaves output by the high-power microwave source are ultimately reduced to a very small power level, allowing for direct measurement using a small power meter.

Since microwave power measurements are based on microwave effects, microwave power meters can be classified based on different effects. They can mainly be divided into thermal effect power meters and non-thermal effect power meters [9–13].

1. **Thermal effect power meters**

Power meters that utilize the thermal effect of microwave absorption materials in the converter, where the absorbed microwave energy of absorption materials causes a temperature rise or changes in electrical characteristics, are known as thermal effect power meters. This power meter also has the following forms.

(1) **Dry-type calorimeter**

In the dry calorimeter, the heat converter for absorbing microwave power is a certain solid material, so it is called the dry calorimeter.

(2) **Liquid calorimeter**

If a power meter is made using a liquid as the heat converter that absorbs microwave power, it is called a liquid calorimeter.

(a) Static calorimeter

If the liquid in the converter of calorimeter remains stationary, then, this calorimeter is called a static calorimeter.

(b) Flow calorimeter

A calorimeter in which the liquid in the transducer flows continuously is called a flow calorimeter.

Whether it is a dry calorimeter, or liquid calorimeter, they both indirectly calculate the magnitude of microwave power by measuring the temperature change caused by the absorption of microwave energy by microwave absorption materials, so they are generally referred to as calorimeters.

(3) **Thermoelectric resistance power meter**
(a) Radiation thermoelectric resistance power meter.
(b) Thermistor power meter.
(c) Resistive sensor power meter.

The converter in the thermoelectric resistance power meter is a resistance material, which causes a change in resistance value when it absorbs microwave, thereby causing a change in output current or voltage, and thus calculating the microwave power energy.

(4) **Thermocouple power meter**

The thermocouple power meter utilizes the temperature rise at one end of the thermocouple by absorbing microwave, resulting in a temperature difference electromotive force between this end and the other end (the cold end) of the thermocouple, and then measures the output electromotive force to calculate the microwave power.

(5) **Ferrite power meter**

Ferrite has the characteristic of resonant absorption of high-frequency electromagnetic energy, causing a change in the resistance value of the ferrite element due to the heating from absorbing microwave energy. This phenomenon allows for microwave power measurement similar to the thermoelectric resistance measurement. However, this measurement method involves complex control and measurement processes, with many factors affecting measurement accuracy, making it difficult to control, and therefore it is rarely widely used.

(6) **Pyroelectric power meter**

The spontaneous polarization strength of piezoelectric materials is a function of temperature (below the Curie temperature). With the absorption of microwave energy, the temperature rise causes a decrease in the spontaneous polarization strength, leading to a change in polarization charge on the crystal's outer surface perpendicular to the direction of spontaneous polarization, thereby generating a current. As a result, an open-circuit voltage that varies with temperature change appears across the crystal's two ends. Utilizing this phenomenon allows for temperature measurement and calculation of power magnitude.

2. **Non-thermal effect power meter**

If the power is not measured using the thermal effect caused by the action of microwave power on the converter, but various electromagnetic effects of the converter are used, such as detection, magnetoresistive effect, Kerr effect, electroacoustic effect, Hall effect, and even mechanical pressure effect to realize power measurement, it is collectively referred to as a non-thermal effect power mete.

(1) **Detection power meter**

The detection power meter is the most convenient and commonly used type of small power meter. It utilizes the square law detection characteristic of the microwave detection diode within a certain range of microwave voltage, making its output directly proportional to the power. Alternatively, even if the detection range of the detector does not strictly conform to the square law detection characteristic, the microwave power can be calculated by calibrating the detector in advance and using the calibration results to convert the detection output into microwave power, so that the calibrated oscilloscope scale can directly indicate the power level.

Since the response time of the microwave crystal detector is less than 1 ns, the oscilloscope can directly display the microwave pulse waveform and pulse duty cycle, thereby allowing measurement of pulse power, peak power, calculation of average power, as well as measurement of pulse width and repetition frequency.

(2) **Pressure power meter**

Electromagnetic waves exert a mechanical effect on the walls of transmission lines and on reflective elements in the interior or exterior space of the transmission line, known as the pressure effect. This indicates that although electromagnetic waves are intangible, they also produce pressure, and the magnitude of the pressure is related to the power of the electromagnetic waves. Therefore, measuring the magnitude of the microwave pressure exerted on the walls of the waveguide wall or on the reflective elements inside or outside the waveguide can be used to calculate the microwave power.

(3) **Hall effect power meter**

A semiconductor Hall chip placed in a magnetic field, when a current passes through the Hall chip in a direction perpendicular to the magnetic field, will produce a potential difference in a third direction that is both perpendicular to the magnetic field and the current. This phenomenon is called the Hall effect, and the generated potential difference is called the Hall electromotive force. The spatial orthogonality between the electric field components and magnetic field components of the electromagnetic field makes it possible to utilize them to obtain the Hall effect in the semiconductor, and the magnitude of the Hall electromotive force is directly related to the power flow of the electromagnetic wave. A power meter made using the semiconductor Hall element based on this property is called a Hall effect power meter.

(4) **Other electromagnetic effect power meters**

As long as electromagnetic effects that cause voltage, current, resistance, polarization plane deflection, and other changes under the action of a microwave field can be generated, they can be used for power measurement.

(a) Magnetoresistiv-effect

When electromagnetic materials undergo a change in resistance under the action of a magnetic field, it is called the magnetoresistiv effect or the Gauss effect, and its output signal will be proportional to the active component of the product of the electric field component and the magnetic field component of the electromagnetic field, that is, the active power. Thus, microwave power measurements can be performed.

(b) Kerr effect (magneto-optical effect)

A beam of linearly polarized light incident on a magnetic medium, after transmission and reflection, will change its polarization direction and transform from linearly polarized light to elliptically polarized light. This phenomenon of transmission light is called the Faraday effect, and the phenomenon of reflection light is called the Kerr effect. These effects can be used to measure the pulse waveform, and after calibration, the power level can be determined.

(c) Electroacoustic effect

When an external alternating electric field acts on a piezoelectric material through an electroacoustic transducer, it will excite material deformation and generate elastic waves in the piezoelectric body, which propagate along the surface of the piezoelectric material. These waves, known as surface acoustic waves, will excite electromagnetic pulses through the output electroacoustic transducer. The piezoelectric deformation of the piezoelectric body is proportional to the electric field strength, making it possible to measure power.

1.2.5 Characteristics of Power Meters

1. **Input impedance and matching**

To enable the power meter to match the connection to the microwave system without causing reflections that would lead to inaccurate measurements and affecting the output of the microwave source, it is necessary to require the impedance of the power meter to be matched with the system as much as possible. This means that the standing wave coefficient should be minimized, generally controlled below 1.5. For higher precision power meters, it should be below 1.2.

2. **Operating frequency range**

The operating frequency range of a power meter, especially the frequency range within which the basic error does not exceed the specified value, is its characteristic index. Of course, by recalibrating the converter, or even replacing the power probe, the operating frequency range of the power meter can be greatly expanded.

3. **Measurement range**

The measurement range refers to the range of microwave power that the power meter can measure, which can be directly expressed in power units, or sometimes in milliwatt decibels, i.e., how many dBm. For the most commonly used thermal effect power meter, it can generally be divided into several types according to the measurement range:

(a) High-power meter: used to measure power greater than 10 W, currently capable of measuring up to 2000 W. Its power probe mostly uses a flow-type liquid calorimetric probe. When the power is below a few hundred watts, a dry calorimetric probe with a heat sink can also be used.
(b) Medium-power meter: used to measure power in the range of 10 W to 10 mW, and its power probe can use a dry calorimetric probe.
(c) Small-power meter: used to measure power in the range of 10 mW to 1 μW, the power probe is generally a thermoelectric resistance or thermocouple converter.

(d) Ultra-small-power meter: used to measure power less than 1 μW, it is a new type of power meter formed by improving the microwave crystal detection diode power meter, with a lower limit of up to 100 pW (− 70 dBm).

4. **Basic error**

The basic error of a power meter refers to the measurement error of the instrument under normal working conditions, including systematic errors and random errors. The basic error is expressed as a positive and negative percentage allowance, such as ± 1%, ± 2.5%, etc.

Based on the allowed size of the basic error, power meters can also be classified into accuracy levels, commonly including 1.0 level, 1.5 level, 2.5 level, 4.0 level, 6.0 level, 10.0 level, and so on. The basic error of the power meter should not exceed its accuracy class value.

1.3 Introduction to the Mode Measurement of High-Power Microwave

1.3.1 Necessity of Mode Measurement of High-Power Microwave

Mode measurement is one of the most important measurement contents for high-power microwaves, mainly because it has the following functions.

(1) **Understanding the mode composition of the output of high-power microwave sources**

Since high-power microwave devices generally operate in high-order modes, it is easy to excite multiple modes during the interaction process between electron beams and high-frequency structures inside the device, as well as in the output structure of the device. This often results in the output modes of high-power microwave sources containing many parasitic modes, which are very detrimental to the transmission of high-power microwaves, making it difficult for the design of various functional components in the transmission system, as well as causing difficulties in the design of certain transmission systems and mode converters, thereby reducing the transmission efficiency of the system and the radiation efficiency of antennas. Therefore, understanding the composition of the output modes of microwave sources is essential for improving the design of devices or implementing measures to suppress parasitic modes in high-power microwave systems.

(2) **Determining the mode composition in high-power microwave multimode transmission systems**

In addition to the impurity of the output modes of microwave sources themselves, parasitic modes are also generated during the transmission of high-power microwaves. Because the mode order and power level of high-power microwaves are both very high, almost without exception, overmoded waveguide systems are used for transmission. Any non-uniformity in the overmoded system, such as changes in the waveguide cross-section, waveguide bends, ellipticity of circular waveguides, waveguide type changes, probes introduced into the waveguide, coupling holes, matching components, etc., will excite parasitic modes in the system, and some of these modes can be transmitted in the overmoded system. The existence of these modes also causes difficulties in the design of functional components and mode converter design, so we need to not only understand the mode composition of the output of high-power microwave sources, but also analyze and master the mode composition in high-power microwave multimode transmission systems.

(3) **Determining the relative or absolute content of each mode component**

It is not enough to just know the composition of the modes; we also need to know the relative or absolute content of each mode component. In this way, we can determine which mode content is within an acceptable range and can be disregarded, and which mode content has reached a level that requires measures to suppress it. If we can not only measure the relative values of each mode, but also determine the absolute content of each mode component, i.e., the power of each mode, then we can directly obtain the total power in the system, thereby completing the power measurement of high-power microwaves at the same time.

(4) **Determining the performance of mode converters**

In order to transform the output mode of a high-power microwave source or the transmission mode in a system into a mode that is more suitable for transmission or that meets the requirements of antenna radiation, a mode converter must be used. The mode converter is a very important component in high-power microwave systems, as it can convert one mode into another and ultimately into the mode we need. The measurement of the mode, including qualitative and quantitative measurements, is necessary to assess the quality of the mode converter (i.e. whether the output is the desired mode) and the conversion efficiency (i.e. how many input mode components are converted into the mode we need).

(5) **Power measurement**

In point (3) above, we have actually pointed out that after measuring the absolute content of each mode component, the total power measurement of high-power microwaves is also completed at the same time. Of course, there are many methods for measuring the power of high-power microwaves, and we will introduce specific methods later. Using mode measurement to achieve power measurement is just one of these methods.

1.3.2 Primary Methods of Mode Measurement for High-Power Microwave

The methods for high-power microwave mode measurement mainly fall into two categories: mode field measurement and mode spectrum measurement.

1. **Mode field measurement**

Different modes have different field structures, so directly measuring the distribution of microwave fields to determine their mode is the simplest and most direct mode measurement method. This field measurement can be performed inside the waveguide, or it can also be radiated into the space using a horn for measurement. The field distribution can be obtained either through field strength or power measurements, or by directly displaying the field distribution image.

(1) **Radiation field or waveguide field measurement method**

In the waveguide for microwave transmission, the waveguide field's electric field distribution is measured using a moving probe, or the microwave field is radiated into space through a horn, and the field distribution in space is measured using a receiving horn. The mode composition of the field distribution can be obtained by analyzing the measuring result. When conducting the radiation field measurement, the receiving horn is generally connected to a standard rectangular waveguide. In the standard rectangular waveguide, there is the TE_{10} base mode, whose electric field is perpendicular to the wide side of the waveguide, and the magnetic field is parallel to the wide side of the waveguide. Therefore, in the spherical coordinate system of the radiation field, the wide side of the receiving horn should be perpendicular to the electric field component of the radiation field and parallel to the magnetic field component.

(2) **Image display method**

When microwaves are irradiated onto a heat-sensitive paper or ordinary paper, or onto an array of light-emitting diodes or an array of thermal probes, a thermal image or light image directly reflecting the electric field distribution is formed. Based on the image, the mode composition can be directly determined. Alternatively, the voltage or current reflecting the field distribution can be output by the pyroelectric effect, and through computer processing, a two-dimensional or three-dimensional image of the mode field can be obtained to determine its mode composition.

2. **Mode Spectrum Measurement**

Rather than directly measuring the distribution of mode fields, different modes are separated in space to form a mode spectrum. In this way, by measuring the power or thermal image at the corresponding position of each mode in space, without measuring the field distribution, it is possible to determine whether the mode exists and even the relative strength of that mode. Alternatively, different modes can be output through different coupling devices to form a mode spectrum. By measuring the output of each mode, it is also possible to determine the existence and relative size of the modes without the need to measure the field distribution. This type of mode measurement method is called mode spectrum measurement.

(1) **Dispersion reflection antenna measurement method**

The dispersion reflection antenna can reflect microwave beams projected onto its surface into space. Different modes have different reflection directions, so different modes can be distinguished in space. By using a microwave absorber plate to receive the reflected field, a thermal image of the mode distribution in space can be obtained. Based on the presence or absence of a thermal image and the temperature level in the corresponding reflection direction of each mode, it is possible to identify the existence and relative power strength of certain modes. This method is called the dispersion reflection antenna measurement method.

(2) **Mode selective coupling method**

The method of mode recognition and analysis based on the mode selective coupling effect is called mode selective coupling. The mode selective coupler only couples a specified mode to output in the secondary waveguide, while other modes are not coupled and do not have output in the secondary waveguide. Therefore, by designing a selective coupling arm for each possible mode in high-power microwave transmission systems, it is possible to determine the mode present in the main waveguide system and its relative content based on the presence or absence of output and the output size of each coupling arm.

(3) **Wavenumber spectrum measurement method**

By opening a series of small holes according to certain rules on the wall of an overmoded circular waveguide used for transmitting high-power microwaves to form a small hole array antenna, microwaves will radiate into space through the small hole array. The radiation direction depends on the wavenumber of the transmission mode, and different modes (different wavenumbers) have different radiation direction angles. Therefore, different modes can be dispersed in space. By measuring the power received by a horn antenna at the corresponding radiation direction of these modes, it is possible to determine which radiation modes exist and their power. This is the wavenumber spectrum measurement method.

1.4 Introduction to the Frequency Measurement of High-Power Microwave

1.4.1 Fundamental Concept of Frequency Measurement

Frequency is one of the most important parameters for characterizing microwave signal characteristics. Therefore, the measurement of frequency f is one of the most fundamental microwave measurement projects. Frequency measurement is often replaced by wavelength λ measurement, as they have a simple relationship:

$$f = \frac{c}{\lambda} \tag{1.10}$$

where, c is the speed of light in free space. Therefore, wavelength measurement and frequency measurement are completely equivalent.

The basic method of measuring frequency is the comparison method, which directly or indirectly compares the frequency being measured with a standard frequency. Frequency and time are conceptually unified, so the time standard serves as the original benchmark for the comparison method.

Comparison methods are divided into two categories: active comparison methods and passive comparison methods. The former uses a standard frequency source as the standard for comparing the measured frequency, that is, directly comparing the signal of the measured frequency with the signal of the known frequency generated internally by the instrument or externally added. Heterodyne method (mixing method) frequency measurement is a representative of this measurement method; the latter uses passive components with known frequency characteristics as the comparison standard, allowing the signal of the measured frequency to pass through the passive component and compare it with the known frequency characteristics of the component, thereby obtaining the measured frequency. In conventional microwave measurements, the most commonly used resonant wavelength meter frequency measurement and the most commonly used dispersion line frequency measurement in high-power microwave frequency measurements are typical representatives of passive comparison methods [11].

(1) **Active Comparison Method**

The commonly used method in the active comparison method is the heterodyne method (mixing method) and the counting methods. The heterodyne method is to input the signal of the frequency to be measured together with a signal of known frequency generated by an external oscillator into a mixer, extracting the intermediate frequency at the output of the mixer, measuring the frequency of the intermediate frequency signal, and inferring the frequency to be measured based on the obtained intermediate frequency. Heterodyne frequency meters used to be the most important precision instruments for measuring high frequencies up to microwave frequencies. However, with the emergence of more accurate and convenient counting frequency meters, the heterodyne frequency meter has been gradually phased out. Nevertheless, in high-power microwave frequency measurements, the heterodyne measurement method remains one of the main measurement methods.

The counting method refers to the use of digital frequency meters as the representative counting frequency measurement method. Essentially, it still compares the frequency to be measured with a standard frequency. Each cycle of frequency of pulse microwave to be measured is converted into a pulse, and an electronic counter is used to count the number of pulses passing through a gate, the opening time of the gate is controlled by a signal with a standard frequency f_b (period T_b). Assuming that the gate opening time τ is equal to m periods T_b of the standard frequency f_b, i.e. $\tau = mT_b$, the number of the frequency signal to be measured passing through the gate and counted during the time of gate opening is n. If the frequency of the signal to be measured is f_s, with a period T_s, then $nT_s = mT_b$, since $T_b = 1/f_b$, $T_s = 1/f_s$, so $f_s = nf_b/m$. In practical measurement systems, all microwave counting frequency meters rely on pre-split frequency technology and various down-conversion techniques to change the measured microwave frequency to a frequency below hundreds of megahertz, and then measure the frequency by the electronic counter, and can be directly converted into the measured frequency display.

(2) **Passive comparison method**

The most familiar device using the passive comparison method for frequency measurement is the resonant wavelength meter. This type of wavelength meter uses a circular waveguide resonant cavity, with its main working modes being the TM_{010} mode, TE_{111} mode, and TE_{011} mode (of which the TE_{011} mode provides the best performance: minimal high-frequency loss, the highest quality factor, and relatively low machining precision requirements). Their resonant wavelength is related to the cavity length, and when a cylindrical cavity with a movable shorting plunger is coupled to the microwave transmission system, adjusting the short-circuit plunger changes the resonant cavity length, thereby changing its resonant wavelength. When the resonant wavelength is adjusted to be equal to the microwave wavelength, resonance occurs, and the resonant cavity absorbs the microwave, causing an absorption dip in the output of the microwave system on the oscilloscope. Alternatively, the resonant cavity will output a resonance peak. Since the resonant wavelength of the resonant cavity can be strictly calculated based on its dimensions (radius and length) and mode, it can be inferred that when the resonant cavity radius and operating mode are given and fixed, only the resonant cavity length is related to the microwave wavelength. This length can be adjusted by tuning the short-circuit piston, so the wavelength or frequency can be directly indicated based on the resonant cavity length through calculation in advance. When the microwave reaches resonance with the resonant cavity, the frequency or wavelength size can be directly read on the outer wall of the resonant cavity. This is the working principle of the resonant wavelength meter. It can be seen that it measures the frequency by comparing the microwave frequency (wavelength) with the resonant frequency (wavelength) of the resonator.

1.4.2 High-Power Microwave Frequency Measurement

1. **Active comparison method for high-power microwave frequency**

Although the counting method has the potential to replace the heterodyne method (mixing method), which has led to the widespread use of digital frequency meters in modern conventional microwave frequency measurements, for high-power microwave pulses, the extremely short pulses of a single (or low repetition frequency) make it difficult for digital frequency meters to measure. At present, there is no mature product that can be used for high-power ultra-short pulse microwave frequency measurement. In addition, high-power microwave signals often contain rich harmonics and parasitic components, which also limit the application of counting methods in high-power microwave frequency measurements, unless the interference signals and noise are suppressed before measurement.

Therefore, many high-power microwave research units still generally use their own assembled mixing systems to measure frequency in the active comparison method. The error of the heterodyne method (mixing method) measurement mainly depends on the accuracy of the externally applied standard signal frequency. When the accuracy of the standard signal frequency is very high, the measurement frequency error is very small, and the sensitivity can also be very high. Since the microwave signal itself contains harmonics, and the ease of generating harmonics when the signal frequency is mixed, it can lead to confusion in determining the measurement results. In addition, the frequency detection circuit of the heterodyne method is more complex, requiring the standard frequency oscillator used to have continuous frequency tuning capabilities. The frequency stability and indication accuracy of the oscillator will affect the accuracy of instantaneous measurement of microwave frequency.

If the sampling rate of the oscilloscope is high enough and the analog bandwidth is wide enough, such an oscilloscope can be used to directly display the waveform of the high-power microwave, and the frequency can be directly calculated, or the frequency can be directly read out in the frequency domain using fast Fourier transform.

2. **Passive comparison method for high-power microwave frequency**

Conventional microwave wavelength measurements generally use the resonant wavelength meter mentioned above, which is a passive component. Although manual tuning is generally required during measurement, it has a simple structure, low cost, and is easy to use, so it is still widely used in frequency measurements that do not require high accuracy. Moreover, the wavelength meter often directly marks the frequency value rather than the wavelength size. The higher the quality factor Q of the resonant cavity of the resonant wavelength meter, the sharper the resonance curve (or the narrower the half power width), and the higher the frequency measurement accuracy. However, for general resonant cavities, it is difficult to achieve a very high Q value, which means that the resonance curve is relatively wide, indicating that resonance can occur over a certain range of frequencies, leading to a reduction in frequency measurement accuracy.

Resonant wavelength meter is no longer suitable for frequency measurement of high-power microwaves, because it is completely impossible to manually tune the resonant wavelength meter to find the resonant frequency within the extremely short pulse duration of a single or low-repetition-frequency pulse of high-power microwaves. Therefore, except for conventional gyrotrons operating in continuous wave or continuous pulse mode, which can use resonant wavelength meters for frequency measurement, other frequency measurements of high-power microwaves cannot use resonant wavelength meters.

Currently, in the frequency measurement of high-power microwave using the passive comparison method, the most widely used approach is to utilize dispersion lines. That is, using the group speed and frequency correlation characteristics of the microwave pulse envelope in the transmission waveguide, i.e. the dispersion characteristics. By measuring the delay time of the microwave pulse traveling through a waveguide with a certain length, the microwave frequency can be calculated. The frequency of high-power microwaves can also be measured using a filter array, which is also a passive comparison method, but it can only determine a certain range of frequencies and cannot accurately determine the frequency value, or in other words, the measurement resolution of this method is relatively low.

Chapter 2
Power Measurements of High-Power Microwave of Continuous Wave and Continuous Pulse Wave

2.1 Introduction

As mentioned earlier, conventional gyrotrons have also been classified in the category of high-power microwaves. Conventional gyrotron mainly output continuous wave or continuous pulse microwaves. Therefore, it is necessary to pay attention to the power measurement of continuous wave or continuous pulse microwaves. This chapter will discuss their measurement issues. It should be pointed out that in ordinary microwave measurements, the power measurement of continuous wave and continuous pulse microwaves is also involved and is not unique to high-power microwaves. However, we will discuss them more from the perspective of high-power microwaves.

1. **Basic methods for measuring power of continuous wave or continuous pulse microwave**

For microwave devices operating in continuous wave or continuous pulse modes, their output power mainly refers to average power (including continuous wave power) and pulse power. The average power can be the average power of continuous waves in a modulated state, or it can also be the average power in a continuous pulse state. In the case of continuous pulse microwaves, the pulses are generally rectangular, and the peak power is equal to the pulse power, so there is no need to measure peak power.

The most common method for measuring the average power of continuous wave and continuous pulse microwaves is to use thermal effect power meters because they are the most basic and widely used microwave power meters. However, due to the thermal inertia of the converter elements themselves for absorbing microwave power and the thermal equilibrium process of the entire converter, the response time of general thermal effect power meters is often much longer than the pulse duration. Therefore, the measured powers can only be the average power and cannot be directly the pulse power. Of course, they are even more impossible to be used for the power measurement of single pulse or low repetition frequency high-power pulse

W. Wang et al., *High-Power Microwave Measurement*,
https://doi.org/10.1007/978-981-96-7763-4_2

microwaves. However, there are a few thermal effect power meters with relatively fast response times, such as resistive sensor power meters, which can be used for pulse power measurement of high-power microwaves.

The thermal effect mentioned here includes two physical phenomena: one is that when an object absorbs external microwave radiation, molecular or lattice vibrations intensify, and the kinetic energy of particles increases, thereby converting microwave energy into heat energy, causing the temperature of the object to rise. Materials that produce this physical phenomenon include dry microwave absorbing loads and microwave water loads. The other is that when an object absorbs external microwave radiation, it causes a change in some electrical or magnetic properties of the object. Materials that produce this effect include thermoelectric resistance and thermistors, which cause changes in their resistance, thermocouples that cause changes in output electromotive force, and pyroelectric materials that produce pyroelectric phenomena. The former detects the radiation power based on temperature changes. Therefore, as long as we directly measure the change in temperature of the object, we can calculate the microwave power. This type of power meter can be called a calorimetric power meter. The latter should measure the change in the physical quantity of the generated electricity or magnetism to detect the strength of microwave radiation and then calculate the microwave power. These two types of power meters are collectively referred to as thermal effect power meters.

2. **Average power and pulse power**

The absorption (terminal) type method is commonly used to measure the continuous wave power or average power of continuous pulse. Typically, calorimetric power meters are used for this purpose. On the other hand, for the pass-type (online) method of measurement, one can use thermoelectric resistance power meters, thermistor power meters, thermocouple power meters etc. for measuring small power, and also directly measure pulse power or peak power using a detection power meter.

In the case of continuous pulse microwave, if one wishes to calculate the pulse power from the average power measured by a thermal power meter, it can be conveniently computed using the Eq. (1.9), and in this case,

$$P_p = \overline{P}Q \tag{2.1}$$

For non-ideal rectangular pulse wave forms, strictly speaking, there is no exact pulse power value. We can define an equivalent pulse power using the waveform coefficient K. A calibrated oscilloscope can indicate the magnitude of the pulse peak power P_{pp}, but it cannot determine the value of the pulse power P_p. In this case, if one wishes to use the peak power of the microwave pulse instead of the average power to obtain the equivalent pulse power, the above formula should be modified to:

$$\mathrm{P_p} = P_{pp}/K; \quad P_{pp} = \overline{P}QK \tag{2.2}$$

Here, K is referred to as the waveform coefficient. The value of K is equal to the ratio of the power of a rectangular pulse to the power of an equivalent rectangular pulse. The rectangular pulse is constructed using the peak power of the actual pulse as the pulse power and has the same pulse durationτ. The equivalent rectangular pulse not only has the same pulse duration τ but also has the same area enclosed by the pulse waveform as the actual pulse, which clearly indicates that $K > 1$. In the Chinese electronic industry standard SJ 20769-1999, the definition of K is the opposite: it is the ratio of the area occupied by the actual pulse waveform to the area of a rectangular pulse constructed with the pulse peak power that has the same pulse duration τ. In this case, $K < 1$, $P_p = P_{pp}K$. In practical measurements, due to the complexity of the actual pulse waveform of non-ideal rectangular pulses, the pulse power does not have a definite value; therefore, the value of K is often estimated. For an ideal rectangular pulse, it is clear that $K = 1$.

In most cases, the measurement of pulse power is more commonly performed using detection power meters. Other measurement methods have been proposed in the past, such as the integration-differentiation method, sample comparison method, notch method, etc., but these methods are now less commonly used.

A detection power meter utilizes the detection characteristics of crystals or vacuum diodes to display the envelope of the microwave pulse on an oscilloscope. By calibrating the indication scale of the oscilloscope using a standard signal source (generally a continuous wave signal source), one can directly read the pulse power (for rectangular pulse cases) or the magnitude of the peak power from the oscilloscope.

2.2 Thermal-Effect Power Meter—Calorimetric Power Meter

The calorimetric power meter is a typical microwave thermal-effect power meter. As the name suggests, it is a device that measures heat quantity or thermal energy, which converts microwave energy into thermal energy to measure microwave power. It uses microwave-absorbing materials in the microwave heating convertor to absorb microwave energy and increase the temperature. So, it is a power meter that obtains the microwave power by directly measuring the temperature change or comparing the temperature with that of a reference absorbing material to obtain the microwave power and commonly known as calorimeter. The microwave power benchmark in the world is established by calorimetry, that is, by using a calorimeter, because the theoretical research of this method is relatively mature and is considered to be the most accurate method to measure microwave power [14, 15].

Calorimetric power meters have two basic types: dry calorimeters and liquid calorimeters. Liquid calorimeters can be further divided into static calorimeters and flow calorimeters based on whether the liquid absorbing microwave energy is stationary or flowing.

Flowing liquid calorimeters are mainly used for medium to high average power (10^{-2} to 10^3 W) measurements, while dry calorimeters are mainly used for medium to small power (10^{-7} to 10^{-2} W) measurements. With good heat dissipation measures, the measurement range of modern dry calorimeters has been expanded to hundreds of watts level.

Calorimetric power meters using the substitution method (see Sect. 2.2.1) have the advantages of high accuracy (up to 0.5% or higher), wide operating bandwidth, low standing wave ratio (generally less than 1.05), and a wide power measurement range, making them a widely used type of power meter. Calorimeters mostly use the terminal absorption method when measuring microwave power.

2.2.1 Dry Calorimeter Power Meter

Since the calorimeter calculates power based on the measurement of temperature changes and some basic electrical quantities such as voltage and resistance, its measurement uncertainty is relatively low. Therefore, dry calorimeters can be used as national metrological standards for measuring microwave small power.

1. **Principle of dry calorimeter**

(1) **Basic structure of the thermal converter**

In dry calorimeters, the thermal converter is a solid-state material, such as structural absorbing materials and coating-type absorbing materials, both of which function to convert electromagnetic energy into thermal energy. Dry calorimeters are generally only suitable for applications of medium to small power due to the power capacity and heat transfer (heat dissipation) limitations of the absorbing materials.

In medium-power meters, our most commonly used structural absorbing materials are primarily ferrites with carbonyl iron powder as the main component and silicon carbide ceramics, as well as sintered ceramics mixed with graphite powder and alumina powder. In small-power meters, thermocouples and thermistor resistors are commonly used. Coating-type absorbing materials are often applied by brushing a layer of carbon powder, such as graphite milk, to surfaces such as bakelite, fiberglass boards, organic glass boards, and ceramic sheets, or by using vacuum deposition to form a layer of metal resistance film, such as titanium film, tantalum-niobium alloy film, or nickel–chromium alloy film. The thickness of the film depends on the surface resistance required for the most effective microwave absorption, generally ranging from tens to hundreds of ohms per unit square (Ω/□, where □ represents unit square area independent of side length). Coating-type absorbing materials are mainly used in small-power meters and widely used as absorbing loads. For example, waveguide-type small-power matching loads are almost all made of coating-type absorbing materials. In microwave electron vacuum devices, resistance films are often deposited on ceramic surfaces to serve as attenuators, capable of withstanding power levels of several tens of watts.

To achieve a good match, the structural absorbing material should generally be made in an oblique plane shape in the direction of microwave transmission. Due to the slow change in the dimensions of the oblique plane, there is minimal reflection of microwaves, resulting in a good match over a wide bandwidth, thus allowing the absorbing material to absorb all the microwave energy and improve the accuracy of power measurements. Based on the same reason, the coating-type absorbing materials should generally be made into the surface resistance gradient distribution form.

(2) **Thermal balance of the thermal converter**

A calorimeter measures power by using the temperature rise of the microwave absorbing material (heat load) in the converter. However, the process of the rising temperatures of the heat load after absorbing microwave energy can accumulate. If the container in which the converter is placed is ideally adiabatic, i.e., there is no heat exchange with the surrounding environment, then with the continuous input of microwave energy, the temperature of the converter will continue to rise. Obviously, this is completely impossible in practice because the temperature rise of the converter will result in the dissipation of heat through conduction, radiation, and air convection. Although various measures have been taken to insulate the container and reduce the heat loss of the converter in order to improve measurement accuracy, it is impossible to completely eliminate heat loss. After a period of time, when the energy added to the converter by the microwaves in a unit of time balances the dissipated heat quantity by the converter in the same unit of time, the temperature will no longer rise, and the microwave power can then be measured.

The thermal balance process of the converter in a dry calorimeter can be described as follows: When the heat load of the converter receives microwave power P, assuming the temperature of the converter is θ_1, with a heat capacity of C_1, and the temperature of the heat insulated container in which the converter is placed is θ_2, with a heat capacity of C_2, and $C_2 \gg C_1$, the thermal conductivity of the heat load (converter) to the surroundings is G (the reciprocal of the thermal resistance R). According to the principle of heat transfer, the thermal balance equation for the system can be written as:

$$\begin{aligned} P &= C_1 \frac{\mathrm{d}\theta_1}{\mathrm{d}t} + G(\theta_1 - \theta_2) \\ C_2 \frac{\mathrm{d}\theta_2}{\mathrm{d}t} &= G(\theta_1 - \theta_2) = -G(\theta_2 - \theta_1) \end{aligned} \tag{2.3}$$

where,

P is in units of W, C in units of J/°C, and G in units of W/°C,

$C_1\mathrm{d}\theta_1$ represents the microwave energy absorbed by the heat load, and $C_1\mathrm{d}\theta_1 / \mathrm{d}t$ is the power it absorbs,

$G(\theta_1 - \theta_2)$ is the power delivered to the container by the thermal load,

$C_2 d\theta_2 / dt$ is the power dissipated by the container to the environment. Since the container itself is not a heat source, it should obviously be equal to the power $G(\theta_1 - \theta_2)$ delivered to the container by the heat load.

Solving Eq. (2.3) and using $C2 \gg C1$, we can obtain

$$(\theta_1 - \theta_2) = \frac{P}{G}\left(1 - e^{-\frac{G}{C_1}t}\right) = \theta_0\left(1 - e^{-\frac{t}{\tau}}\right) \tag{2.4}$$

where,

$$\begin{aligned} \theta_0 &= P/G \\ \tau &= C_1/G \end{aligned} \tag{2.5}$$

θ_0 is the steady-state temperature rise, τ is the thermal time constant of the system, which is a constant that characterizes how quickly the system reaches a steady state. If the system is already in a thermal equilibrium state before the microwave power P is applied, i.e., in $\theta_1 = \theta_2$ state, then after the microwave power is applied, θ_1 will no longer be equal to θ_2, then Eq. (2.4) represents the temperature change of the load after absorbing microwave power P. Once the calorimeter reaches thermal equilibrium, C and G can be considered constants. After calibration, using Eq. (2.3), the steady-state temperature rise can be used as a measure of the power being measured. When the added power is constant, the better the adiabaticity of the calorimeter, i.e., the smaller the thermal conductivity G, the larger θ_0, the higher the power sensitivity. However, the smaller the thermal conductivity, the larger the time constant, and the longer the equilibrium time of the calorimeter, which can be inconvenient for use.

It is not difficult to see from Eq. (2.4) that for a stable power, the temperature rise follows an exponential law, as shown in Fig. 2.1. When measurements are made at 4, 5, or 6 times the thermal time constant, the temperature rise will be 98.2, 99.3, and 99.8% of the final stable temperature. Since the thermal time constant is relatively short compared to the measurement personnel's operation process of the calorimeter, in general, for continuous wave or continuous pulse microwaves, what we measure will be the power after the temperature rise is fully stable.

In actual measurements, we do not need to calculate the power based on Eqs. (2.4) and (2.3). but simply calibrate the measuring power meter with a standard power meter beforehand, or compare the temperature rise caused by the heating power of a direct current (or alternating current) with measured temperature rise, in order to indicate the magnitude of the microwave power.

2. **Measurement method of dry calorimeter**

(1) **Direct measurement method**

The principle schematic diagram of the direct measurement method is shown in Fig. 2.2, where a thermocouple is installed between the working load and the reference load.

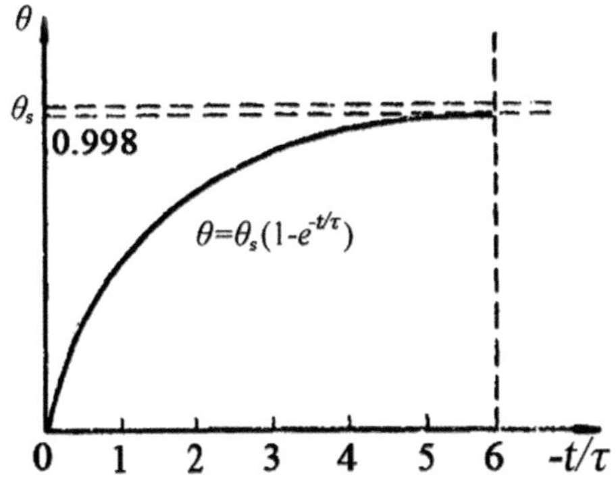

Fig. 2.1 The relationship between the temperature rise of the calorimeter and time at a stable microwave power *P*

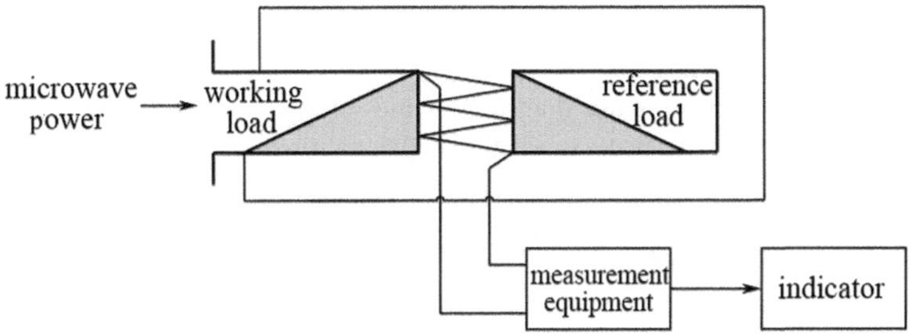

Fig. 2.2 Direct measurement method of microwave power by a dry calorimeter

In Fig. 2.2, the material, size, and surrounding environment of the working load and the reference load should be as identical as possible. The entire calorimeter should also be kept in adiabatic conditions as much as possible, that is, to minimize the heat exchange between the entire thermal converter and its surrounding environment.

Under the condition that the thermal performance of the working load and the reference load is completely the same as described above, we can think that before the microwave power is introduced, the temperatures of the two loads are completely the same, and this temperature is called the background temperature of the load. After introducing the microwave power, the working load will increase in temperature due to absorbing microwave energy, while the reference load does not absorb microwave energy, so, the temperature remains unchanged. As a result, there will be a temperature difference between the two loads, and this temperature difference causes the thermocouple to generate a temperature difference electromotive force, which is output through the lead connected to the thermocouple. If, during this measurement process, the background temperature of the load changes due to other factors, such as changes in the surrounding environment temperature, it is obvious that this change will be completely the same for both loads and will not affect the temperature difference between them, that is, it will not affect the magnitude of the output temperature difference electromotive force. Therefore, we can fully consider that this temperature difference is only caused by the absorption of microwave power by the working load. So, by analyzing the output of the thermocouple and performing the necessary calculations, we can determine the magnitude of the microwave power. This conversion method should be based on the relationship between the output of the calibrated thermocouple and the microwave power.

However, the thermocouple output and microwave power calibration, in addition to calibration using a standard power meter, a more convenient method is to use direct current or alternating current instead of microwave to heat the working load of the converter, and by comparing the output of the thermocouple with the direct current or alternating current power, we can obtain the corresponding microwave power size of the thermocouple output. In essence, this method is the alternative measurement method that we will introduce below, so the direct measurement method is almost only a theoretical method, and the alternative measurement method is more commonly used in practical applications.

(2) **Alternative measurement method**

Figure 2.3 shows the principle diagram of the alternative measurement method. Most of the calorimeters used to establish the national small power meter standard also adopt this scheme, which is based on the substitution of direct current (or low-frequency alternating current) power for microwave power. Its basic structure consists of a working load and a reference load with completely identical thermal performance, which are placed together in a heat-insulating container. Unlike the structure used for direct measurement, a direct current (or alternating current) power heater for substituting microwave power is also placed in the working load.

When using the alternative method for microwave power measurement, first use a direct current or alternating current power source to calibrate the indication of the microwave power indicator. The method is to use a direct current or alternating current power source to heat the working load instead of microwave power. This heating power can be conveniently calculated based on voltage and current. In fact, this power can be directly indicated on the meter. After the working load is heated, the thermocouple output will cause the microwave power indicator to have a certain

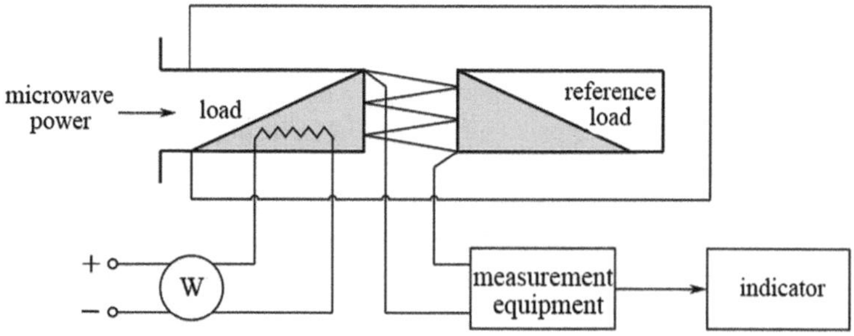

Fig. 2.3 Alternative measurement method of microwave power by a dry calorimeter

indication. At this time, the adjustment knob of the indicator should be adjusted so that the reading of the microwave power indicator is consistent with the reading of the heating power meter. By changing the size of DC (or AC) heating power, the reading of the microwave power indicator is adjusted again to make it consistent with the reading of the heating power meter. By repeatedly calibrating several times, the reading of the microwave power indicator under various heating power levels can be fully consistent with the size of the heating power, which completes the calibration of the indicator reading. However, if the heating power meter or microwave power indicator changes its range, calibration should be performed again. After calibrating the reading of the microwave power indicator to be the same as that of the heating power indicator, turn off the heating power supply to perform microwave power measurement. At this time, the output of the thermocouple will directly indicate the size of microwave power on the indicator.

The working principle of the alternative measurement method can be understood as follows:

Let C_1 and C_2 be the heat capacity of the working load and the reference load respectively, and θ_1 and θ_2 be their temperatures respectively; G_1 and G_2 be their thermal conductivities to the heat-insulated container; G_m be the thermal conductivity between the two loads; θ_c be the temperature of the container. When a microwave power P is added to the working load, the thermal equilibrium equation of the system can be written as:

$$\begin{aligned} P &= G_1(\theta_1 - \theta_c) + C_1\frac{\mathrm{d}\theta_1}{\mathrm{d}t} + G_m(\theta_1 - \theta_2) \\ C_2\frac{\mathrm{d}\theta_2}{\mathrm{d}t} &= G_m(\theta_1 - \theta_2) - G_2(\theta_2 - \theta_c) \end{aligned} \tag{2.6}$$

Solve the Eq. (2.6) to obtain

$$\frac{P}{C_1} = \frac{\mathrm{d}(\theta_1 - \theta_2)}{\mathrm{d}t} + \left(\frac{G_m + G_1}{C_1} + \frac{G_m}{C_2}\right)(\theta_1 - \theta_2) + \left(\frac{G_1}{C_1} - \frac{G_2}{C_2}\right)(\theta_2 - \theta_c) \tag{2.7}$$

If the calorimeter satisfies the thermal symmetry condition, i.e., $G_1/C_1 = G_2/C_2$, the temperature difference between the two loads can be determined from Eq. (2.7)

$$\theta_1 - \theta_2 = \frac{P}{C_1}\tau\left(1 - \mathrm{e}^{-\frac{t}{\tau}}\right) \tag{2.8}$$

The steady-state temperature difference θ_0 after thermal stabilization is

$$\theta_0 = \theta_1 - \theta_2 = \frac{P}{C_1}\tau \tag{2.9}$$

where, the time constant τ is

$$\tau = \frac{1}{\frac{G_m+G_1}{C_1} + \frac{G_m}{C_2}} \tag{2.10}$$

As seen from Eq. (2.9), the steady-state temperature difference θ_0 is proportional to the microwave power applied, and thus can be used as a measure of the measured power. When the microwave power is constant, the smaller the thermal conductivity G_m, the larger the steady-state temperature difference θ_0. However, the decrease in thermal conductivity, that is, the increase in thermal resistance, also increases the time required for the calorimeter to reach thermal stability.

When calibrating the calorimeter with direct current power instead of microwave power, after the direct current power P_{dc} is added to the working load, the steady-state electromotive force E_{dc} generated by the thermocouple is proportional to the direct current power P_{dc} absorbed by the load. That is,

$$E_{dc} = K_{dc}P_{dc} \tag{2.11}$$

where, K_{dc} is the response coefficient of the thermocouple to the direct current power P_{dc}.

When the direct current power is disconnected for microwave power measurement, the working load is heated by microwave power, and the steady-state electromotive force E_{rf} generated at both ends of the thermocouple is:

$$E_{rf} = K_{rf}P \tag{2.12}$$

where, K_{rf} is the response coefficient of the thermocouple to the microwave power P.

In general, it can be considered that the thermocouple has the same response to direct current calibration power and microwave power, $K_{dc} = K_{rf} = K$, thus, from Eqs. (2.11) and (2.12) can be obtained

$$P = P_{dc}\frac{E_{rf}}{E_{dc}} \tag{2.13}$$

Equation (2.13) is the fundamental formula for using an alternative method to measure microwave power in a calorimeter. The calibration we initially performed on the microwave power indicator is essentially to make the measurement using the alternative method in the calorimeter satisfy the above equation, thereby enabling the microwave power indicator to directly display the magnitude of the microwave power.

The advantage of the alternative measurement is that it is not necessary to convert the relationship between the output of the thermocouple and the microwave power, reducing the error caused by the conversion, so its measurement error is small. In addition, it has strong overload capacity, large dynamic range, which can reach 30–40 dB, and good impedance matching.

Alternative method measurement is also called comparative measurement, it greatly improves the convenience of measurement and has high measurement accuracy. Therefore, it is not only widely used in dry calorimeter measurement, but also in liquid calorimeter measurement. Therefore, modern calorimetric power meters all use alternative method to achieve power measurement. Its disadvantages are complex structure, long time constant, high environmental temperature requirements, and it is more suitable for high-frequency range power measurement above centimeter wave band, because the higher the frequency, the smaller the heat capacity of the load, the easier it is to reach thermal equilibrium.

2.2.2 *Liquid Calorimeter Power Meter*

1. **Basic principles and converter**

(1) **Working principle of liquid calorimeter**

The main disadvantage of dry calorimeters is that they have a large time constant and require a certain amount of time to reach thermal equilibrium. More importantly, their power capacity is limited, and too large microwave power can easily cause damage to the heat load, making them unsuitable for high-power measurements. If a certain liquid is used as the microwave absorption load to make a power meter, it is called a liquid calorimeter. Generally speaking, liquids have higher microwave absorption efficiency than solid microwave absorption materials, but liquids have fluidity, which is both an advantage and a disadvantage. The advantage is that the flowing liquid can quickly take away the heat generated by microwave heating, thereby improving the power capacity of the power meter. The disadvantage is that the liquid must have a special container to hold it, and this container must be able to allow the microwave to pass through the container wall into the liquid without significant attenuation, and be absorbed by the liquid as much as possible, and it also must have sufficient strength and not be easily damaged.

For the measurement of medium and high microwave power, flow-type liquid calorimeters are commonly used, and the liquid used is generally water. This is not only because the relative dielectric constant of water reaches 81 (at a water temperature of 20 °C), and the absorption efficiency is very high for microwaves, but also because the calculation formula of the absorption power of water is simple and relatively accurate.

When water is used as the microwave absorption material, if the water is flowing, the power absorbed by the water is obviously related to the temperature rise of the water after absorbing the microwave, the flow rate of the water, the specific heat capacity of the water, the density of the water, etc. It can be simply expressed as:

$$P = 4.18 c_p \rho V \Delta T (\text{W}) \tag{2.14}$$

where,

P is the power absorbed by water (W).

c_p is the specific heat capacity of water (Cal/g·°C).

ρ is the density of water (g/cm^3).

V is the flow rate of water (cm^3/s).

ΔT is the temperature rise of water (°C).

For water, under a standard atmospheric pressure, we can approximately take

$$\rho = 1\,\mathrm{g/cm^3}$$

$$c = 1\,\mathrm{Cal/g\,{}^\circ C}$$

Thus, the power calculation Eq. (2.14) can be greatly simplified to

$$P = 4.18\,V\Delta T\,(\mathrm{W})(\text{when the unit of } V \text{ is } \mathrm{cm^3/s}) \quad (2.15)$$

where, the number 4.18 is the thermal work equivalent of water, that is, the Joule value (J) of the work corresponding to the unit heat Calorie (Cal), i.e.

$$1\mathrm{Cal} = 4.18\,\mathrm{J}$$

Therefore, what is calculated by Eq. (2.14) is the size of J/s, but because

$$1\mathrm{W} = 1\,\mathrm{J/s}$$

So the result of Eq. (2.14) is the size of power W.

However, we must point out that under a standard atmospheric pressure (101.325 kPa), the specific heat capacity and density of water are functions of temperature, and their values at different water temperatures are not always equal to 1. Therefore, more rigorous calculations should take into account the changes in specific heat capacity and density at different temperatures. In engineering calculations, they can be approximated as 1.

(2) **The converter of the liquid calorimeter**

The converter of the liquid calorimeter, as the name suggests, should be some kind of liquid, usually water, and therefore it is also habitually called a water load. The requirements for the water load are, first and foremost, that it must be able to withstand sufficient power capacity, which mainly depends on the volume of the water load and the flow rate of the water. Secondly, the water load should match the microwave transmission system as much as possible to prevent the microwave power from being reflected and causing measurement errors. For this reason, the water load is always tipsily placed relative to the microwave transmission direction, such as the container wall forming a conical tilt, or the entire water load being tiltedly, or the water load being conical distribution or curve distribution. Thirdly, the material of the water container that makes up the water load must have sufficient transparency to microwaves, that is, it must allow the microwaves to be absorbed by the water without being reflected or attenuated as much as possible. Fourthly, the container material should have a certain structural strength and bonding strength (when the water load needs to be bonded to the waveguide wall to form a water container), and should not be easily damaged or detached. The container of the water load is usually made of quartz glass, but can also be made of ordinary hard glass, glass fiber-epoxy resin material, organic glass, polytetrafluoroethylene, etc., depending on the specific usage and conditions.

Figure 2.4 shows several common water loads used in flow-type liquid power meters, where Figure 2.4c, d are water loads that need to be bonded to the waveguide wall.

(a) Absorption water load

The water container of this water load is directly placed at the strongest point of the electric field in the waveguide (Figs. 2.4a, 2.5, 2.6, 2.7 and 2.8e), and the microwave power is directly absorbed by the water load.

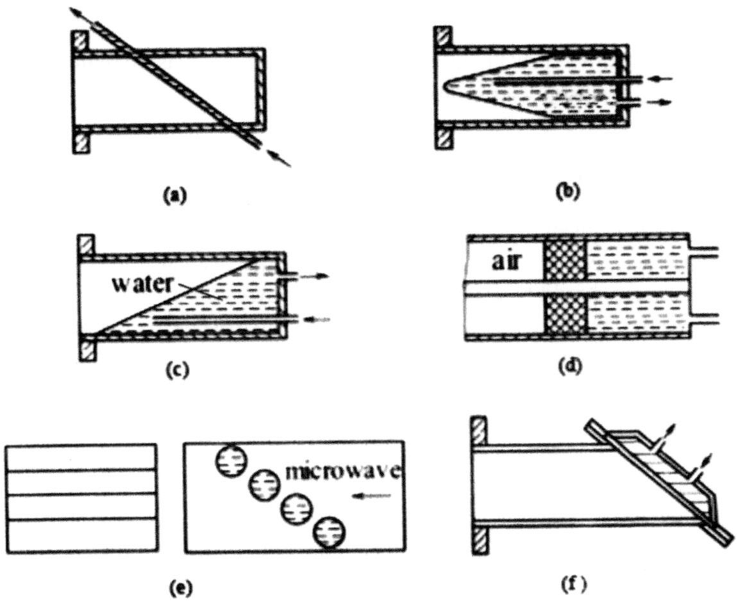

Fig. 2.4 Several common water loads for flow-type liquid power meters **a** plug-in water load, **b** cone water load, **c** slant water load, **d** coaxial water load, **e** multi-tube distributed water load, **f** radiative water load

(b) Radiation water load

The water container of the radiation water load is not directly placed inside the waveguide, but is placed on the waveguide port, so the microwave power is radiated through the port into the water load and absorbed (Fig. 2.4f).

Of course, there are various types of water loads, which have different shapes of water containers, different arrangements, different power capacities, different container materials, different water inlet and outlet methods, and different placement methods in the waveguide, and so on.

2. **Working mode of liquid calorimeter**

Liquid calorimeters can be divided into static and flow types.

(1) **Static calorimeter**

If the liquid in the liquid calorimeter does not flow, it is actually completely similar to the dry calorimeter. Obviously, the microwave energy it can absorb is limited and will be limited by the volume of the liquid. At the same time, the microwave heat absorbed by the liquid cannot be taken away by the liquid, so the power capacity is not high. Generally, it is only made as a small-power meter and used for power measurement in medium and small power systems as through-type measurement, or in large power systems as radiative measurement. During measurement, similar to dry calorimeters, either direct measurement or alternative measurement methods can be employed. This calorimeter is rarely used in practice, and the capillary calorimeter introduced in Sect. 5.1 is a static calorimeter.

(2) **Flow calorimeter**

If the microwave power is large enough, especially in the case of large average power, which exceeds the measurement range of the dry calorimeter, a flow-type liquid calorimeter should be used to measure the power. Due to the continuous flow of the liquid, the heat generated by the absorbed microwave can be taken away at any time and cooled in the calorimeter cooling chamber for recycling, or directly discharged as waste water. Therefore, it can measure very high power levels. The schematic diagram of the flow-type liquid calorimeter is shown in Fig. 2.5. It is mainly composed of a water load, a heating resistor wire placed in the water load outlet pipe, a heating power source and its power indicator, a thermocouple and a temperature difference electromotive force indicator (whose scale has been calibrated to directly indicate microwave power, so it is usually called a microwave power meter), a water flow regulating valve, a water pump and a cooling container.

When using a flow-type liquid power meter for power measurement, first start the power meter to make the water in the water load start to flow, and turn off the microwave power source to calibrate the power meter indication. During calibration, turn on the heating power source, use DC or AC current to heat the water flow, adjust the heating power size so as to make the heating power reach a certain reading on the heating power indicator. Due to the heating power, the thermocouple has an output,

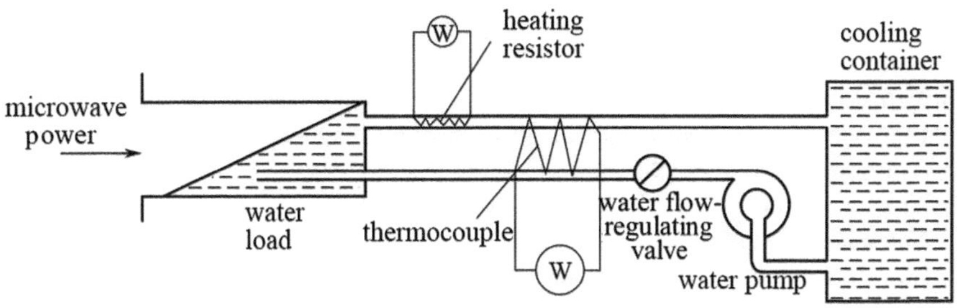

Fig. 2.5 Schematic diagram of flow calorimeter measurement

and the microwave power meter also has an indication. At this time, adjust the valve of the water flow rate, which will cause different temperature rises of the water due to the different water flow rates, which will change the output size of the thermocouple, that is, the size of the microwave power indication. Therefore, adjusting the water flow rate can change the microwave power indication, so that the power indication of the microwave power meter can be the same as the indication of the heating power meter. This process should be repeated several times, and each time it should be calibrated at different heating power levels until the microwave power indication and the heating power indication can be consistent at different power levels. After the calibration is completed, fix the water flow regulating valve and no longer change it, then turn off the heating power source and turn on the microwave power source for microwave power measurement.

Flow-type liquid calorimeter is currently the most widely used power meter for measuring continuous wave and continuous pulse microwave power. It can measure continuous wave power or average power of 5–2000 W and continuous pulse microwave power of 1–2 MW.

2.3 Thermal-Effect Small-Power Meter

For the power measurement of continuous wave and continuous pulse microwaves, besides the most commonly used calorimetric power meters, there are also some thermal effect power meters with small power capacity. The thermal conversion elements and devices in these power meters can only withstand very small power, so they are usually used for measuring small power microwaves. However, they are also widely used in large-power and high-power continuous wave or continuous pulse microwave power measurements. This is because, when using a through-type or radiative-type measurement method, as long as the coupling mechanism makes the power received by the receiving system small enough to reach the power range allowed by the converter of the small-power meter, the small-power meter can be used for measurement. Therefore, in many large-power and high-power microwave power measurements, especially when using absorption-type measurements is practically difficult, small-power meters are often used to indicate power, and then the actual high power in the system is calculated by the coupling coefficient (attenuation) of the coupling element or radiation-receiving system. Therefore, it is still necessary for us to understand the basic working principle of this type of power meter.

The resistive sensor power meter is a new type of microwave power meter that has developed in the past twenty or thirty years. Because its converter response time can reach the nanosecond level, and it can withstand pulse power of kilowatts or more. Therefore, it is more suitable for high-power microwave power measurement. It is seldom used in conventional continuous wave or continuous pulse microwave power measurement. Therefore, we will introduce it in the power measurement of other electromagnetic effects of high-power microwaves of Chap. 5. Here, we will mainly introduce several conventional thermal effect small-power meters that are widely used in microwave measurement [14, 15].

2.3.1 Thermoelectric Resistance Power Meter

1. **Types of thermoelectric resistance**

Thermoelectric resistance, also known as thermal resistance, radiation thermal resistance, is an element whose power measurement based on the change in resistance value of metal conductors or semiconductor materials with temperature. When the thermoelectric resistance receives microwave irradiation, it will increase in temperature due to the absorption of microwave energy, and cause a change in resistance value. By measuring this change with an electric bridge, after calibration, the size of the microwave power can be obtained. Thermoelectric resistance can be divided into three types: metal thermoelectric resistance, semiconductor thermoelectric resistance, and ferrite thermoelectric resistance.

(1) **Ballast resistor-metal thermoelectric resistance**

Metal thermoelectric resistance can also be called ballast resistor, which is mostly made of pure metal materials. Platinum and copper are currently the most widely used materials. Now, materials such as palladium, nickel, manganese, and rhodium are also used to make thermoelectric resistance. Among them, the measurement accuracy of platinum thermoelectric resistance is the highest. Thermoelectric resistance usually needs to transmit the resistance signal through a lead to a computer control device or other secondary instrument (measurement device).

The metal thermoelectric resistance used to measure microwave power can also be vacuum-sprayed onto mica sheets and used in the form of thin film resistors.

Metal thermoelectric resistance is generally suitable for temperature measurement in the range of − 200–800 °C, with a power sensitivity of 3–12 Ω/mW. Its characteristics are accurate measurement, good stability, reliable performance, and extremely wide application in engineering control.

(2) **Thermistor-semiconductor thermoelectric resistance**

Thermistors are generally made of semiconductor materials. Their typical feature is that they are very sensitive to temperature, and their temperature coefficient of resistance is more than 10–100 times that of metals. The power sensitivity can reach 50–100 Ω/mW, and they can detect temperature changes of 10^{-6} °C. At room temperature, its resistance value is very high (usually above several thousand ohms), but the interchangeability is poor, the nonlinearity is serious, and the temperature measurement range is only − 50–300 °C. According to the temperature coefficient, it can be divided into positive temperature coefficient thermistor (PTC) and negative temperature coefficient thermistor (NTC).

The thermistor used for microwave power measurement is generally a metal oxide, such as $CuO + Mn_3O_4$, $Mn_3O_4 + NiO$, etc., and is often made into a spherical or thin film shape. They belong to negative temperature coefficient thermistors.

(3) **Ferrite thermoelectric resistance**

When ferrite materials magnetized by a DC magnetic field interact with a microwave field, they absorb microwave energy when ferromagnetic resonance occurs in the ferrite. This energy absorption is achieved within a certain frequency range, and the bandwidth depends on the half-power point width of the ferromagnetic resonance curve of this type of ferrite.

Since the ferrite thermoelectric resistance is interacting with the microwave magnetic field, it can be placed on the wall of a waveguide or a resonant cavity, because the electric field on the wall is zero while the magnetic field is generally non-zero.

Ferrite thermoelectric resistance is a frequency-selective element. Therefore, by using a ferrite with a very narrow ferromagnetic resonance absorption curve, a microwave power meter with power distributed according to the spectrum can be manufactured, and the spectrum can be adjusted by changing the intensity of the ferrite magnetization.

2. **The measurement method of thermoelectric resistance power meter**

(1) **Converter–power probe**

The converter used in the thermoelectric resistance power meter, that is, the power probe that converts the absorbed microwave power into heat and outputs it as an electrical signal through a specific circuit, can be a waveguide type, but more are coaxial types. The thermistor used has a body resistance type and a thin film resistance type. Figures 2.6 and 2.7 show schematic diagrams of coaxial power probes and waveguide power probes for some body resistance type thermistors.

(2) Measurement circuit–electric bridge

The output circuit of the thermoelectric resistance power meter uses a bridge circuit. Figure 2.8 shows two principle diagrams of the bridge circuit [14].

In Fig. 2.8, the role of the thermoelectric resistance R_t in the power probe is to be placed as an absorbing load of microwave power into the waveguide or coaxial line, as shown in Figs. 2.6 and 2.7. When measuring, the leads at both ends of the thermoelectric resistance Rt are connected to the electric bridge, so that the thermoelectric resistance becomes one arm of the bridge, and the resistors R_1, R_2, and R_3 of the other arms of the bridge are selected so that the bridge can reach balance without microwave power. Generally speaking, the thermoelectric resistance will have a certain DC resistance value, and we can adjust the variable resistor R4 and the size of the power supply current of the bridge (equivalent to changing the thermal state of the thermoelectric resistance, that is, the resistance value) to make the bridge reach balance, that is, the current of the bridge is zero when there is no microwave power input.

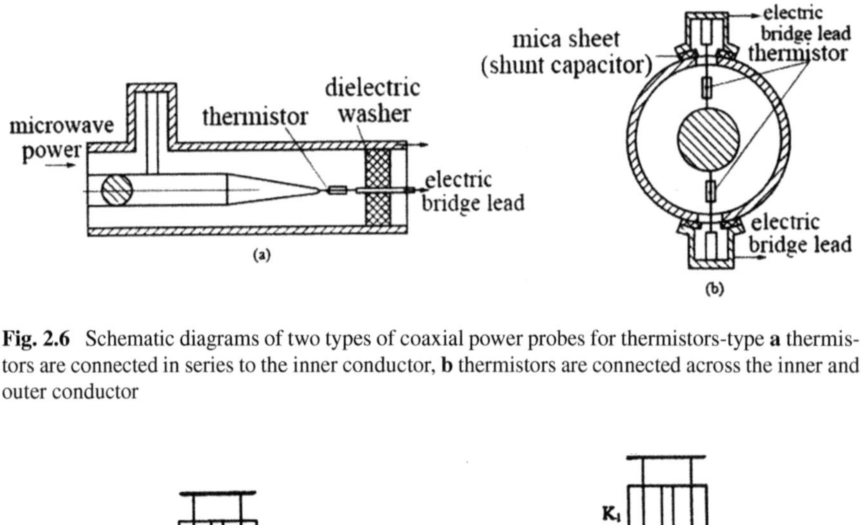

Fig. 2.6 Schematic diagrams of two types of coaxial power probes for thermistors-type **a** thermistors are connected in series to the inner conductor, **b** thermistors are connected across the inner and outer conductor

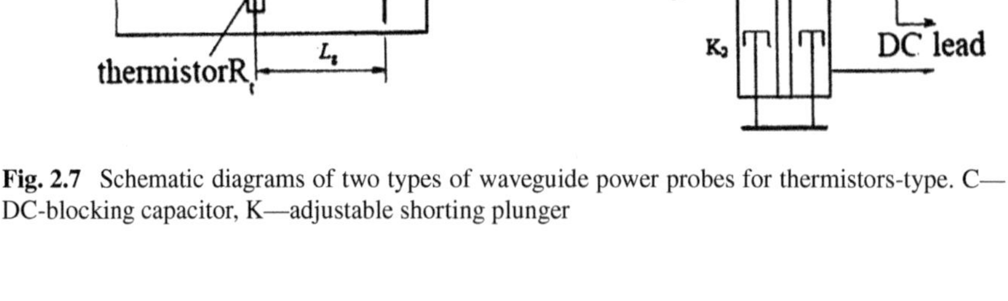

Fig. 2.7 Schematic diagrams of two types of waveguide power probes for thermistors-type. C—DC-blocking capacitor, K—adjustable shorting plunger

When microwave power is applied to the thermistor, the thermistor will heat up due to the absorption of microwaves, resulting in an additional temperature increase that causes its resistance value to change. Assume the change in resistance isΔR. At this point, the balance of the bridge is disrupted, and the galvanometer located on the diagonal of the bridge will indicate the magnitude of the current caused by the imbalance. WhenΔR is not large, meaning the imbalance of the bridge is minor, the imbalance current is linearly related to the magnitude of ΔR, which corresponds to the level of microwave power.

$$P = kI \tag{2.16}$$

where, P is the microwave power, I is the bridge indication current, k is a proportionality coefficient, which should be calibrated in advance. After calibrating k, the size of the microwave power can be calculated based on the bridge indication current value.

Figure 2.8a shows the simplest bridge circuit. However, it has some drawbacks in practical use: for example, changes in the surrounding medium and environment, as well as the replacement of the thermoelectric resistance will affect the coefficient k, which means k must be calibrated constantly, an inconvenience in practical use. In addition, as the measured power changes, the resistance value of the thermoelectric resistance, which acts as a microwave system load, also changes, making its matching degree with the system change, which not only affects the measurement accuracy but also limits the dynamic range of the measurement. Therefore, overall, the accuracy of this bridge measurement is not high. In practical work, the alternative bridge circuit is commonly used. The so-called alternative method is to compensate for the additional heating of the thermoelectric resistance caused by microwave power by reducing the DC heating power of the thermoelectric resistance, so that the temperature of the thermoelectric resistance remains unchanged during the measurement process, that is, the resistance value remains unchanged, and the change in DC power represents the size of the microwave power, that is, the microwave power is obtained by measuring the alternative power.

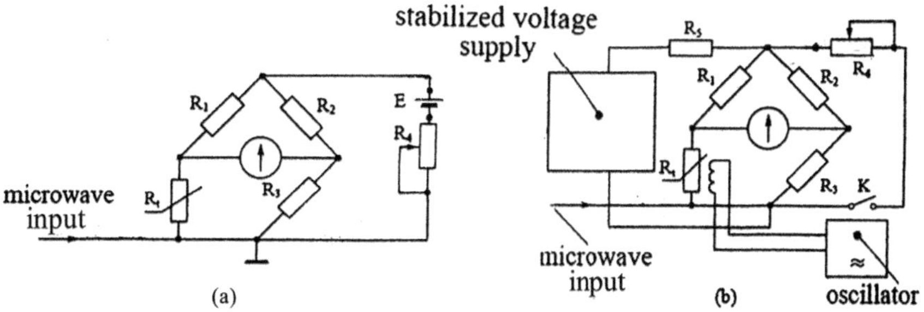

Fig. 2.8 Thermoelectric resistance electric bridge circuit for measuring microwave power **a** principle circuit of electric bridge, **b** schematic diagram of electric bridge for power measurement by DC voltage alternative method

Figure 2.8b shows a practical alternative bridge circuit. The thermoelectric resistance R_t is powered by a DC power supply with a stable output voltage through a series resistor R_5. The internal resistance of the DC power supply is much greater than the bridge resistance. Before adding microwave power, the switch is disconnected, and the bridge is brought to the initial balance with an audio AC power supply (oscillator in the figure). At this time, the current flowing through the thermoelectric resistance is the known current provided by the DC power supply. After the switch is closed and microwave power is added, the bridge loses balance. Adjust the shunt resistor R_4 to change the current flowing through the thermal resistance to restore the balance of the bridge. Since the voltage supplied to the thermoelectric resistance is stable, the current that is shunted directly corresponds to the microwave power. If a precision potentiometer is used as the shunt, the power size can be directly engraved on the potentiometer.

The alternative measurement method maintains the temperature and resistance value of the thermoelectric resistance unchanged, and the matching with the microwave system can also be kept unchanged. Therefore, the measurement accuracy is greatly improved. In practical applications, the bridge circuit of the alternative method is even more complicated to further improve the measurement accuracy.

3. **Microcalorimeter**

The microcalorimeter consists of a thermal resistance, a thermocouple, and a bridge, as shown in Fig. 2.9. A and B are two completely identical thermoelectric resistance probes. The measuring heat body A is used to absorb microwave power and DC calibration power, and B is the reference measuring heat body. The working principle of the microcalorimeter is actually a combination of the thermoelectric resistance power meter and the calorimeter, which places the thermoelectric resistance in the measuring heat body A that absorbs microwave energy. Like the thermoelectric thermal resistance power meter, the thermoelectric resistance serves as a heat converter. It also operates in a dual-load working mode of the working load and the reference load like in the calorimeter, and outputs the microwave power by the thermocouple. It not only uses the high accuracy of the calorimeter, but also uses the advantages of fast response time and convenient measurement of the thermoelectric resistance power meter [16, 17].

After adopting very careful heat insulation measures, the microcalorimeter is often used as a standard power meter. Its measurement uncertainty can generally reach less than 0.5–1%, and its operating frequency can reach above 100 GHz.

2.3.2 *Thermocouple Power Meter*

In the previous section discussing calorimetric power meters, we already mentioned thermocouples. In calorimetric power meters, the microwave power is actually obtained by measuring the thermoelectric potential output of the thermocouple.

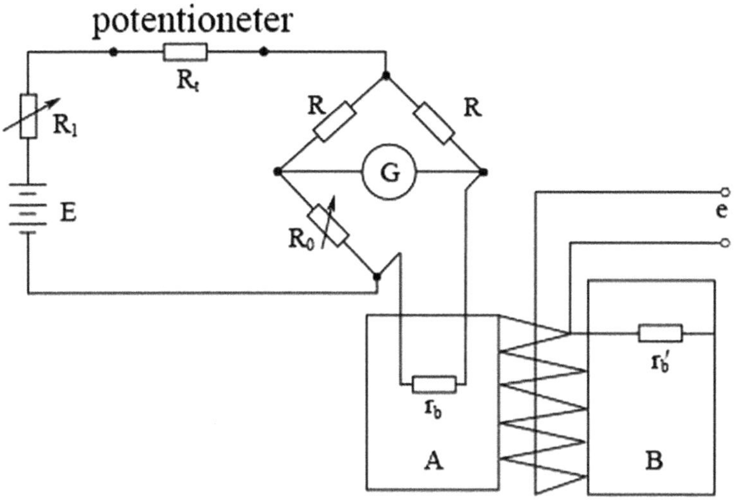

Fig. 2.9 Principle diagram of microcalorimeter

However, in calorimetric power meters, the thermocouple itself is only used as a measuring device rather than a power converter, that is, it is not a load that absorbs microwave power, and its hot end is placed in the actual thermal load, which absorbs microwave power to generate heat and then transfer it to the thermocouple to raise its temperature, thus creating a temperature difference electromotive force between its hot end and its cold end, or in other words, the heating of the thermocouple is indirect. Now we will introduce a power meter made by directly using a thermocouple as a converter to absorb microwaves-thermocouple power meter. In this type of power meter, the thermocouple is directly heated by microwaves.

1. **Basic principle**

By utilizing the Seebeck effect of the thermocouple, the microwave power directly heats one end of the thermocouple to generate a thermoelectric potential. The thermoelectric potential is correlated to the microwave power absorbed by the thermocouple, so that the microwave power can be measured.

The basic principle of thermocouple temperature measurement is: when two different conductors or semiconductors A and B are connected to each other to form a loop, as long as the temperature at the two junctions is different, one end is at a temperature T, called the working end or hot end, and the other end is at a temperature T_0, called the free end (also called the reference end) or cold end, an electromotive force will be generated in the loop. This phenomenon is called the thermoelectric effect, also known as the Seebeck effect. The loop composed of two conductors (or semiconductors) is called a thermocouple, and the electromotive force generated is called a thermoelectric potential or a temperature difference electromotive force. The direction and magnitude of the electromotive force are related to the material of the conductor (or semiconductor) and the temperature of the two junctions, and it is independent of the shape and size of the thermocouple circuit. After the two electrode materials that make up the thermocouple are fixed, the thermoelectric potential is a function of the temperatures T and T_0 of the two junctions. If the cold end temperature T_0 is constant, the thermoelectric potential generated by the thermocouple will only change with the temperature change of the hot end (measurement end), that is, a certain thermoelectric potential corresponds to a certain temperature. We only need to measure the size of the thermoelectric potential to achieve the purpose of temperature measurement.

When a third metal material is connected to the thermocouple circuit, as long as the temperature at the two connection points of this material is the same, the thermoelectric potential generated by the thermocouple will remain unchanged, that is, it will not be affected by the third metal connected to the circuit. Therefore, by using this characteristic, we can connect the thermocouple circuit's cold end to a measuring instrument through leads to measure the thermoelectric potential, and determine the temperature of the measuring end (hot end). When measuring the temperature with a thermocouple, it is required that the temperature of its cold end remains unchanged, and the magnitude of its thermoelectric potential can be in a certain proportion to the measured temperature. If the temperature of the cold end (environment) changes, it will seriously affect the measurement accuracy. A certain measure is taken to compensate for the influence of the cold end temperature change on the thermocouple, which is called the cold end compensation of the thermocouple.

Thermocouples have a wide temperature measurement range and relatively stable performance; high measurement accuracy, direct contact between the thermocouple and the measured object, no intermediate medium influence, fast thermal response time, and sensitive response to temperature changes. The measurement range is large, and the temperature can be measured continuously from $-$ 40 °C to $+$ 1600 °C, reliable performance, good mechanical strength; long service life, easy installation.

1. **Measurement method**

(1) **Converter–power probe**

In the microwave band, thermocouple power meters generally use thin-film thermocouples as microwave loads. They are also divided into coaxial and waveguide types. Coaxial power probes are more commonly seen in daily life. The structure of the thin-film thermocouple used in the coaxial probe is shown in Fig. 2.10.

A circular metal film that serves as the center electrode and a half-moon-shaped metal film on each side that serves as the outer electrode are deposited on a mica sheet or a polyester film substrate. During the operation of the power probe, they will make close contact with the inner and outer conductors of the coaxial line, respectively. A bismuth, antimony thin-film thermocouple is cross-coated on the gap between the center electrode and each outer electrode, as shown by points a_1 and a_2 in Fig. 2.10. One end of the two thermocouples is connected in the middle of the center electrode, such as point b_2 in the figure, and the other end is respectively connected to the outer electrode, such as points b_1 and b_3. In this way, for the thermoelectric potential output of the thermocouple, the two thermocouples output in series, for the

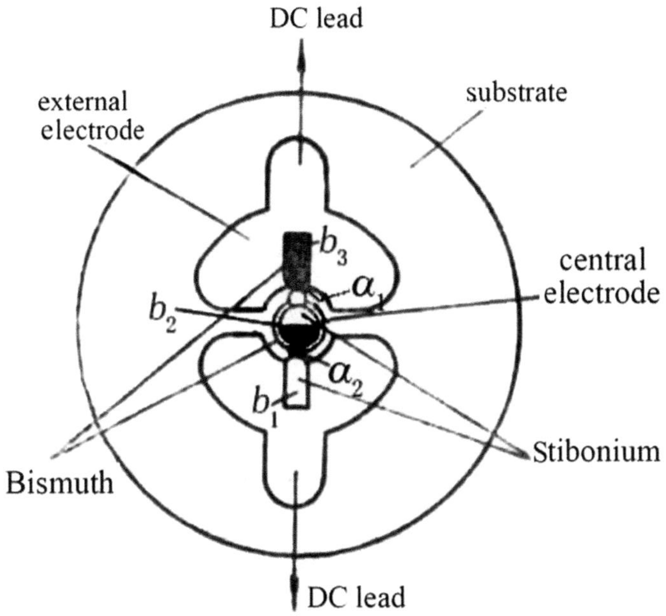

Fig. 2.10 Schematic diagram of the thermocouple structure in the coaxial power probe

microwave input signal, the two thermocouples in parallel, regardless of whether the electric field direction is from the center electrode to the outer electrode or from the outer electrode to the center electrode. If the internal resistance of each thermocouple is 100 Ω, the parallel resistance is 50 Ω, which matches the 50 Ω coaxial line. A layer of insulating film is then coated on the bismuth and antimony thin films to make the thermocouple capacitively coupled with the inner and outer conductors of the coaxial line to become an absorbing load of the microwave signal, while ensuring that the center electrode, the outer electrode, and the inner and outer conductors of the coaxial line are in good contact.

When there is no microwave signal input, the two ends of each thermocouple have no temperature difference because the inner and outer conductors of the coaxial line are at the same ambient temperature, so there is no output voltage. When a microwave signal is input, the thermocouple absorbs microwave power through capacitive coupling, so that the contact points al and a2 of the thermocouple bismuth and antimony thermocouples are heated, and the other contact points b2, b1, and b3 will not increase in temperature due to their close contact with the inner and outer conductors of the coaxial line. Therefore, a temperature difference is generated between a1and b2, b3 and between a2 and b2, b1, causing a thermoelectric potential output. The magnitude of the thermoelectric potential reflects the magnitude of the microwave power, but the thermoelectric potential generated by the thermocouple is very small, so it needs to be amplified by a low-noise, high-gain amplifier before it can be indicated by a microampere meter. After calibration, the dial can be directly engraved with the power value.

(2) **Power indication**

The output signal of the thermocouple converter is a DC voltage, so a digital or pointer millivoltmeter or microvoltmeter can be used as a measurement indicator. The relationship between the measured microwave power and the thermoelectric potential output by the thermocouple can be expressed as

$$P = KU \tag{2.17}$$

where, P is the microwave power, U is the thermoelectric potential output by the thermocouple; K is a proportionality coefficient. K is influenced by two factors: one factor is the conversion coefficient of the thermocouple, which indicates how many volts of thermoelectric potential are generated per unit microwave power. This coefficient is a nonlinear function of temperature, because when the microwave power being converted causes the thermocouple to heat up, its heat dissipation coefficient will also change; The other factor is the efficiency of the converter, which represents the percentage of microwave power that can be absorbed and converted into electromotive force by the thermocouple. This is because there are always some factors in the converter that cause microwave power loss or reflection, such as microwave power reflection caused by the mismatch between the thermocouple itself and the transmission line, gaps between components, joints, sockets, dielectric supports, structural capacitors, and other non-uniformities, which result in some power not being directly transferred to the thermocouple, thereby reducing the efficiency of the converter. Particularly as the frequency increases, this nonlinear behavior of K will become more serious.

Therefore, the thermocouple power meter must be calibrated in advance to give a calibration curve. Some thermocouple power meters have a special calibrator set up in the instrument to automatically calibrate the entire instrument.

Thermocouple power meters can be used for power measurement from the meter-wave band to the millimeter wave band, and can measure power from microwatts to milliwatts. Attenuators and directional couplers can be used to further expand the range.

2.3.3 *Ferrite Power Meter*

1. **Resonance absorption of ferrite**

(1) **Ferrite**

Ferrite has a wide range of applications in the field of microwave engineering. It is both a ceramic material and a magnetic material, commonly known as black porcelain or magnetic porcelain. It has a black-brown appearance and a hard and brittle texture.

The main raw material of ferrite is $XO{\cdot}Fe_2O_3$, where X represents a divalent metal ion. Ferrite used in the microwave band, X is often a mixture of Ni-Zn, Ni-Mg, or Mn-Mg.

Like ceramics, microwave ferrite is a good insulator with a high resistivity greater than 10^6 Ω cm, so the electrical loss is extremely small, and the relative dielectric constant is 10–20. Ferrite is also a ferromagnetic material, and its relative permeability can change with an external magnetic field. Under a constant magnetic field bias, its permeability in different directions is different, that is, it has anisotropy. It is precisely based on the anisotropy of ferrite that different effects will occur when electromagnetic waves pass through the ferrite from different directions.

(2) **Electromagnetic waves in ferrite**

We know that a linearly polarized wave can be decomposed into two circularly polarized waves with opposite rotation directions. When such an electromagnetic wave passes through a ferrite, if there is an external DC magnetic field H_0, then under the combined action of the DC magnetic field and the microwave magnetic field, the magnetic permeability of ferrite for circularly polarized waves of different rotation directions will be different, as shown in Eq. (2.18).

$$\left.\begin{aligned} \mu_r^+ &= 1 + \frac{\omega_M}{\omega_0 - \omega} \\ \mu_r^- &= 1 + \frac{\omega_M}{\omega_0 + \omega} \end{aligned}\right\} \tag{2.18}$$

where,

$\mu_r{}^+$ is the relative magnetic permeability of ferrite material for right-handed polarized waves,

$\mu_r{}^-$ is the relative magnetic permeability for left-handed polarized waves,

ω is the microwave angular frequency,

$\omega_M = \gamma M_0$ is called the intrinsic angular frequency of ferrite, where M_0 is the macroscopic spin magnetic moment of the ferrite material,

$\omega_0 = \gamma H_0$ is the precession angular frequency of the spin electrons of the ferrite, where H_0 is the external DC magnetic field,

$\gamma = e/mc$ is called the gyromagnetic ratio, e and m are the electron charge and mass, and c is the speed of light.

In magnetized ferrites, we define the direction of rotation for the circularly polarized waves based on the direction of the external applied DC magnetic field H_0. When you make a fist with your right hand, with your thumb pointing in the direction of the applied direct current magnetic field, if the direction of the four fingers of a fist is counterclockwise and coincides with the rotation direction of the polarized wave, then the circularly polarized wave is right-handed circular polarization. Conversely, if the direction of the four fingers is opposite to the rotation direction of the polarized wave, then the circularly polarized wave is left-handed circular polarization. When you make a fist with your left hand, with your thumb pointing in the direction of the applied DC magnetic field, if the direction of the four fingers is clockwise and coincides with the rotation direction of the polarized wave, then the circularly polarized wave is left-handed circular polarization. Conversely, if the direction of the four fingers is opposite to the rotation direction of the polarized wave, then the circularly polarized wave is right-handed circular polarization.

(3) **The phenomenon of cyclotron resonance in ferrite**

It can be seen from Eq. (2.18) that when the microwave frequency is equal to the precession frequency ω_0 of the electron, that is, $\omega = \omega_0$, $\mu_r{}^+$ tends to infinity. This means that ω_0 is a resonance point for $\mu_r{}^+$, and $\mu_r{}^-$ has no resonance point. This phenomenon is called the cyclotron resonance of ferrite. When ferrite has actual loss (cyclotron loss) for electromagnetic waves, the right-handed polarized wave will have a large attenuation near the resonance point, while the left-handed polarized wave does not have a resonance phenomenon, so the attenuation is small. Because at the same position in the rectangle waveguide, the circularly polarized waves formed by the transverse and longitudinal components of the TE_{10} mode magnetic field have polarization rotation directions that are exactly opposite for forward and backward waves, so the effect of the ferrite element on forward and backward waves is different. Figure 2.11 shows the image of the TE_{10} mode magnetic field observed from the wide side of the rectangular waveguide. If the direction of an external DC magnetic field H_0 is from the inside to the outside of the paper, it can be clearly seen that when the electromagnetic wave propagates from left to right (assuming it corresponds to the forward wave), the microwave magnetic field observed at point A rotates counterclockwise with time (from top to bottom in the figure, from $t = 0$ to $t = T/2$), indicating that it is a right-handed polarized wave; while the microwave magnetic field at point B rotates clockwise, indicating that it should be a left-handed polarized wave. If the electromagnetic wave propagates from right to left (corresponding to the backward wave), then with the passage of time (from bottom to top in the figure, from $t = 0$ to $t = T/2$), the microwave magnetic field at point A becomes a left-handed polarized wave, and the microwave magnetic field at point B becomes a right-handed polarized wave, which is opposite to the forward wave. That is to say, for the forward wave, the ferrite placed at point A will cause a large loss to the electromagnetic field during cyclotron resonance, while there is no cyclotron resonance for the backward wave, so there will be no loss. If the ferrite is placed at point B, the situation will be reversed, with a large loss for the backward wave and no loss for the forward wave.

The DC magnetic field of the cyclotron resonance microwave element should be perpendicular to the microwave magnetic field. For the rectangular waveguide that propagates the TE_{10} wave, the external magnetic field should be perpendicular to the wide side of the waveguide.

2. **Characteristics of ferrite power meter**

Ferrite components have the characteristic of generating resonance absorption of high-frequency electromagnetic energy, which causes the resistance value of the

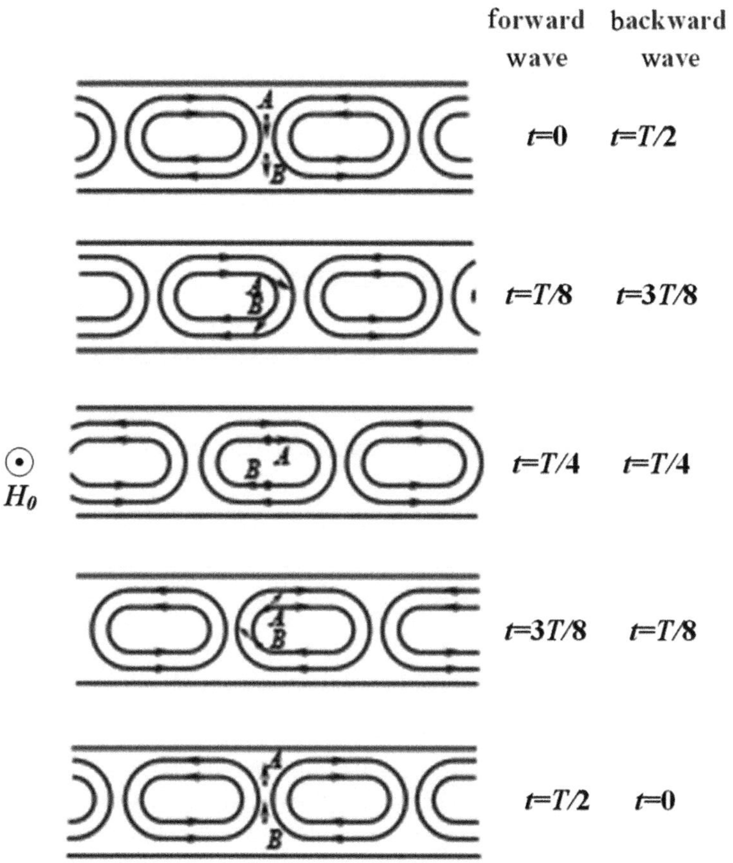

Fig. 2.11 Circularly polarized wave of TE_{10} mode magnetic field in rectangular waveguide

ferrite component to change due to the heating caused by the loss of microwave power when $\omega \approx \omega_0$. Therefore, when using a ferrite component to measure microwave power, the ferrite transducer is similar to the thermoelectric resistance converter, and its temperature rise is measured. After calibration, the measured microwave power can be obtained. In actual work, thermocouples or thermistors are often used to measure the temperature of ferrite components.

Characteristics of ferrite power meter:

(a) By changing the external magnetization field, the sensitivity of ferrite temperature change and the resonant absorption frequency can be adjusted over a wide range.
(b) The electrical strength of ferrite resistance is high, so it has strong overload resistance and long-term stability of parameters, and can be used for large-power measurement.

(c) The resonant absorption of ferrite has frequency selectivity, so it has strong anti-interference ability, and can also be made into a power meter according to the spectral distribution.
(d) Ferrite components have different effects on forward and backward waves, so they have directivity, and the incident wave power and reflected wave power can be measured separately.

2.4 Non-thermal Effect Power Meter

In both through-type and radiative-type microwave power measurements, for measuring small powers coupled out from waveguides or radiation fields, in addition to commonly used thermal effect small power meters, some non-thermal effect small power meters are also commonly used measuring instruments, especially the detector power meter is widely applied.

2.4.1 Detector Small-Power Meter

In microwave measurement, crystal detectors are not only often used to display microwave waveforms, especially pulse envelope waveforms, pulse width, repetition frequency, etc., but also the most commonly used power probe. It is often used as a signal level indicator, which can not only indicate the presence or absence of microwave power, indicate the relative size of power, but also directly indicate the size of microwave power after necessary calibration. In conventional microwave power measurement, it can be made into a power meter for measuring ultra-small power. In the power measurement of short pulse high-power microwaves, it is widely used to indicate the size of the received radiation power, so as to calculate the total power of the microwave system.

1. **Crystal detector**

(1) **Characteristics of detector diodes**

Crystal diodes, usually called semiconductor diodes, have detection characteristics, so they are directly called detector diodes. The structure of the detector diode involves knowledge related to semiconductor materials, which can be found in many materials, so it is not introduced here.

The semiconductor detector diode has unidirectional conductivity, and its volt-ampere characteristic is shown in Fig. 2.12. Obviously, it is nonlinear. The volt-ampere characteristic can be expressed as:

$$I = I_s\left(e^{\alpha V} - 1\right) \tag{2.19}$$

and,

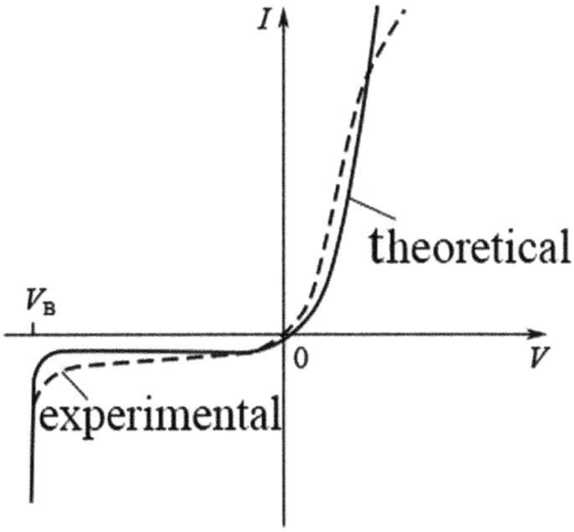

Fig. 2.12 Detector diode volt-ampere characteristic

$$\alpha = \frac{e}{nkT} \tag{2.20}$$

where,

I_s is the reverse current saturation value of the diode,

V is the voltage at both ends of the diode,

k is the Boltzmann constant,

T is the absolute temperature,

e is the electron charge,

n is the slope parameter, which depends on the manufacturing technics of the diode, generally between 1–2.

As can be seen from Fig. 2.12, the forward current of the diode increases rapidly after the inflection point, current can be very large, while the reverse current tends to be very small saturation value I_s, and when the reverse voltage reaches a certain value V_B, the reverse current suddenly increases, indicating that the diode has been broken down. This volt-ampere characteristic indicates that the semiconductor diode has variable resistance characteristics. Under forward voltage, its resistance is very small, allowing large current to flow, while under reverse voltage, its resistance is very large, so the reverse current is very small.

(2) Detection principle of detector diode

When a microwave signal is applied to a diode, the voltage generated on the tube is

$$V(t) = V_0 + V \sin \omega t \tag{2.21}$$

where, V_0 is the DC bias voltage of the diode, that is, the voltage of the operating point of the diode, V is the amplitude of the microwave signal. At this voltage, the detection current obtained after detection is

$$I(t) = f(V_0 + V \sin \omega t) \tag{2.22}$$

Expand the equation into a Taylor series at the V_0 point

$$\begin{aligned} I(t) &= f(V_0) + f'(V_0) V \sin \omega t + \frac{1}{2} f''(V_0)(V \sin \omega t)^2 + \cdots \\ &= \left[f(V_0) + \frac{1}{4} f''(V_0) V^2 + \cdots \right] + \left[f'(V_0) V \sin \omega t - \frac{1}{4} f''(V_0) V^2 \cos 2\omega t + \cdots \right] \end{aligned} \tag{2.23}$$

Under the small signal assumption, the higher-order terms above the quadratic term are omitted, and all high-frequency alternating currents are filtered out. Therefore, the detection current we obtain is

$$I(t) \approx f(V_0) + \frac{1}{4} f''(V_0) V^2 \tag{2.24}$$

where, $f\ (V_0)$ is the contribution of the DC bias voltage to the detection current, and the detection current generated by the microwave signal voltage is

$$I_0 \approx \frac{1}{4} f''(V_0) V^2 \propto V^2 \propto E^2 \tag{2.25}$$

It can be seen that the detection current is proportional to the square of the microwave signal electric field amplitude E. This means that the detector exhibits the characteristic of square-law detection under small signals.

The actual measurement of the volt-ampere characteristics of the detector diode shows that when the detection current is less than the microampere level, it conforms to the square-law detection, and when the detection current is greater than the microampere level, the square-law no longer holds. At this time, the relationship between the detection current and the microwave signal electric field amplitude can be expressed as

$$I_0 = kE^{\alpha} \quad (\alpha \le 2) \tag{2.26}$$

where, k is a constant, and the exponent a should be determined by experiment. Moreover, α is not a constant, and that it is different for different detector diodes and for different detection currents.

2. **Detector working mode**

(1) **Crystal detector power probe**

In order to connect the detector diode to the microwave system and achieve wideband matching applications as much as possible, it is often necessary to make a kind of special waveguide element-crystal detector, which is what we call the power probe. Figure 2.13 shows the structure of waveguide-type and coaxial-type crystal detector probes, respectively.

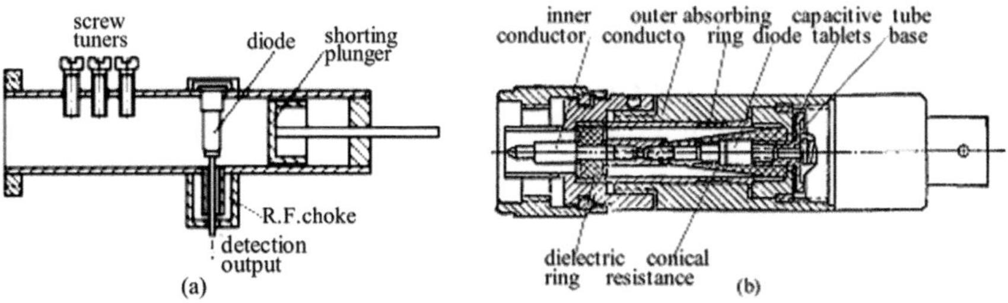

Fig. 2.13 The structure of crystal detector probes **a** waveguide-type, **b** coaxial-type

Figure 2.13a is a waveguide type crystal detector probe with an adjustable shorting plunger, and three screw tuners at the input end, which can be adjusted to achieve broadband matching between the detector and the high-frequency circuit, that is, to maximize the output detection current of the detector. The detection diode should be bridged at the center of the wide side of the rectangular waveguide so that it is exactly at the position of the maximum electric field of the TE_{10} wave. The output terminal of the detection current is equipped with a choke structure to prevent the leakage of microwave power, while also serving as a high-frequency bypass capacitor to prevent the high-frequency components of the detection current from entering the DC or low-frequency circuit, thereby affecting the correct indication of the indicator.

Figure 2.13b is a broadband coaxial crystal detector probe. In order to prevent the leakage of microwave, a ferrite absorption ring is added to the inner wall of the outer conductor of the coaxial line, and a conical absorption resistor is connected in parallel between the inner and outer conductors. The diode is connected in series on the inner conductor of the coaxial line, and a layer of dielectric film is sandwiched between the tube base for fixed diode and the outer conductor of the coaxial line to form a high-frequency bypass capacitor.

(2) **Detection way of detector**

When the input microwave of the crystal detector is a continuous wave with equal amplitude, due to the nonlinear characteristics of the detection diode, the positive half cycle of the microwave oscillation will have an output from the detector, while the negative half cycle will not have an output. In this way, the detection current is no longer a sinusoidal oscillation, and it can be decomposed into a DC component and various order high-frequency harmonic components. Due to the high-pass filtering effect of the high-frequency bypass capacitor of the detection device outlet, only the DC component l_0 is output from the detector output, which can be directly indicated by a DC instrument. This detection process is shown in Fig. 2.14a.

If the modulated microwave signal is input to the crystal detector, the detection process at this time is as shown in Fig. 2.14b. In addition to the DC and high-frequency components in the detection current, there is also a modulated wave, that is, the signal envelope waveform. What we need to obtain from the detector is the signal envelope rather than its DC component. Therefore, at this time, the DC component can be

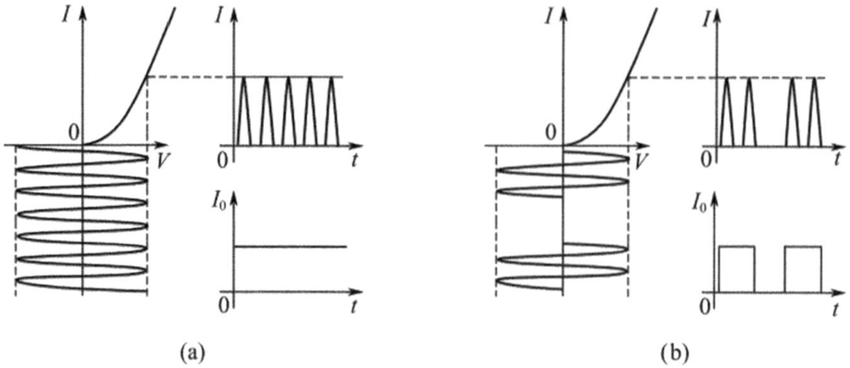

Fig. 2.14 The working way of the detector **a** detection of equal-amplitude continuous wave microwave signal, **b** detection of square wave pulse modulated microwave signal

removed by a DC-blocking capacitor, the high-frequency bypass capacitor filters out the high-frequency component, and the low-frequency modulated signal is output. After being amplified by a dedicated measurement amplifier—a frequency-selective amplifier, it is indicated by an oscilloscope, or directly input to the oscilloscope for display.

In Fig. 2.14a, b, the lower left part shows the microwave signal waveform imported to the detector. In (a), the lower left part of the figure shows a continuous wave microwave signal with constant amplitude, while in (b), the lower left part of the figure shows a square wave pulse-modulated microwave signal. The upper left part also gives the voltage-current characteristic curve of the detector. The upper right graph shows the signal waveform after detection, and the lower right graph shows the actual output signal waveform of the detector. It can be seen that for a continuous wave input with constant amplitude, the high-frequency components are filtered out by the high-frequency shunt capacitor, resulting in a DC waveform at the output (Fig. 2.14a lower right graph). For a pulse-modulated signal input, the high-frequency components are filtered out by the high-frequency shunt capacitor, and the DC component is blocked by a DC-stoppage capacitor, resulting in a pulse envelope waveform at the output (Fig. 2.14b lower right graph).

(3) **Indicator**

In general, for equal amplitude continuous wave input signals, the detector output signal can be displayed using a microammeter or galvanometer. Alternatively, it can also be indicated using an oscilloscope. The I_0 reaching microampere level can be indicated using a DC microammeter, while values I_0 less than a microampere can be indicated using a light spot galvanometer. When I_0 is converted to voltage, its magnitude can be displayed on an oscilloscope. For modulated wave input signals, the detector output signal can be amplified using a frequency-selective amplifier and then indicated using an oscilloscope. Whether it is a continuous wave signal or a pulse-modulated wave signal, it is more common to directly use an oscilloscope for detector output indication. This is because, with proper calibration of the oscilloscope, not only can the output signal amplitude be read, but the oscilloscope can also display the waveform of the output signal, including the DC signal waveform or the pulse envelope waveform, as well as the pulse microwave's pulse width, period, and repetition frequency.

3. **Detection small power meters**

(1) **Detection diode small power meter**

The detection efficiency of the crystal detector is high, with high sensitivity, short response time, and ease of use. However, when used as a power meter, it is difficult to calibrate and is not very stable, requiring frequent recalibration.

In the square law detection region at small signal levels, the detection output voltage of the diode is directly proportional to the input power, thus, allowing for direct power measurement. When the microwave power is above − 20 dBm, the output characteristics of the diode transition to the linear region, the square law relationship no longer holds. Modern crystal diode small power meters use extended diode power sensors to expand the square law detection region of the diode. The common method is to use calibration compensation, i.e., using a continuous wave power source to calibrate the output characteristics of the diode, thereby compensating for the deviation from the square law in the transition region (approximately – 20 to 0 dBm) and the linear region (above 0 dBm), so using such a sensor can accurately measure the continuous wave power of constant amplitude from − 70 to + 20 dBm, for noncontinuous wave and non-constant amplitude signal, the application of this sensor technology will affect the measurement accuracy to some extent.

It can be seen from this that when used as a power meter, the detection sensitivity of the detector diode should be calibrated, that is, the relationship between the known microwave power applied to the detector tube and the output voltage displayed on the oscilloscope should be calibrated. Since the detection sensitivity not only varies with the change in input power, but also changes with the change in microwave frequency, the standard signal source used for calibration should have the same frequency and modulation parameters as the output signal of the microwave source to be tested in the future. By changing the output power of the standard signal source and measuring the detector output voltage point by point within the estimated power range of the microwave signal to be tested in the future, the sensitivity curve can be obtained. This calibration must be performed again before each measurement and every time the detector tube is replaced.

Surface contact low barrier Schottky diodes have good square law characteristics and are more suitable for use in small-power meter.

In order to suppress the influence of higher-order harmonics, dual diode structures are currently used in diode power meters, which can effectively suppress even-order harmonics such as the second and higher harmonics. In order to achieve better linearity and a wider power measurement range, the latest diode power meters also use diode cascading, dynamic channel switching, and automatic correction technology.

(2) **Comparison of diode small power meters and thermocouple small power meters**

(a) The dynamic range of thermocouple power meters is only 50 dB, while diode power meters with calibration compensation generally have a dynamic range greater than 80 dB.

(b) Thermocouples respond slower to changes in input power, while diodes respond faster.

(c) Thermocouples can withstand high pulse power. The maximum continuous wave power they can withstand is 300 mW, and for pulses, it is 30 W per microsecond, while diodes can withstand lower power, with a maximum continuous pulse power of 200 mW. Once a test signal exceeds the peak power that the diode can withstand, the diode will be damaged.

2.4.2 Hall-Effect Power Meter

1. **Hall effect**

(1) **Working principle**

The Hall effect is a kind of electromagnetic effect, discovered by American physicist Edwin Hall (1855–1938) in 1879 while studying the electric conduction mechanism of metals. When a conductor carrying current is subjected to a magnetic field and the current is perpendicular to the external magnetic field, the charge carriers are deflected, creating an additional electric field in the direction perpendicular to both the current and the magnetic field, thereby creating a potential difference at both ends of the conductor. This phenomenon is the Hall effect, and the resulting electric potential difference is also known as the Hall electromotive force.

Materials with relatively high carrier mobility and the ability for electromagnetic fields to penetrate significantly at microwave frequencies, such as semiconductor materials like germanium, indium antimonide, etc., are commonly used as Hall pieces. These materials can measure microwave power across a wide frequency band.

As shown in Fig. 2.15, current Is flows through the N-type or P-type semiconductor Hall element, with the magnetic field B perpendicular to the current I_s direction, and the magnetic field direction is from the inside to the outside. For the N-type and P-type semiconductors, they produce Hall electric fields E_h in opposite directions as shown in Fig. 2.15a, b respectively. The direction of the Hall electric field can be determined using the left-hand rule. For an N-type semiconductor, when the magnetic field passes through the palm of the left hand and the four fingers point in the direction of the current, the thumb points in the direction of the Hall electric field. The direction of the Hall electric field for a P-type semiconductor is opposite to the direction of the thumb (or can be determined using the right hand). Based on this, the properties of the Hall element, whether N-type or P-type, can be determined.

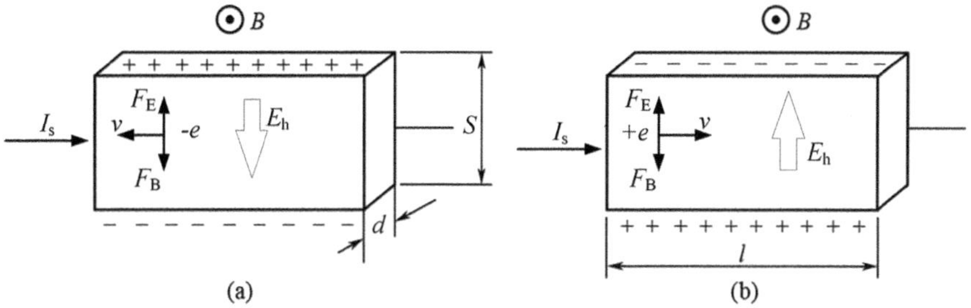

Fig. 2.15 The Hall effect of semiconductors **a** N-type semiconductors. **b** P-type semiconductors

On different types of semiconductors, adding a magnetic field perpendicular to the direction of the current will cause the electrons and holes in the semiconductor to experience Lorentz forces in different directions, causing them to accumulate in different directions. The accumulated electrons and holes will generate an electric field between them. When the electric field force (F_E) balances the Lorentz force (F_B), the electrons and holes will no longer accumulate. At this time, the electric field force will balance out the magnetic field's Lorentz force on the subsequent electrons and holes, allowing them to pass through smoothly without deviation. This phenomenon is the physical essence of the Hall effect.

(2) **Hall electromotive force** [14]

The Hall electromotive force E_h can be expressed as:

$$E_h = \frac{R_h I_s B}{d} \tag{2.27}$$

where,

R_h is the Hall constant.

B is the magnetic induction intensity.

I_s is the current through the Hall element.

d is the thickness of the Hall element in the direction of the magnetic field.

The relationship between the Hall electromotive force and microwave power is as follows: When measuring microwave power, the Hall element should be placed in such a way that the high-frequency electric field of the electromagnetic wave establishes a current in the Hall element, while the high-frequency magnetic field passes through the Hall element in a direction perpendicular to the electric field. For a Hall piece, this is generally perpendicular to the surface of the Hall piece.

When the force of the magnetic field component H of the electromagnetic field acting on the charge carriers in the Hall element is equal to the force of the Hall electric field acting on the charge carriers, equilibrium is reached. At this time

$$eE_h = evB = e\mu vH = e\mu uEH \tag{2.28}$$

and,

$$v = uq \tag{2.29}$$

where,

q is the carrier charge (Coulomb),

v is the velocity of carrier charge motion (m/s),

u is the carrier charge mobility ($m^2/(V \cdot s)$),

μ is the magnetic permeability of the medium (H/m),

E and H are the electric field and magnetic field components of the electromagnetic field (V/m, A/m).

Equation (2.28) can also be written as

$$\begin{aligned} eE_h &= e\mu up \\ E_h &= \mu up \end{aligned} \tag{2.30}$$

where, p is Poynting vector. The average value of Poynting vector over one period is

$$p = \frac{1}{2}E_m H_m \tag{2.31}$$

where, E_m, H_m is the amplitude of the field.

It can be seen that there is a completely deterministic relationship between the energy flow density at a given point in an electromagnetic field and the Hall potential on the semiconductor element placed at that point, regardless of whether the energy is being transmitted in free space or in a waveguide. The Hall electromotive force is directly proportional to the Poynting vector of the electromagnetic field, so it can be used to measure power. The product of $E_m H_m$ at any point on an ideal transmission line remains constant, so as long as the Hall element doesn't disrupt the original structure of the electromagnetic field, it won't affect standing waves on the transmission line, and the Hall element can be placed at any distance from the load. If the microwave power source outputs modulated power or pulse power, the Hall electromotive force will also be modulated electromotive force or pulse electromotive force, but in any case, the average value of the electromotive force over one period is directly proportional to the average power.

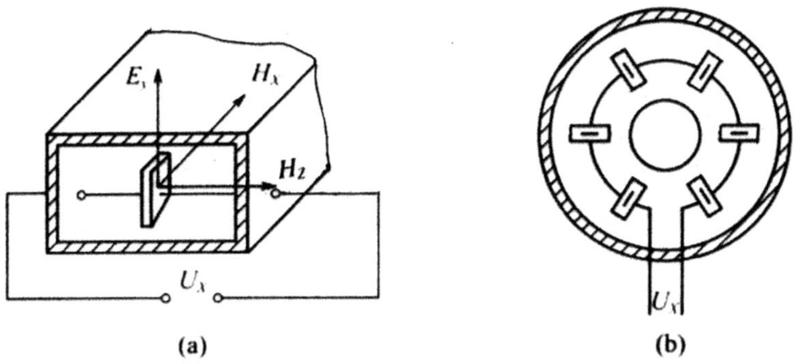

Fig. 2.16 Placement of the Hall piece in the power probe of Hall effect power meter **a** waveguide-type, **b** coaxial-type

2. **Hall effect power meter**

(1) **Power probe**

Figure 2.16 shows the schematic diagram of the placement of the Hall piece in the waveguide-type power probe and coaxial-type power probe of the Hall effect power meter.

When measuring, the electric field of the electromagnetic wave generates a current in the Hall element, and the magnetic field is perpendicular to the surface of the element. If the phase difference between the current and the magnetic field is equal to the phase difference on the load connected to the transmission line, the average value of the Hall potential is directly proportional to the active power transmitted on the transmission line.

(2) Characteristics of the Hall effect power meter

(a) When using a Hall effect power meter, the maximum allowable power transmitted along the waveguide depends on the maximum allowable operating temperature of the Hall element, the heat dissipation conditions, and the size of the element. For example, for the 3 cm band, when the Hall piece with dimensions of $(4 \times 0.15 \times 2)\ \text{mm}^3$ is placed at the center of a rectangular waveguide, it can withstand a maximum continuous wave microwave power of 12W at a working temperature of 100 °C, or a maximum pulsed power of 12 kW with a duty cycle of one thousandth.

Placing the Hall element near one narrow side of the waveguide can increase the upper limit of power measurement range of the power meter.

(b) The Hall electromotive force depends on the active power, so it is independent of the impedance of the transmission line as well as the position of the Hall element relative to the standing wave waveform, that is, it is not affected by the mismatch on the transmission line.

(c) The Hall effect has almost no inertia, so it can be used to measure continuous wave microwave power, continuous pulse power, and can be also used to measure single pulse power. The frequency range can reach several tens of gigahertz. This means that it is fully applicable for power measurement of high-power microwave nanosecond-level ultra-short pulses.

(d) Low sensitivity (1 ~ 10 μV/W), low accuracy, strong temperature dependence, and the necessity of performing temperature compensation are the disadvantages of the Hall effect power meter. The Hall electromotive force has a strong dependence on temperature because the carrier mobility in the semiconductor material that makes up the Hall element varies with temperature. Therefore, without a temperature compensation circuit or temperature-stable circuit, the Hall element cannot be applied, otherwise it will cause significant errors in power measurement.

2.4.3 Electromagnetic Wave Pressure Effect Power Meter

1. **Working principle** [18]

As early as 1873, Maxwell predicted that electromagnetic waves exert pressure on reflective surfaces. In 1900, Russian scientist Lebedev first experimentally confirmed the existence of electromagnetic wave pressure.

The power meter made from the mechanical action effect of the electromagnetic field on the sidewall of the transmission line or the reflecting element inside the transmission line is called an electromagnetic wave pressure effect power meter.

(1) **Electromagnetic wave pressure in free space**

(a) The line size of the force-receiving surface is much larger than the wavelength of the electromagnetic wave.

In this case, when the electromagnetic wave is irradiated on the force-receiving surface, the edge diffraction effect can be ignored, which we call the shortwave limit situation. At this time, the pressure generated by the electromagnetic field incident perpendicularly onto the ideal conductor plane is

$$F = \frac{1}{2}\mu_0 H_x^2 S_e \tag{2.32}$$

where, H_x is the instantaneous value of the transverse magnetic field intensity on the force-receiving surface, S_e is the effective force-receiving surface area.

Because

$$P_{in} = \eta_0 H_{xm}^2 S_e = p_{in} S_e \tag{2.33}$$

where, P_{in} is the incident microwave power; p_{in} is the power surface density of the incident electromagnetic wave; η_0 is the wave impedance of the space in front of the reflecting surface, and for free space, it is

$$\eta_0 = \sqrt{\frac{\mu_0}{\varepsilon_0}} \tag{2.34}$$

In the case of total reflection, the reflection coefficient $\Gamma = 1$, $H_x = 2H_{xm}$, where H_{xm} is the amplitude of the incident wave magnetic field. From Eq. (2.32),we obtain

$$F = 2\frac{P_{in}}{c} = 2\frac{p_{in}S_e}{c} \tag{2.35}$$

here, $c = 1/\sqrt{\mu_0\varepsilon_0}$ is the speed of light in free space.

If it's not an ideal reflection, then $\Gamma \neq 1$, in this case:

$$F = \frac{P_{in}(1 + |\Gamma|^2)}{c} = \frac{p_{in}S_e(1 + |\Gamma|^2)}{c} \tag{2.36}$$

When $P = 1$ W, the pressure F it generates is about 0.677×10^{-8} N in this case. To detect such a tiny force, must use a highly sensitive indicator, such as a torsion scales, a pressure sensor, etc.

If the force-receiving surface is a circular plane with a radius r, then in Eqs. (2.35) and (2.36),

$$S_e = \pi r^2 \tag{2.37}$$

(b) The line size of the force-receiving surface is smaller than or approximately equal to the wavelength of the electromagnetic wave.

In this case, the edge diffraction effect when the electromagnetic wave is irradiated on the force surface must be considered, and the effective radiation pressure Eqs. (2.35) or (2.36) should be corrected to

$$F_e = FK_d \tag{2.38}$$

where, F is the pressure when not considering the edge effect, K_d is the correction coefficient after considering the edge effect. For a circular fully reflecting plane with a radius r, K_d is

$$K_d = \frac{64r}{3\lambda} \tag{2.39}$$

λ is the wavelength of the incident wave.

(2) **The electromagnetic waves transmitted in the waveguide exert pressure on the reflection surface in the transmission direction**

When using a waveguide, as long as we know the power flow of the transmitted microwaves, we can accurately calculate the wave pressure on the reflecting surface placed transversely in the waveguide, because the diffraction problem of electromagnetic waves in free space can be ignored in the waveguide, and the size of the reflecting surface can be restricted only by the waveguide.

In the waveguide, the power flow of the microwave can be expressed as

$$P_{in} = \eta \int_S (H)^2 dS \tag{2.40}$$

where, η is the wave impedance of the known wave type in the waveguide.

Assuming the microwave source matches the waveguide, it can be considered that when the ideal reflecting plate shorts the waveguide, P_{in} will not change, and at this time, the transverse component of the magnetic field on the reflecting surface doubles, and the electric field component becomes zero. The pressure acting on the reflecting surface according to Eq. (2.35) is

$$F = 2\mu_0 \int_S (H)^2 dS \tag{2.41}$$

For TE waves in the waveguide, we have:

$$\begin{aligned} \eta &= \eta_0 \frac{\lambda_g}{\lambda} \\ F &= \frac{2P_{in}}{c} \frac{\lambda}{\lambda_g} \end{aligned} \tag{2.42}$$

And for TM waves, we have:

$$\begin{aligned} \eta &= \eta_0 \frac{\lambda}{\lambda_g} \\ F &= \frac{2P_{in}}{c} \frac{\lambda_g}{\lambda} \end{aligned} \tag{2.43}$$

where, λ is the wavelength of the microwave in free space, λ_g is the wavelength of the waveguide.

In the ideal situation described above, because the reflecting surface has short-circuited the entire waveguide, the S in Eqs. (2.40) and (2.41) should be the entire waveguide cross-section. Of course, this is an idealized scenario. In reality, when there is an ideal short circuit of the reflecting plate in the waveguide, total reflection occurs, making it impossible to match the microwave source.

(3) **The pressure of the electromagnetic waves transmitted in the waveguide on the waveguide sidewall**

According to the concept of partial waves in the waveguide, TE waves in the waveguide can be considered as a unique field structure formed by the TEM plane wave incident to the narrow side of the waveguide at a certain angle, and reflecting back and forth between the waveguide sidewalls (narrow side), due to wave interference. The angle at which the plane wave is incident on the sidewall is related to the frequency of wave and size of the waveguide. Therefore, it is clear that TE waves also exert pressure on the waveguide sidewall, and this wave pressure is also related to the frequency of wave and size of the waveguide. For example, as the frequency increases, the incident angle of the plane wave tends to $\pi/2$, and the electromagnetic wave will be incident along the longitudinal direction of the waveguide, reducing the wave pressure on the waveguide sidewall to zero.

It can be proven that the average pressure exerted on the unit surface area of the waveguide sidewall in one microwave period is given by

$$F = \frac{P_{in}}{c}\left(\frac{\lambda}{\lambda_c}\right)^2 \frac{\lambda_g}{\lambda} \frac{1}{ab} \tag{2.44}$$

where, λ_c is the cutoff wavelength.

This pressure is uniform and constant in the height direction of the sidewall. It is not difficult to see from Eq. (2.44) that when the incident power P_{in} is constant, the pressure on the waveguide sidewall is related to the microwave frequency. As the frequency decreases and approaches the cutoff frequency, the $(\lambda/\lambda_c)^2$ term approaches 1, and the (λ_g/λ) term increases, thus the pressure becomes greater. In other words, the lower the frequency, the smaller the incident angle of the plane wave, which means the plane wave is closer to perpendicular incidence on the narrow side of the waveguide, and therefore the pressure on the sidewall is greater. Conversely, as the frequency increases, the $(\lambda/\lambda_c)^2$ term becomes smaller and approaches zero, while the (λ_g/λ) term approaches 1, therefore the pressure F tends zero.

2. **Electromagnetic wave pressure effect power meter**
(1) **Pressure sensor power meter**

If a metal elastic thin plate is used instead of a portion of the sidewall (narrow side) of the waveguide, the elastic thin plate will bend under the action of wave pressure, and the power of the electromagnetic wave can be measured based on the degree of bending. Piezoelectric elements or capacitive elements can be used as pressure indicators. If the time constant of these pressure sensors is small enough, then this type of pressure effect power meter can also measure pulse power.

Soviet scientists have reported the results of measuring microwave power using the above method. They used a quartz piezoelectric sensor and metallized one surface of the quartz wafer, making it a part of the narrow side of the waveguide. In a waveguide containing such a piezoelectric quartz body, when the input pulse power is 300 kW and $f/f_c = 1.5$, the output voltage of the quartz piezoelectric sensor is 150 µV.

It should be noted that the time constant of the piezoelectric body wafer is very small, so it can measure pulse power with a width of less than 2 µs. A power meter using a quartz piezoelectric sensor can measure power lower limit for pulse power of 40–50 kW and for average power of 40–50 W. When using potassium sodium tartrate double crystal state as the sensor, the lower limit of power measurement can be extended to 1W or even a fraction of a watt. The upper limit of the measured power is actually limited by the breakdown strength of the waveguide itself.

(2) **Torsion balance power meter**
(a) Spatial torsion balance power meter

The pressure generated by microwave radiation is very small, on the order of 10^{-8}N, so a device capable of measuring tiny pressure must be selected. Considering the sensitivity, accuracy, and convenience of measurement, a torsion scales force measurement device is a good choice.

As shown in Fig. 2.17, one end of the suspension wire is fixed, and the other end is connected to a crossbar. One end of the crossbar is fixed with a thin metal circular slices reflector as the force-receiving surface. The circular slice should be placed in a vertical direction, parallel to the suspension wire. A balancing object is placed at the other end of the crossbar to balance with the reflecting surface, so that the crossbar can remain horizontal. A reflecting mirror is fixed at an appropriate position on the suspension wire, and the mirror surface is preferably perpendicular to the reflecting surface. A laser beam is vertically projected onto the mirror surface from the laser source. When the mirror surface is in a balanced state, the laser beam will be reflected back to the starting point. If the suspension wire twists, the mirror surface will follow the twist, causing the reflected laser beam to deviate from the starting point at a certain angle. This deviation can be read on the scaleplate.

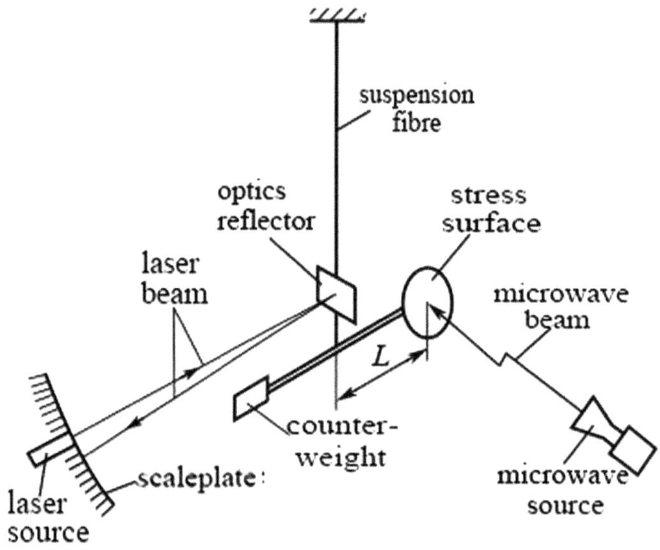

Fig. 2.17 Schematic diagram of the spatial torsion scale power meter

The microwave power generated by the microwave source is vertically directed towards the force-receiving surface through the radiation horn. Under the electromagnetic wave pressure, the force-receiving surface undergoes rotation around the suspension wire axis. When the torsional torque of the suspension wire is equal to the rotational torque of the force-receiving surface, the force-receiving surface will stop rotating. For the continuous waves, the force-receiving surface will maintain its position after rotating a certain angle, while for pulse waves, the force-receiving surface will return after rotating to a certain maximum angle as the pulse ends.

The equation of motion for the above torsion pendulum can be represented as follows:

$$N(t) = I\frac{\mathrm{d}^2\varphi}{\mathrm{d}t^2} + b\frac{\mathrm{d}\varphi}{\mathrm{d}t} + C\varphi \tag{2.45}$$

where,

φ is the torsional angle.

I is the rotational inertia of the system.

b is the damping coefficient of the system.

C is the torsional elastic modulus.

t is time.

$N(t)$ is the moment of force.

For continuous waves, whether $t = 0$ or $t > 0$, $N(t)$ is equal to N_0, and there will no longer be a situation where it is equal to 0.

For pulse waves, the expression of $N(t)$ can be written as:

$$N(t) = F(t)L = \begin{cases} N_0 & \mathrm{t} = 0 \\ 0 & \mathrm{t} > 0 \end{cases} \tag{2.46}$$

here, L is the force arm, i.e., the perpendicular distance from the center of the force-receiving surface to the suspension wire. For continuous waves, regardless of $t = 0$ or $t > 0$ in the above equation, they are all equal to N_0, and there will no longer be any cases where they are equal to 0. Thus, according to the Eq. (2.46), the maximum torsion angle for pulse response can be obtained

$$\varphi_m = \frac{N_0}{\sqrt{CI}} \tag{2.47}$$

For a suspension wire with a circular cross-section, we have

$$C = \frac{\pi d^4 G}{32l} \tag{2.48}$$

where, G is the shear elastic modulus of the material, d is the diameter of the suspension wire, and l is its length. Substituting Eq. (2.48) into Eq. (2.47), we get

$$\varphi_m = \frac{4N_0}{d^2}\sqrt{\frac{2l}{\pi GI}} \tag{2.49}$$

Based on the size of the system's C and l, after obtaining φ_m by measurement, N_0 can be calculated, and F can be obtained from Eq. (2.46), and then the power magnitude can be calculated using Eq. (2.35).

Because the electromagnetic wave pressure on the force-receiving surface is not applied to the center of mass of the system, along with the rotation of the force-receiving surface, there will also be swinging of the suspension wire acting like a pendulum. Although the swinging amplitude is much smaller than the rotation amplitude, this swinging in the vertical plane is still detrimental to the measurement accuracy. We place the reflecting mirror and the force-receiving surface vertically relative to each other, so that the forward and backward lateral swinging of the suspension wire does not affect the distance between the mirror surface and the scale, and thus does not affect the measurement of the rotation angle of the force-receiving surface, greatly reducing the decrease in measurement accuracy caused by lateral swinging.

Tsinghua University of China once used the above method to experimentally test the magnetron of a household microwave oven [18]. The output power of the magnetron was 700 W. The radiation main lobe axis of the microwave radiation horn was aligned with the torsion scale reflecting surface (i.e., the force-receiving surface). The distance between the torsion scale reflecting surface and the mouth of the microwave radiation horn was 0.8 m. A semiconductor laser beam was projected onto the reflecting mirror of the torsion scale, and the laser spot was reflected onto the scaleplate of the observation screen after reflection on the mirror surface. The distance between the torsion scale reflecting mirror and the observation screen was 2.8 m. The parameters of the torsion scale are as follows:

Suspension wire diameter d: 0.1 mm.
Suspension wire length l: 130 mm.
Reflection surface area, i.e., force-receiving area S_e: 600 mm^2
Lateral distance between force-receiving surface and suspension wire L: 25 mm.
Suspension system mass: 1.5 g.
Suspension system rotational inertia I: 1.5 g.cm^2
Torsion pendulum period T: 5 s.
Swing period: 0.7 s.
Maximum torsional angle 0.5° under the action of 10 W/cm^2, 0.1 s.

The torsional elastic modulus C is deduced from the torsion pendulum's torsional period.

$$\omega = \sqrt{\frac{C}{I}} = \frac{2\pi}{T}\ C = \frac{4\pi^2 I}{T^2} \tag{2.50}$$

The theoretically calculated data shows that the power density on the force-receiving surface is 7W/cm^2, the corresponding radiation pressure is $7 \times 0.667 \times 10^{-8}$N, and the maximum torsional angle is $\varphi_m = 0.03$ rad, the maximum displacement of the light spot should be $r_\text{m} = R\theta_\text{m} = 8.4$ cm. Experimental results show that in 4 trials, the maximum swing observed on the observation screen ranged from 10 to 12 cm, averaging at 10 ± 1 cm. The main sources of error between the experimental and calculated values are due to the experiment not being conducted in a microwave darkroom, where stray reflections are relatively large. Additionally, microwave diffraction errors are also a contributing factor. However, the consistency in order of magnitude between the experimental and calculated results proves the feasibility of this measurement method. As an estimate, at a pulse peak power of 1GW (peak power density of about 5 MW/m^2) and a pulse duration of 100 ns, the aforementioned torsion scale can achieve a maximum swing angle of $1°–2°$. Thus, the torsion scale system is fully capable of measuring the power of high-power narrow pulsed microwaves.

(b) Waveguide torsion scale power meter

If one wishes to use a torsion scale pressure power meter in a waveguide for power measurement, in practical measurements, it is impossible to achieve ideal reflection of the reflection element, i.e., to allow the reflection element to occupy the entire internal space of the waveguide. Doing so would cause the microwave power generator to operate under a short-circuit load, affecting the output of the microwave source and even damaging the microwave source. Therefore, the waveguide pressure effect power meter must consider allowing most of the microwave power to pass through the waveguide, with its end connected to a matching load or a real load.

The schematic diagram of the torsion scale power meter placed in the waveguide is shown in Fig. 2.18. Using an elastic quartz wire 3 (diameter 10 μm), the force-receiving thin plate 4 is suspended in a segment of rectangular waveguide 2. Electromagnetic waves propagating along the waveguide produce pressure on the thin plate, causing it to twist to a certain angle, which is indicated by the light beam reflected on the scaleplate. For a given thin plate, waveguide, and microwave wavelength, the torsional angle of the force-receiving thin plate is determined by the electromagnetic wave power and the torsional torque of the suspension wire. The inductive film 5 is used to compensate for the capacitive shunt conductance introduced by the thin plate [13].

The relationship between electromagnetic wave power and the quartz wire torsion angle can be expressed as follows,

$$P = \frac{M\Delta\varphi}{\xi} \tag{2.51}$$

where,

P represents electromagnetic wave power,

M is the unit torsional torque of the suspension wire,

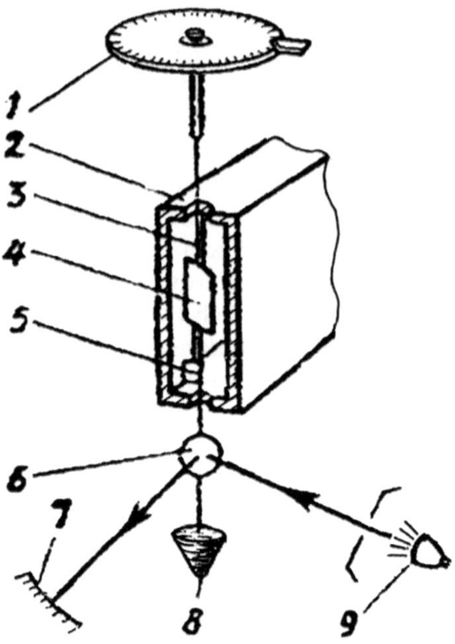

Fig. 2.18 Schematic diagram of structure of the torsion scale power meter in the waveguide. 1—calibration disc, 2—waveguide, 3—suspension wire (quartz wire), 4—force-receiving thin plate (reflection surface), 5—inductive film, 6—flat mirror, 7—scale, 8—shock absorber, 9—light source

$\Delta\phi$ is the torsional angle of the suspension wire,

$\xi$is the proportional relationship between the suspension wire torque and the electromagnetic wave power.

If M and ξ are ~~is~~ known, the power can be determined based on the magnitude of $\Delta\phi$. Generally, M and $\xi$are determined experimentally.

The force-receiving thin plate is a capacitive susceptance. So, in order to compensate for the inhomogeneity caused by the force-receiving thin plate on the waveguide, two symmetric inductive film pieces should be added on both sides of the suspension wire that fixes the force-receiving thin plate and the reflecting mirror (in the actual instrument, this section of the suspension wire may be replaced by a section of a rod). The force-receiving thin plate should be placed at a 45° angle to the waveguide cross-section, allowing it to bear the maximum torque.

The calibration of the instrument is to determine the dependence between power and torque (electrical calibration) and determining the unit torsional torque of the suspension wire (mechanical calibration). Once calibrated, the instrument can be used for absolute measurements.

(3) **Features of electromagnetic wave pressure effect power meter**

The electromagnetic wave pressure effect power meter has several advantages compared to calorimetric power meters and thermoelectric resistance power meters:

(a) It is possible to be an absolute measurement method, meaning that the measurement of electricity quantity (power) may be attributed to the measurements of mass, length, and time, and thus can be used as a standard instrument.
(b) It is possible to make a direct reading instrument for measuring power, with an accuracy of up to ± 2%. However, the accuracy of measuring small power (< 20 W) will be reduced to a certain extent.
(c) No power consumption is needed, which is particularly important when measuring high power.
(d) It has a strong overload capacity. The instrument will not damage when measuring the measured power far beyond the rated power range, and can measure the continuous wave power at the level of several hundred kilowatts or even higher.
(e) It can measure pulse power, and the pulse power can reach the GW level or above.
(f) The time constant of the force-receiving thin plate is small enough to measure pulse power with a pulse width of less than 2 μs, and can even measure a single pulse.
(g) The upper limit of the electromagnetic wave pressure effect power meter's power measurement is actually limited only by the breakdown strength of the waveguide, while the lower limit is limited by external vibrations, generally around 0.1 W. Improving the device's anti-vibration measures will extend the lower limit to 100–10 mW.

The shortcomings of the electromagnetic wave pressure effect power meter are: it is relatively cumbersome when measuring small power (< 1 W), and the transmission system must be carefully matched. Due to the high matching requirements for the power meter's transmission line segment, the processing accuracy of the mechanical structure must be strictly controlled. The pressure effect power meter is very sensitive to external vibrations, resulting in low reliability. Additionally, its measurement results are frequency-dependent, requiring the use of frequency calibration curves, which can be inconvenient. Although most of these drawbacks are not decisive and can be mitigated during design and manufacturing. Due to limitations of many practical factors, such as the high sensitivity of the measurement device to external interference, the difficulty in ensuring the reliability and repeatability of measurements, the challenges in device debugging, matching, and maintaining stability, as well as the high requirements for the testing environment, the application of the pressure effect power meter is greatly limited.

2.4.4 Other Electromagnetic-Effect Power Meters

The previously introduced Hall effect power meter and electromagnetic wave pressure effect power meter are both based on electromagnetic effects in the microwave field. There are many other microwave electromagnetic effects. Although various electromagnetic effects can in principle be used to measure microwave power, their reliability and stability are not satisfactory, so they are rarely used in practice. We will only briefly introduce their basic concepts without going into detail.

1. **Magnetoresistive sensor power measurement—magnetoresistive effect**

(1) **Principle of magnetoresistive effect**

The phenomenon in which the resistance value of certain metals or semiconductors that carry electric current changes with the size or direction of an applied magnetic field is called the magnetoresistive effect or Gauss effect. Its output signal will be proportional to the active component of the product of the electric field component and the magnetic field component of the electromagnetic field, similar to the Hall effect, so it is collectively referred to as the magneto-electric effect (Galvanomagnetic Effect). Like the Hall effect, the mechanism of the magnetoresistive effect is also caused by the deflection of charge carriers by the Lorentz force in the magnetic field, resulting in the accumulation of charges at both ends and the generation of a Hall electric field. If the Hall electric field force experienced by charge carriers at a certain velocity is exactly equal to the Lorentz force, then charge carriers slower than that velocity will deflect in the direction of the Hall electric field force, while those faster will deflect in the direction of the Lorentz force. This deflection causes the charge carriers to move in an arc trajectory, increasing the drift path and decreasing the mobility, resulting in an increase in resistance. In other words, the number of charge carriers moving in the direction of the applied electric field decreases, leading to an increase in resistance, which is the magnetoresistance effect.

As shown in Fig. 2.19, where B is the applied magnetic field, I is the current, and Hall electromotive force is generated at both ends of a and b, the resistance of the material changes with the change of B. When two points of a and b are short circuit, the magnetoresistance effect is more obvious.

The electromagnetic materials that can produce the magnetoresistive effect can be semiconductor materials such as germanium (Ge), PIN (P^+-I-N^+ type) semiconductors, indium stibide. and metal ferromagnetic films, such as nickel–cobalt

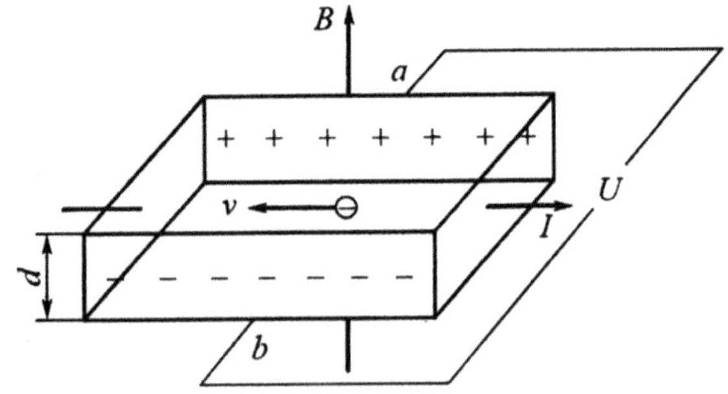

Fig. 2.19 Principle of magnetoresistive effect

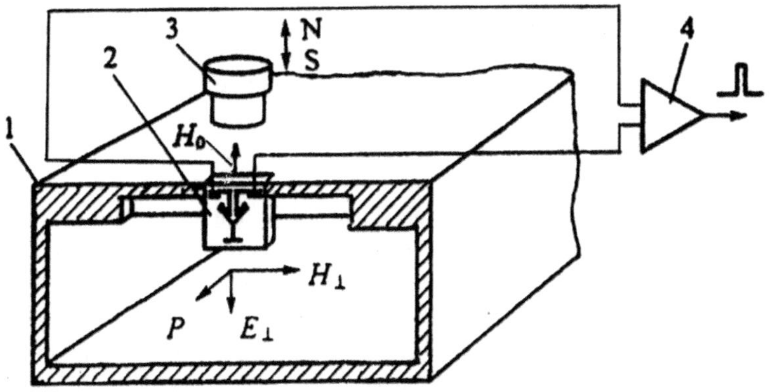

Fig. 2.20 Schematic diagram of power measurement using a magnetic resistance sensor 1—waveguide, 2—magnetoresistancesensor element, 3—magnetic field adjustment element, 4—pulse amplifier

alloy. The most typical indium stibide (lnSb) sensor is a low-cost, highly sensitive magnetoresistance device with very important practical value.

(2) **Using magnetoresistance sensor for power measurement**

Figure 2.20 shows a schematic diagram of an experimental setup for power measurement using a magnetic resistance sensor. The magnetic resistance sensor is made of a metal ferromagnetic thin film, which can not only reduce reflection but also have the ability to resist parasitic interference. For example, the ability to resist temperature effects can be improved by 2–3 orders of magnitude. In this setup, at a maximum pulse power of 42 kW and an average power of 50W, a linear dynamic range of 40 dB can be achieved. Its output signal is proportional to the instantaneous value of the active power of the input signal. The input signal pulse width can be 0.2–500 ps, and the S_{11} parameter can reach − 30 dB within a 2% bandwidth and − 20 dB within an 8% bandwidth [19].

Due to its high sensitivity and strong resistance to interference, magnetoresistance devices are widely used in industry, transportation, instrumentation, medical equipment, and mining, such as digital compasses, traffic vehicle detection, navigation systems, counterfeit money inspection, and position measurement.

Just like the Hall-effect power measurement, the output signal of magnetoresistance devices depends on the active power of electromagnetic wave, so it doesn't matter how well the load matches. The power capacity of the magnetoresistance thin film material is small, so it should be made into a through-type power meter.

2. **Kerr-effect power measurement—electro-optic effect**

Isotropic media such as glass, paraffin, water, and nitrobenzene, under the action of a strong electric field, their molecules are subjected to the force of the electric field and undergo orientation (deflection) and will present the anisotropic optical properties, resulting in the birefringence phenomenon, namely the refraction ability of light in the two different directions is different. Experiments show that under the action of an electric field, the difference between the two refractive indices is proportional to the square of the electric field intensity. This phenomenon is called the electro-optic Kerr effect, or simply the Kerr effect [20].

A glass box containing a certain liquid (such as nitrobenzene or nitrotoluene) is called a Kerr cell, and it contains a plate capacitor. When a high voltage is applied between the two parallel plates, the medium exhibits optical properties similar to uniaxial crystals due to the regular arrangement of molecules under the influence of the electric field, with the direction of optical axis corresponding to the direction of the electric field. When linearly polarized light passes through the medium perpendicular to the electric field direction, it decomposes into two beams of linearly polarized light, one vector follows the direction of the electric field, and the other vector is perpendicular to the electric field.

$$n_{\parallel} - n_{\perp} \propto E^2 \tag{2.52}$$

where, $n_{\parallel}$is the refractive index of light in the direction of the parallel electric field, $n_{\perp}$is the refractive index of light in the direction perpendicular to the field, E is the amplitude of the field.

As the electromagnetic waves propagating in the waveguide are the fundamental mode TE_{10} mode, it only has one electric field component E_y, so E^2 is directly proportional to the power flow P in the z direction. Thus, Eq. (2.52) can be written as

$$\phi(t) \propto P_z \tag{2.53}$$

where, $\phi(t)$ is the phase shift between the two polarized light beams of the parallel and perpendicular electric field directions.

By calibration, the size of the pulse power can be measured, and the pulse waveform can also be displayed. It can perform power measurements of hundreds of kilowatts or more.

When a liquid is subjected to an electric field, it produces polarization, which is the reason for birefringence. The polarization effect of the electric field is very rapid. The polarization process can be completed in less than 10^{-9} s after applying the electric field, and it returns to isotropy in the same short time after the electric field is removed. This rapid response property of the Kerr effect can be used to make almost inertial-less optical switches, which have important applications in high-speed photography, speed measurement of light, laser technology, and can also be used for measuring the power of nanosecond pulses.

3. **Pyroelectric detector power measurement—pyroelectric effect** [21]

Polarized crystal in piezoelectric material, such as triglycerol sulfate (TGS), barium titanate ($BaTiO_3$), and lithium tantalate ($LiTaO_3$), they spontaneous polarization intensity is a function of temperature (below the Curie temperature), and decreases as the temperature increases. This change causes a variation in the polarized charges on the outer surface of the crystal perpendicular to the spontaneous polarization intensity direction, thereby generating a current. As a result, an open-circuit voltage that changes with temperature appears across the two ends of the crystal. This type of polarized crystal is called a thermoelectric body, also commonly known as a pyroelectric body or pyroelectric element, and this phenomenon produced by the thermoelectric body is called the pyroelectric effect, which is a type of thermoelectric effect. If a resistor is connected across the two ends of the pyroelectric element, a current will flow through the resistor when the temperature of the pyroelectric body changes, and a voltage signal can also be obtained across the resistor.

At a constant temperature T, the spontaneous polarization of the thermoelectric body is neutralized by the charges inside the body and the surface adsorbed charges (suspended charges). If the pyroelectric material is made into a parallel thin slice perpendicular to the polarization direction, when a signal is irradiated onto the surface of the thin slice, the temperature of the thin slice changes ($T + \varDelta T$) due to radiation absorption, causing a change in the polarization intensity. The neutralization of the charges cannot keep up with this change due to the high resistivity of the material, resulting in a transient voltage appearing between the two surfaces of the thin slice. If an external resistor is connected between the two surfaces, the charge is released through the external circuit. The magnitude of the current is proportional to the pyroelectric coefficient and also to the rate of temperature change of the thin slice, which can be used to measure the strength of the incident radiation waves. The formation process of this pyroelectric effect can be represented by Fig. 2.21.

The biggest difference between pyroelectric materials and thermosensitive materials (such as thermocouples and thermistors) is that the voltage generated by the pyroelectric material is only related to the change in material temperature, regardless of the actual temperature of the material itself. In contrast, the characteristics of thermosensitive materials depend on the actual temperature of the material itself.

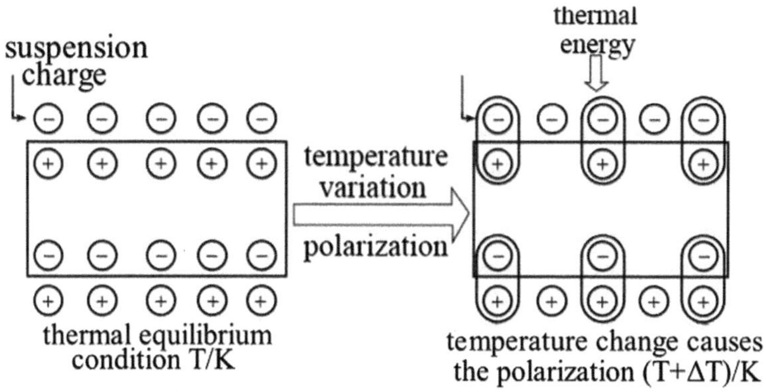

Fig. 2.21 Formation principle of pyroelectric effect

The latter uses the temperature value of the responsive component for measurement, and the response time depends on the establishment process of the new temperature equilibrium, which takes a relatively long time and cannot measure rapidly changing signals. On the other hand, the former utilizes the rate—of temperature change, making it capable of detecting rapidly changing signals. Therefore, it is a type of AC or pulse instantaneous response device and does not respond to stable radiation.

In theory, pyroelectric detectors can detect energy radiation of any frequency, but their primary applications are for infrared, soft X-rays, and laser measurements. Microwave power can also cause temperature changes in pyroelectric detectors, so there have been proposals in recent years using pyroelectric detectors to measure the pulsed microwave. Currently, pyroelectric detectors are widely used in radiation and non-contact temperature measurements, infrared spectroscopy measurements, laser parameter measurements, industrial automatic control, space technology, and infrared imaging. When used for CO_2 laser pulse measurements, a pyroelectric detector can detect pulses with a width of 10 ns and withstand a pulse power density of up to 2 MW/cm^2. This indicates that it should be entirely capable of measuring the power of high-power microwave.

Pyroelectric detectors have extremely high sensitivity, with a temperature response rate of 4–5 μA/°C. They can detect temperature changes of 10^{-6} °C, and their temperature resolution is less than 0.2 °C.

2.5 Power Measurement of Continuous Pulse Microwave

Continuous pulse microwave refers to periodically repeated pulse microwaves, which are the most commonly encountered microwave signals in our applications, also known as conventional pulse microwaves. It's different from single or low-repetition frequency pulse microwaves in high-power microwaves. Apart from the difference in pulse repetition frequency, the pulse width of regular pulse microwaves is much wider compared to high-power microwaves, generally in the range of microseconds, up to milliseconds, while the latter is mainly in the range of nanoseconds. The waveform of regular microwave pulses is basically rectangular or close to rectangular, while the pulse output waveform of high-power microwaves is often irregular. Currently, the output waveform of some high-power microwave sources is also close to rectangular. The former has stable pulse output and good repeatability, while the latter often has unstable output and poor pulse repeatability.

For an ideal rectangular pulse waveform in the continuous pulse state, its peak power is obviously the pulse power (see Eq. (1.6)), that is,

$$P_p = P_{pp} \tag{2.54}$$

Since the microwave pulse signals we usually encounter are often rectangular pulse waveforms, in many cases, pulse power is equated with peak power. This should be noted, that is, in many literature and materials, the peak power mentioned actually refers to pulse power, or it could imply both peak power and pulse power.

The most commonly used method for measuring regular pulse power is the pulse power meter, also often referred to as a peak power meter. Of course, there are many other methods for measuring pulse power, such as the integration and differentiation method, the sample comparison method, and the notch technology method, but relatively speaking, these methods are less commonly used now.

The measurement of peak power mentioned in this chapter generally refers to pulse power measurement. Of course, some methods can also measure peak power, but in general, peak power is often measured after detecting the pulse microwave through a previously calibrated detector and displaying the envelope waveform on an oscilloscope.

2.5.1 Average Power Method

1. **Basic principle**

For microwaves working in a continuous pulse state, because the time constant of the thermal effect power meter we introduced above for responding to microwave pulses is much larger than the microwave pulse width, its response cannot keep up with the change of the pulse envelope, so it cannot show the pulse power, but can only indicate the average power of repeated pulses at steady state. If one wants to calculate the pulse power using the average power measured by the thermal power meter, the simplest way is to use the pulse width and pulse period (or repetition frequency), that is, the reciprocal of the pulse duty cycle, to calculate the microwave pulse power through the average power, as given in Eq. (2.55).

$$P_p = \overline{P} Q \tag{2.55}$$

$$Q = \frac{T}{\tau} = \frac{1}{F\tau}$$

where, Q is the reciprocal of the pulse duty cycle. For non-ideal rectangular pulse waves, Eq. (2.2) has given the correction result using the waveform coefficient K to the above equation.

2. **Measurement method**

(1) **Absorption-type**

The absorption-type measurement system for pulse power is shown in Fig. 2.22. If a thermal-effect power meter is used as the terminal of the measurement system to measure the average power output of the pulse microwave source, the oscilloscope will display the envelope waveform of the pulse and the repetition frequency of the pulse. The pulse width can be read from this, and according to Eq. (2.55), the pulse

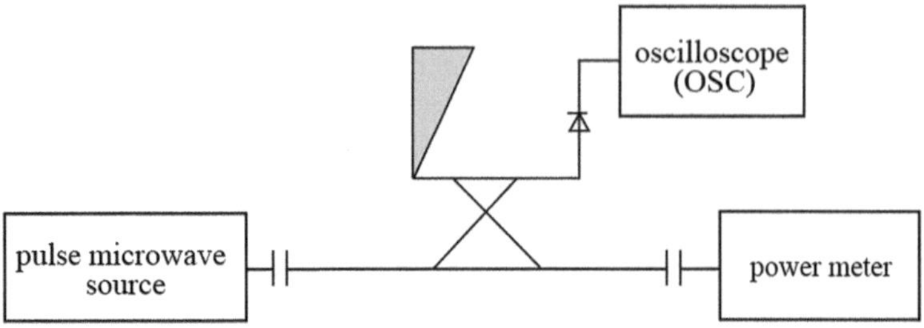

Fig. 2.22 System diagram for absorption-type measuring microwave pulse power using average power method

power can be calculated. If a peak power meter is used as the terminal, the pulse power can be directly measured.

(2) **Pass-through type**

When measuring the microwave pulse power using the pass-through type, the measurement system is generally as shown in Fig. 2.23. At this time, the power meter only measures a part of the power coupled out by the directional coupler, so after calculating the pulse power (when using a thermal power meter) or directly reading out the pulse power (when using a peak power meter), the output pulse power of the pulse microwave source should be converted according to the coupling coefficient of the directional coupler.

(3) **Characteristics**

The average power $\overline{P}$ can be measured with a conventional thermal-effect power meter, and then the pulse power can be determined based on τ, T (or F) and K. However, the measurement of τ and T also requires the assistance of a non-thermal type detector, such as a crystal detector, to display the microwave pulse envelope on the oscilloscope and measure τ and T, while K can only be estimated in general.

When using the pass-through type measurement, it is also necessary to calibrate the total attenuation of the power coupled to the power meter (or coupling coefficient A (dB)). When A is a positive value, the pulse power output of the pulse power source can be calculated according to Eq. (2.56).

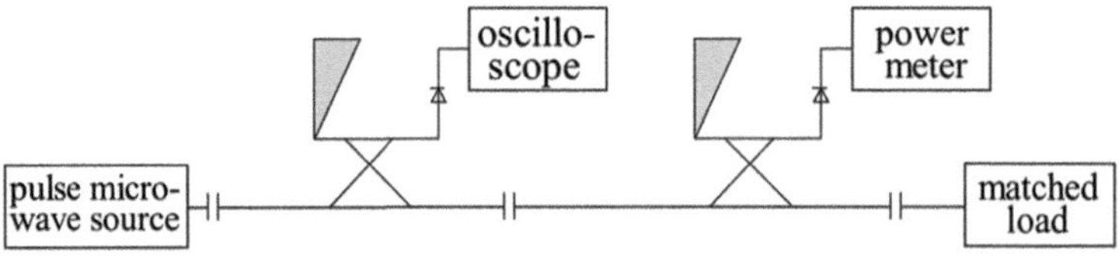

Fig. 2.23 System diagram for pass-through type measuring microwave pulse power using average power method

$$P_p = \overline{P}Q\ 10^{A/10} \tag{2.56}$$

However, when using Eq. (2.56), there are two problems:

(a) When the duty cycle is not constant, the pulse power cannot be calculated;
(b) When the pulse waveform is not an ideal rectangle, because there is no strictly defined pulse power, the calculated result is actually only the pulse power of an equivalent rectangular pulse.

2.5.2 Peak Detection Method

In order to overcome the drawback of slow thermal equilibrium time of the thermal-effect power meter, which can only display the average power of the microwave pulse over several periods, and to overcome the uncertainty caused by irregular pulse waveforms when calculating pulse power through the average power using the duty cycle, in most cases, the measurement of pulse power is more commonly carried out using a peak detection power meter method [22].

Calibrating the relationship between the indicated value of oscilloscope and the microwave power allows using the oscilloscope to measure the microwave pulse power, which is the most commonly used method for measuring pulse power. But the problem is that we require that the output of the detector should directly reflect the size of the pulse power. The essential point of the peak detection method is that the output voltage of the peak detector is proportional to the pulse power of the microwave signal applied to the detector. If this can be achieved, the detector can be considered as peak detector.

When the charging and discharging time constants of the detection circuit are the same, its detection output will be proportional to the average power, hence it is called the average detection, which is particularly suitable for the measurements of continuous wave. If the charging time constant of the detection circuit is very small, even a very narrow pulse can be quickly charged to a stable value. At the same time, it is required that the discharging time constant of the detection circuit is very large, so that after the pulse ends, the output voltage of the detection can remain at the peak value for a long time. This is known as peak detection.

1. **The principle of peak detection**

The peak detection circuit is generally formed by a voltage follower (composed of an operational amplifier), diode and capacitor. When the input signal is in the positive half-cycle, the diode conducts and charges the capacitor until it reaches the peak, i.e., the maximum value. When the input voltage is in the negative half-cycle, the diode is cut off. However, at this time, the capacitor does not discharge quickly and can maintain the maximum voltage (peak voltage) for a long time, thus detecting the signal peak. The foundational peak detection circuit is shown in Fig. 2.24.

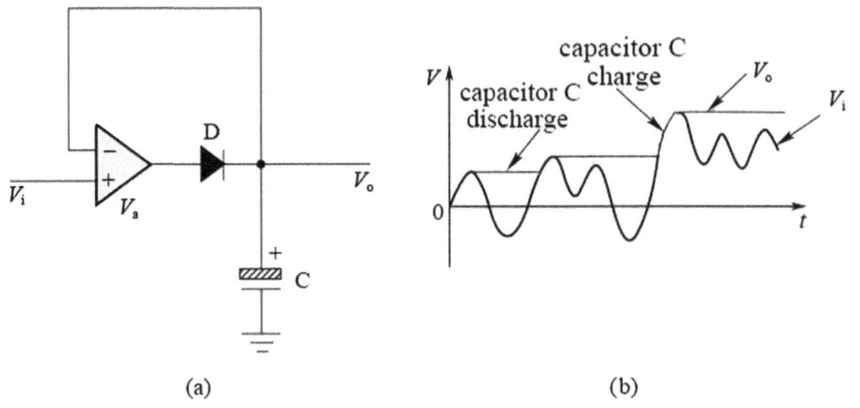

Fig. 2.24 Principle of peak detection **a** The principle circuit of simple peak detection, **b** the waveform of peak detection

In Fig. 2.24, Vi is the input signal voltage of the peak detector, V_o is the voltage on the capacitor.

The operational amplifier has two input ports, one is the anti-phase (reverse direction) input port, indicated by "–", and the other is the in-phase (same direction) input port, indicated by "+". Note that the "+" and "–" here do not refer to the positive and negative polarity of the voltage.

When $V_i > V_o$, the signal is input from the "+" terminal, and the output voltage V_a of the operational amplifier is a positive bias voltage. The diode D conducts, so the output current charges the capacitor C through D, continuously charging until it reaches the same voltage as V_i. Choosing the suitable capacitance value so that the charging speed of the charging and discharging circuit containing capacitor C is much faster than the discharging speed. The voltage across the capacitor can then be maintained at the maximum value of V_i, which is V_o, thus achieving peak detection.

When $V_i < V_o$, the operational amplifier output V_a is a reverse-biased voltage, and the diode *D* is cut off, which is equivalent to an open circuit. The capacitor does not charge or discharge, maintaining the maximum value of the input voltage, i.e., the peak value.

2. **Peak detection power meter**

(1) **Crystal detection diode power meter**

Using a detector for microwave power measurement, there are now many different models of commercial power meters available that can measure both continuous wave power and pulse average power and pulse peak power. In general, peak detection power measurement mainly refers to the measurement of pulse power, but as long as the response time of the detection diode is short enough, it can also measure the peak power of high-power microwave pulses at the nanosecond level. The application of high-performance microwave diode detection technology and multidimensional calibration compensation technology has made crystal detection diode microwave power meters have the characteristics of wide frequency coverage, high measurement accuracy, large dynamic range, and fast measurement speed, greatly facilitating microwave power measurement. However, these power meters are all small-power meters, and the power level that a single power probe can withstand is generally below 20dBm, so they are only suitable for measuring the output power of small-power signal sources. In most cases, especially for measuring the output power of electric vacuum microwave tubes, a through-type, i.e. coupling-type measurement should be used to ensure that the power level entering the power meter is less than the rated power capacity of the power meter, and then the actual power of the microwave source can be calculated. Figure 2.23 has already provided the measurement system diagram for this situation.

Chinese microwave power meters can accurately measure the average power and peak power of microwave signals with different probes. The maximum dynamic range of a single probe can reach − 67 to + 20 dBm, and the pulse power measurement range is − 40 to + 20 dBm. The minimum measurable pulse width is less than 30 ns.

(2) **Vacuum detection diode power meter**

Crystal diodes can withstand very small power capacities, while the power capacity of vacuum diodes is at least 4–5 orders of magnitude higher than that of crystal diodes, and in some cases, even much higher, so they can be directly used for detecting and measuring high-power microwave pulses. However, because vacuum diodes not only require high voltage on the anode but also need the filament to be preheated, the entire power supply for the vacuum diode is more complex, and the necessary filament preheating also extends the normal working time, so they are now less frequently applied.

2.5.3 Comparative Measurement Method [7]

As we have pointed out, when not using a pulse power meter, measuring the peak power of pulses using an oscilloscope or digital display, it is necessary to first calibrate the oscilloscope scale or indicator reading. This calibration is generally done using a continuous wave microwave signal source, and the calibration process will inevitably introduce errors. If, after obtaining the oscilloscope scale or indicator reading for measuring the microwave pulse power, a continuous wave microwave signal is directly used to replace the microwave pulse signal for measurement, aligning the readings on the oscilloscope scale to equal those on the indicator for both measurements, the output power reading of the continuous wave signal source then represents the magnitude of the microwave pulse power. This is the comparative measurement method for microwave pulse power, also often referred to as the substitution measurement method. Obviously, since there is no longer a need to calibrate the oscilloscope scale or display beforehand, the power indication on the continuous wave signal source represents the pulse power, thus improving the measurement accuracy.

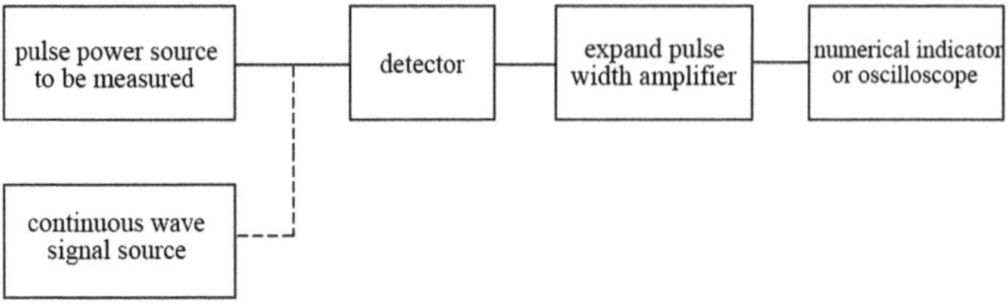

Fig. 2.25 System diagram for measuring microwave pulse power using pulse broadening method

1. **Pulse broadening method**

The peak detection method extends the discharge time of the crystal diode in the circuit, essentially acting to broaden the pulse width. The pulse broadening method further develops on the basis of the peak detection method, directly broadening the pulse envelope waveform obtained by the detector using a broadening amplifier into a direct current. The magnitude (amplitude) of the direct current voltage displayed on the voltage digital display or oscilloscope is directly proportional to the original pulse voltage magnitude (amplitude). By calibrating the voltage digital display or oscilloscope indication using a continuous wave signal, the original pulse power can be measured. The measurement system diagram is shown in Fig. 2.25.

During the measurement, the pulse power source is connected to the input port of the detector, and the microwave pulse signal is input into the detector. After the signal is detected, the pulse envelope waveform is obtained and then broadened into a direct current voltage by the broadening amplifier. The magnitude of this voltage is displayed on the digital display, or the voltage is displayed on the oscilloscope by the size of scale. Record the reading on the display or the scale on the oscilloscope. Then, replace the pulse power source with a continuous wave microwave signal source connected to the input port of the detector, and let it output a continuous wave signal. Its microwave frequency should be the same as that of the pulse power source. Adjust the output power of the continuous wave signal source so that the displayed value on the digital display or the indicated scale on the oscilloscope matches the value or scale recorded when measuring the pulse power. At this point, the continuous wave power output from the continuous wave microwave signal source is the measured pulse power.

Since the measured pulse power is determined based on the output power indication of the continuous wave signal source, the accuracy of the power indication of the continuous wave signal source directly affects the measurement accuracy of the pulse power, so the continuous wave signal source must have sufficient accuracy. Of course, errors indicated by the display or oscilloscope will also affect the measurement accuracy.

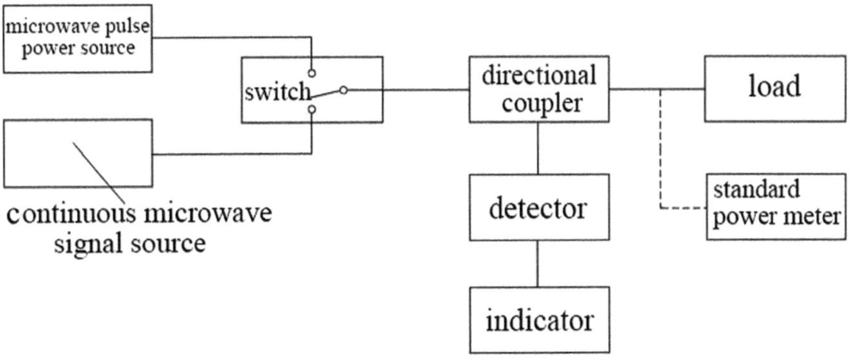

Fig. 2.26 System diagram for measuring high-power microwave pulse power by substitution method

2. **High-power pulse substitution method**

The pulse broadening method is mostly used for measuring small-power pulses. When the pulse power is relatively high, it is necessary to couple out a small portion of the power using a directional coupler for measurement. In this case, the block diagram of a measurement method is shown in Fig. 2.26.

The measurement proceeds as follows:

(a) Sample the output of the microwave pulse power source. The switch connects the microwave pulse power source, and a small portion of the output pulse signal is coupled into the detector through a directional coupler. After detection, the indicator displays its peak amplitude value, which can be the peak scale displayed on the oscilloscope, or the peak amplitude can also be converted through an AD converter (converting analog signal to digital signal) and displayed on the computer.

The vast majority of the output power from the pulse power source is absorbed by the load through the direct pass output port of the directional coupler.

(b) Replacing the pulse power source with a continuous wave signal source. At this point, the switch is connected to the continuous wave signal source, and its output signal size is adjusted to match the scale or amplitude displayed on the indicator with the peak value of the pulse power source output.
(c) Measurement. Keep the output of the continuous wave signal source unchanged, and connect a standard power meter to the direct pass output port of the directional coupler. Read the power size P_s indicated on the standard power meter. Since the output of the continuous wave signal source is determined based on the peak scale of the pulse power source, the power P_s is the size of the pulse power P_p.

When using this method to measure pulse power, attention should be paid to the following: the microwave frequency of the continuous wave signal source should be the same as that of the pulse power source, and its output power should be adjustable to the level of the pulse power source output, with stability better than the requirement of pulse power measurement accuracy. The isolation between the two paths of the microwave switch should be greater than 30 dB. The measurement error of the standard power meter will directly affect the final measurement accuracy of the pulse power, so it must have sufficient precision.

Using a standard power meter to measure power can omit the calibration of the directional coupler coupling coefficient and the calibration of the indicator scale, improving measurement accuracy. If power is measured directly without using a standard power meter, the power can also be obtained from a calibrated indicator, and then the output power of the pulse power source can be calculated using the coupling coefficient of the directional coupler. For sufficient measurement accuracy, the directivity of the directional coupler should be greater than 30 dB and the coupling coefficient should be at least 30 dB.

2.5.4 Notch Method [7]

The notch method is to modulate a continuous wave signal into a pulse signal whose repetition period is the same as the repetition period of the measured microwave pulse signal, but the pulse duration and the gap time between pulses are exactly opposite to the measured microwave pulse signal. In other words, the interpulse gaps between the pulses of the pulse signal modulated from the continuous wave signal match the duration of the measured pulse, and its pulse duration matches the gap between pulses of the measured pulse. This way, when the two pulses are added together, the measured pulse is just right for filling in the interpulse gap of the pulse modulated from the continuous wave signal, and the pulse modulated from the continuous wave signal is just right for filling in the interpulse gap of the measured pulse, effectively combining the two pulses into a continuous wave. By adjusting the output of the original continuous wave signal source to match the amplitude of the two pulses, the two pulses are combined into an equally-amplitude continuous wave. Since the amplitude of the measured pulse and the amplitude of the pulse modulated from the continuous wave signal are now completely equal, measuring the output power of the continuous wave signal source obviously represents the pulse power source's output pulse power.

The measurement system of the notch method is shown in Fig. 2.27, where the microwave frequency of the RF signal source and the continuous wave signal source should be the same, with a frequency error of less than 1×10^{-4} and the stability of the two outputs must be sufficiently high. The on/off ratio of the PIN modulator should be greater than 30 dB, and the directivity of all directional couplers should be greater than 30 dB, and the coupling coefficient of directional couplers 1 and 2 should ideally be strictly the same. The measurement error of the indicated power meter and the standard power meter should be better than $\pm 1\%$.

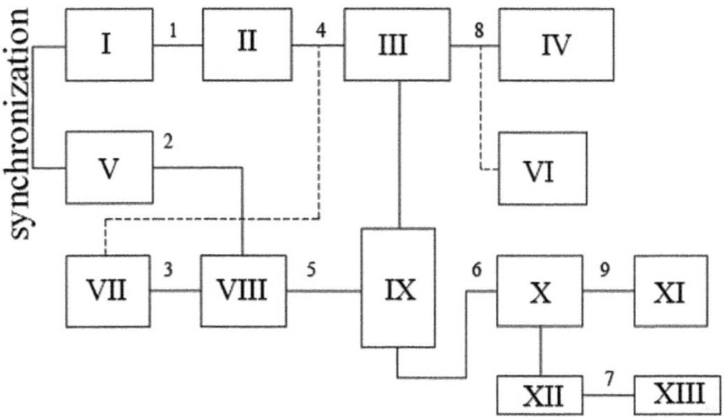

Fig. 2.27 Block diagram of the pulse power system measured by the notch method. I-pulse signal source 1, II-R. F. signal source, III-directional coupler1, IV-load (tested pulse power meter), V-pulse signal source 2, VI-standard power meter, VII-continuous wave signal source, VIII-PIN modulator, IX-adder (directional coupler2), X-directional coupler3. XI-Indicating power meter, XII-detector, XIII-oscilloscope

The measurement method of the notch method is as follows:

Pulse signal source 1 outputs a video pulse signal, the waveform of which is shown in Fig. 2.28 as waveform 1. This pulse is input into the RF (microwave) signal source and is modulated by the microwave signal, thereby outputting the desired microwave pulse signal, as shown in the signal of waveform 4 in Fig. 2.28. Of course, in reality, the output of the microwave signal source is generally already the desired pulse signal, i.e., the signal shown in waveform 4, so at this point, pulse signal source 1 is often redundant.

Pulse signal source 2 outputs a video pulse signal synchronized with pulse signal source 1whose pulse repetition period is the same as the measured microwave pulse signal (waveform 4), but the pulse duration and the gap time between pulses are exactly opposite to the output pulse of pulse signal source 1, as shown in waveform 2 of Fig. 2.28. This pulse controls the on/off of the PIN (P-I-N type semiconductor) diode modulator, thereby modulating the continuous wave signal source's output continuous microwave (waveform 3) to turn it into a pulse, resulting in a wide pulse signal as shown in waveform 5 of Fig. 2.28, which is the pulse signal used for comparison. It essentially removes the pulse width part of the measured microwave pulse signal (waveform 4) from the continuous wave.

The measured microwave pulse signal output from directional coupler 1 enters the adder, while the comparison microwave pulse signal output from the continuous wave signal source through the PIN modulator also enters the adder. The waveform obtained by adding the two signals together becomes the waveform shown in Fig. 2.28 as waveform 6, which is displayed by the oscilloscope after detection, as shown in waveform 7 of Fig. 2.28.

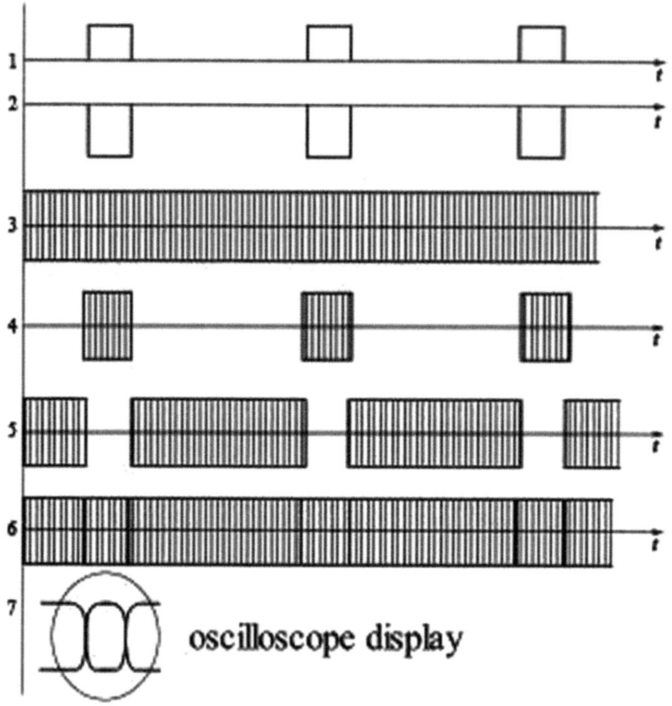

Fig. 2.28 The waveform of each port when the notch method measures the pulse power

The adder can be constructed using a directional coupler, where the two original output ports of the directional coupler (i.e., the direct pass port and the coupled output port) become input ports, inputting the measured microwave pulse signal and the comparison microwave pulse signal, respectively, and the original input port of the directional coupler becomes the output port, outputting the combined signal of the two signals above.

Carefully adjust the comparison pulse signal source so that its notch time is exactly the same as the width of the microwave pulse signal to be measured, and the microwave pulse to be measured falls exactly within the notch time. Then carefully adjust the output amplitude of the continuous wave signal source so that its amplitude is the same as the amplitude of the measured microwave pulse signal on the oscilloscope, i.e., ensuring that the two in the oscilloscope as far as possible highly consistent.

At this point, connect the continuous wave signal source to port 4, and the standard power meter to port 8. With the output amplitude of the continuous wave signal source fixed and unchanged after the previous adjustment, use the standard power meter to measure its power, and after appropriate calibration, will obtain the pulse power of the microwave pulse signal to be measured.

If the coupling coefficients of directional couplers 1 and 2 are strictly the same, their directivity is also very high. The microwave pulse signal to be measured is input from the input port of directional coupler 1 and output from the coupled port, entering the input port of directional coupler 2 and output from the direct pass port, and the comparison microwave pulse signal is directly input from the coupled port of directional coupler 2 and output from its direct pass port. According to the reciprocity principle of linear passive components, inputting from the input port of the directional coupler and outputting from the coupled port in the same direction is equivalent to inputting from the coupled port and outputting from the direct pass port in the same direction, meaning the coupling is the same. This indicates that under the same coupling coefficient, the amplitude reduction caused by the measured microwave pulse signal through directional coupler 1 and the comparison microwave pulse signal through directional coupler 2 is also the same, or in other words, the ratio of the coupled power to the input power is the same for both signals. Both signals are output from the direct pass port of directional coupler 2. When the measured microwave pulse signal passes through directional coupler 2, it can be considered that its power is not affected by directional coupler 2 because it is directly output from the direct pass port after entering from the input port, so we can ignore the power coupled to the secondary waveguide, therefore, we can consider that its power is not affected by the directional coupler 2. In this way, if the power coupled out by directional couplers 1 and 2 for the measured microwave pulse signal and the comparison microwave pulse signal is the same, and the amplitude displayed on the oscilloscope after output from directional coupler 2 is completely consistent, it means that their original input powers are also the same. In this case, we only need to connect the continuous wave microwave signal source to port 4 and connect the standard power meter to port 8. Use the standard power meter to measure the power of the unmodulated output of the continuous wave signal source, which has been adjusted and fixed during the measurement, this power represents the pulse power of the measured microwave pulse signal. At this point, the indicator power meter is no longer needed and can be replaced with a matching load.

If one wants to directly read the power of the high-power microwave pulse on the indicator power meter, considering that there may be some differences in the coupling coefficient of directional couplers 1 and 2, and the measurement accuracy of the indicator power meter is lower than that of the standard power meter, so the obtained measurement results should be calibrated to obtain a more accurate power level. The calibration method is: move the last step of the above measurement process to the first step. First, connect the continuous wave signal source to port 4 and connect the standard power meter to port 8. Output the continuous wave signal and read the power indication Ps on the standard power meter. Then, keep the output of the continuous wave signal source unchanged and return it to port 5, and read the power indication P1 on the indicator power meter. The ratio between the power obtained at the two ports of 8,9 is as follows:

$$L = \frac{P_s}{P_1} \tag{2.57}$$

The subsequent measurement steps are the same as those described above, until the step that the two pulse waveforms are completely highly consistent on the oscilloscope, because at this point, in order to make the amplitudes of the two pulses equal on the oscilloscope, we have adjusted the output of the continuous wave signal source, so the reading on the indicator power meter is no longer P1 but has become P_1'. Thus, the pulse power (i.e., peak power) Pp of the measured microwave pulse signal converted to port 8 should be,

$$P_p = L \times P_1' = \frac{P_S}{P_1} P_1' \tag{2.58}$$

To ensure that the modulation pulse amplitude of the continuous wave signal source is equal to the amplitude of the microwave pulse signal to be measured on the oscilloscope, truly achieving the high consistent, the amplitude of waveforms on the oscilloscope should be adjusted to be greater than 5 cm to better observe the amplitude levels of the two signals.

The measurement accuracy of the notch method can reach 3–5%.

Chapter 3
Radiation Field Measurement of High-Power Microwaves Power

3.1 Introduction

We have pointed out that the most important feature of high-power microwave is extremely high power and extremely short pulse, which determines that its power measurement will be different from conventional microwave power measurement. So far, there are few converters (power probes) that can directly withstand power of hundreds of megawatts and respond to nanosecond-level pulse microwave. Therefore, the current measurement of single (or low repetition frequency) pulse power of high-power microwave generally adopts radiation measurement or coupling measurement, and less directly adopts absorption measurement.

The extremely short pulse of high-power microwave determines that the power converter must have extremely high response speed, that is, it must be able to respond to nanosecond-level pulses. Therefore, among the many pulse power meters we introduced in the previous chapter, any power meters with a response speed of nanoseconds can be used in the power measurement of high-power microwave in the radiation field (radiation-type) or coupling field (pass-type), such as crystal detector power meters, Hall-effect power meters, etc. In addition, the electro-acoustic method, resistive sensor, etc. which we will introduce later can also respond to nanosecond pulses, of course, the most widely used at present is still the crystal detector power meter.

There are also a few power meters that can withstand high power, and they can directly measure high-power microwave without coupling or radiation methods, such as the electromagnetic wave pressure effect power meter we have introduced, and the resistive sensor we will introduce, but the power level of the resistive sensor cannot yet reach the upper limit of the power of high-power microwave.

According to the characteristics of high-power microwave, various practical high-power microwave power measurement methods have been proposed, and we will introduce these methods starting from this chapter.

Radiation field measurement is one of the main methods for measuring high-power microwave power. The basic idea is to use a radiation horn to radiate high-power

W. Wang et al., *High-Power Microwave Measurement*,
https://doi.org/10.1007/978-981-96-7763-4_3

microwave to space (microwave dark room), and use a receiving horn in the far field to receive a very small part of the radiation field power. After detection by the crystal detector serving as a converter, the detected signal is sent to an indicator (usually a storage oscilloscope) that has been calibrated in advance to indicate the received power, and then by using the total attenuation calibrated from the radiation horn to the oscilloscope, the total power of the entire radiation field can be calculated. Radiation-type measurement can actually be attributed to a special type of coupling measurement. Because the aperture size of the receiving horn is very small compared to the entire radiation space, the coupled power is also very small compared to the original radiation power, or in other words, its coupling coefficient can be very large (when the coupling coefficient is positive), so it is very suitable for high power microwave power measurement.

There are many specific measurement methods for radiation field measurement, and their basic principles are the same [23–25].

3.2 Radiation Field Measurement of Microwave Power

3.2.1 Overview

The basic principle of radiation field measurement of high-power microwave power is to radiate high-power microwaves to space (microwave dark room), receive a very small part of the radiation field power in the radiation far field, directly measure its power with a small power meter or indicate it with a calibrated oscilloscope after detection. Then calculate the total radiation power through conversion [26].

The far-field condition of the radiation field is

$$R \geq \frac{2D^2}{\lambda} \tag{3.1}$$

where,

R is the distance from the test point to the radiation port,

D can be taken as the sum of the maximum dimensions of the aperture of the radiating antenna and the receiving antenna (the aperture is defined as the area which is perpendicular to the direction of the incident wave, and effectively captures the incident wave energy),

λ is the wavelength of the radiation microwave.

The schematic diagram of the radiation field power measurement system is shown in Fig. 3.1. It consists of a radiation horn of high-power microwave, a receiving horn used as a receiving antenna, a detector, and an oscilloscope used to indicate the receiving power. The detector and oscilloscope can also be directly replaced with a small-power meter. The measurement should be carried out in a microwave dark

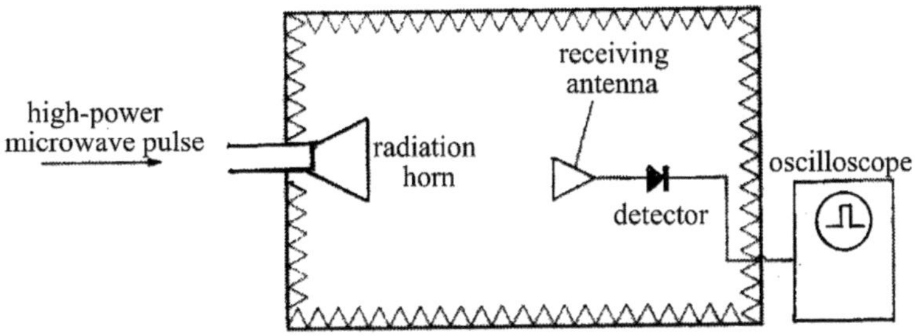

Fig. 3.1 Schematic diagram of the radiation field power measurement system

room to improve the measurement accuracy and reduce the influence of indoor wall reflections and external interference on the measurement.

All detectors used in high-power microwave measurements should be crystal detectors that can work in the microwave band and respond to extremely short pulses, and special attention should be paid to the matching problem when the detector is connected to the system. The oscilloscope used should be a storage oscilloscope, so that we can still observe the recorded waveform on the oscilloscope after the microwave pulse ends.

Not only that, all components and transmission lines of the entire receiving system should be matched as far as possible and prevent microwave leakage, especially when soft coaxial cables are used for connection, the quality of the cable joint will bring serious mismatch errors to the system, as well as microwave leakage, which leads to inaccurate measurements.

3.2.2 Microwave Radiation Field

We investigate the field components of microwave radiation far-field in the spherical coordinate system (R, θ, φ) as shown in Fig. 3.2.

1. **TE_{mn}, TM_{mn} mode radiation field**

The output waveguide of high-power microwave sources is generally overmoded circular waveguide, with only a few using overmoded rectangular waveguides. If the microwave energy radiates directly to free space through the opened terminal of the waveguide, low-power measurement shows that the terminal reflection of such overmoded waveguide can be ignored. Due to the polarization degeneracy in the circular waveguide, that is, the initial angle of the field is random, so when writing out the radiation field, the direction of the electric field or magnetic field must be specified first. If the relationship between the direction of the microwave field and the coordinate system is: the electric vector of the TE_{11} mode is along the x-axis at the center of the waveguide end face, that is, the electric field in the center of the radiation field is located on the *xoz* plane, then the *xoz* plane is called the E plane.

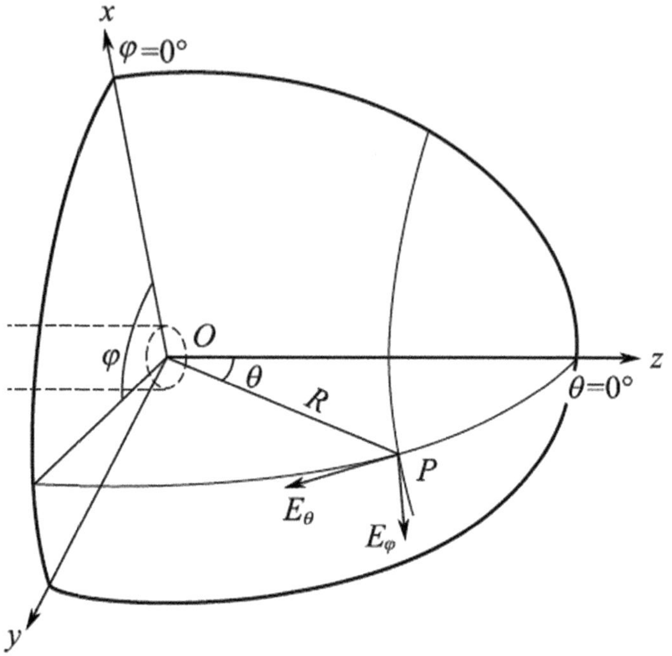

Fig. 3.2 The spherical coordinate system for power measurement of high-power microwave radiation field

At this time, the magnetic field in the center of the waveguide end face is located in the *yoz* plane, and the *yoz* plane is called the H plane.

(1) **Radiation field in the far field of TE$_{mn}$ mode**

Under the above specified relationship between the direction of the electric field and the magnetic field and the coordinates, the radiation field in the far field region of the TE$_{mn}$ mode can be expressed as

$$
\begin{aligned}
E_R &= 0\\
E_\theta &= \mathrm{j}^{m+1}\frac{m\omega\mu}{2R}F_{mn}(\theta)\cos\left[m\phi+(m-1)\frac{\pi}{2}\right]\mathrm{e}^{-\mathrm{j}kR}\\
E_\phi &= -\mathrm{j}^{m+1}\frac{ka\,\omega\mu}{2R}f_{mn}(\theta)\sin\left[m\phi+(m-1)\frac{\pi}{2}\right]\mathrm{e}^{-\mathrm{j}kR}
\end{aligned}
\tag{3.2}
$$

where,

$$
\begin{aligned}
F_{mn}(\theta) &= \left[1+\frac{\beta_{mn}}{k}\cos\theta+\Gamma(1-\frac{\beta_{mn}}{k}\cos\theta)\right]J_m(k_{mn}a)\frac{J_m(ka\sin\theta)}{\sin\theta}\\
f_{mn}(\theta) &= \left[\frac{\beta_{mn}}{k}+\cos\theta-\Gamma(\frac{\beta_{mn}}{k}-\cos\theta)\right]\frac{J_m(k_{mn}a)J_m'(ka\sin\theta)}{\left[1-(k\sin\theta/k_{mn})^2\right]}
\end{aligned}
\tag{3.3}
$$

where,

k is the wave number of free space,

a is the radius of the overmoded circular waveguide,

(R, θ, φ) is the coordinate of the inspection point,

$k_{mn} = \mu'_{mn}/a$ is the cutoff wave number,

μ'_{mn} is the nth zero point of the derivative of the first kind Bessel function with order m, i.e. $j'_m(\mu'_{mn}) = 0$, $\beta_{mn} = k\sqrt{1 - (k_{mn}/k)^2}$ is the phase constant.

Γ is the reflection coefficient of the electromagnetic wave at the waveguide terminal.

(2) **TM$_{mn}$ mode far-field radiation field**

The radiation field in the far field region of the TM$_{mn}$ mode can be expressed as

$$\begin{aligned} E_\phi &= 0 \\ E_R &= 0 \\ E_\theta &= -\mathrm{j}^{m+1}\frac{ka}{2R}Q_{mn}(\theta)\cos(m\phi)\mathrm{e}^{-\mathrm{j}kR} \end{aligned} \tag{3.4}$$

where,

$$Q_{mn}(\theta) = k_{mn}\left[\frac{\beta_{mn}}{k} + \cos\theta + \Gamma(\frac{\beta_{mn}}{k} - \cos\theta)\right]\frac{J_m(ka\sin\theta)J'_m(k_{mn}a)}{\sin\theta\left[1 - (k_{mn}/k\sin\theta)^2\right]} \tag{3.5}$$

here $k_{mn} = \mu_{mn}/a$, μ_{mn} is the nth zero point of the first kind Bessel function with order m.

When the reflection at the waveguide terminal is negligible, the terms containing Γ in the above equations can be omitted.

2. **TM$_{0n}$, TE$_{0n}$ mode radiation field and its characteristics**

It is not difficult to see from the above radiation field equations that if the output of the high-power microwave source is the TM$_{0n}$ mode or the TE$_{0n}$ mode, since $m = 0$, their radiation fields will be expressed as follows:

(1) **TM$_{0n}$, TE$_{0n}$ mode radiation field**

For the TM$_{0n}$ mode, $E_\varphi = 0$, $E_R = 0$, and E_θ in Eq. (3.4) is simplified to

$$E_\theta = -\mathrm{j}\frac{ka}{2R}Q_{0n}(\theta)\mathrm{e}^{-\mathrm{j}kR} \tag{3.6}$$

For the TE$_{0n}$ mode, $E_R = 0$, and because $m = 0$, E_θ is also zero, only the E_φ component in Eq. (3.2) exists.

$$E_\phi = \mathrm{j}\frac{ka\omega\mu}{2R}f_{0n}(\theta)\mathrm{e}^{-\mathrm{j}kR} \tag{3.7}$$

where $Q_{0n}(\theta), f_{0n}(\theta)$ can be obtained by substituting $m = 0$ into Eqs. (3.5) and (3.3).

(2) **Characteristics of TM_{0n}, TE_{0n} mode radiation field**

From Eqs. (3.6) and (3.7) we can immediately see that the radiation field of TM_{0n}, TE_{0n} modes are axisymmetric (z axis), i.e., the field distribution depends only on θ but not φ, in other words, the field changes only with θ and is uniformly distributed in φ direction. The vircator (virtual cathode oscillator) generally outputs the TM_{0n} mode, and others such as multi-wave Cerenkov generator, relativistic backward wave tube, magnetically insulated line oscillator, relativistic magnetron tube, etc. may also output the TM_{0n} mode, while the output of conventional gyrotron is mostly TE_{0n} mode. The rotationally symmetric distribution of their radiation field brings convenience to the power measurement of the radiation field, which will be discussed in the following multi-point measurement method.

3. **TE_{1n}, TM_{1n} mode radiation field**

When high-power microwave radiates through antenna, in order to obtain good radiation directivity, it is generally hoped to use the nearly linearly polarized HE_{11} hybrid mode, or the TE_{11} mode for radiation. For HE_{11} mode, its radiation field can be regarded as the superposition of the TE_{1n} field and the TM_{1n} field, and the TE_{11} mode is the TE_{1n} mode at $n = 1$, so we should also understand the radiation field distribution of the TE_{1n} mode and the TM_{1n} mode.

TE_{1n} mode ($E_R = 0$)

$$\begin{aligned} E_\theta &= -\frac{\omega\mu}{2R} F_{1n}(\theta)\cos\phi\, \mathrm{e}^{-\mathrm{j}kR} \\ E_\phi &= \frac{ka\omega\mu}{2R} f_{1n}(\theta)\sin\phi\, \mathrm{e}^{-\mathrm{j}kR} \end{aligned} \tag{3.8}$$

TM_{1n} mode ($E_\varphi = 0$, $E_R = 0$)

$$E_\theta = \frac{ka}{2R} Q_{1n}(\theta)\cos\phi\, \mathrm{e}^{-\mathrm{j}kR} \tag{3.9}$$

where $F_{1n}(\theta), f_{1n}(\theta)$, $Q_{1n}(\theta)$ can be obtained according to Eq. (3.3) and Eq. (3.5).

It can be seen that the radiation field of the TE_{11} mode and the TM_{11} mode is not only related to θ, but also related to φ, namely that it is no longer has the characteristic of axial symmetry.

3.3 Total Attenuation Measurement Method

3.3.1 Overview

1. **Experimental calibration method**

In order to obtain the total power of radiation, when using the radiation field power measurement, the measured small power must be converted, that is, the transmission attenuation of the receiving system must be considered, including the percentage of the power received by the receiving horn in the total power of the radiation, which is also called the coupling coefficient, and including various attenuation losses introduced on the receiving transmission line, from which the total power of the radiated microwave is inferred,

$$P = P_0 \times 10^{A/10} \tag{3.10}$$

where,

P is the total radiation power to be sought,

P_0 is the power received by the receiving horn measured by the small power meter or oscilloscope,

A is the total attenuation of the entire measurement system, expressed in positive decibels, and should include the coupling coefficient of the receiving horn, the losses caused by the entire transmission path of the receiving system, the attenuation losses introduced by any attenuation components in the path, and the sensitivity of the power indicator (including the oscilloscope).

The total attenuation A should be calibrated in advance. Theoretically, the most straightforward way of calibration is to determine it experimentally: make the continuous microwave or continuous pulse microwave whose power has been measured by other power meters (such as calorimetric power meters) radiate through the horn. The frequency and operating mode of this radiating wave should be identical to those of the high-power microwave source to be measured. The total attenuation A of the entire radiation, reception and transmission system is calibrated by measuring the power received by the receiving antenna, that is,

$$A = 10\lg\frac{P'}{P_0} \tag{3.11}$$

where P' is the total power of the microwave source used for calibration.

It should be noted that due to the spatial distribution of the radiation field, the power received by the receiving horn in different positions of the radiation field is different, so the calibrated value of A is also not the same. Thus, when formally measuring the radiation power of the high-power source, the position of the receiving horn must be same as the position of calibration, and cannot be changed.

Because the field distribution of the microwave radiation field is highly related to the mode of radiating wave, different modes have different radiation field distribution, therefore, during experimental calibration, it must be carried out under the same mode as the radiation field to be tested, otherwise, the calibration result will be meaningless. Thus, it can be seen that although the experimental calibration method is very intuitive and relatively accurate, but it requires a calibration source with known power and frequency and operating mode that are exactly the same as the microwave source to be tested, which becomes the biggest obstacle to the realization of this method.

Generally, it is difficult to find such calibration sources that can fully meet these requirements. For the high-power microwave source to be tested, its output mode is generally a high-order mode. A calibration source with such mode and known power, except for the fundamental mode and several simple modes, is almost impossible to find. Therefore, this experimental calibration method is currently rarely used.

2. **Network analyzer calibration method**

When experimental calibration is almost impossible to achieve, an alternative method can be used with a network analyzer. From the port of radiation horn to the oscilloscope (or small power meter) input port is regarded as a specific two-port network to be measured. The total attenuation of this network can be expressed as

$$A' = 10\lg\frac{1 - |S_{11}|^2}{|S_{21}|^2} \tag{3.12}$$

The obtained A' is not the same as the total attenuation A of the entire radiation system. This is because, when calibrating with the experimental method, the entire system from the radiation horn to the oscilloscope can be included, that is, in Eq. (3.11), P' is the total radiation power during calibration, and P_0 is already the final power measured by the oscilloscope, so A has already included total the attenuation of all components and transmission lines and radiation space in the measurement system. But when calibrating by a network analyzer, the input and output of the network analyzer are both microwave signals. The signal which is output from the crystal detector and fed into the oscilloscope is already the video signal after detection, so only from the port of radiation antenna to the input port of the crystal detector can be regarded as a specific system for network analyzer testing. Thus, A' does not include the attenuation caused by the detection sensitivity of the crystal detector itself, and the loss of the video cable from the detection signal input to the oscilloscope. However, A' will be consistent with the results of experimental calibration with a small power meter, because the input of the small power meter is also a microwave signal, which is equivalent to the input signal of the crystal detector.

The biggest uncertainty in network analyzer calibration is that the output mode of the network analyzer is either the TEM mode in the coaxial line or the fundamental mode TE_{10} mode in the rectangular waveguide. When it enters the waveguide in front of the radiation horn (note that the output waveguide of the high-power microwave source is all overmoded waveguide), the output transmission line of the network analyzer should be transitioned to the overmoded waveguide in front of the radiation horn, and at the same time, a special mode excitation device is needed to excite the same mode in the overmoded waveguide as the high-power microwave source to be measured. The transition of the transmission line will cause additional reflections, and the inconsistency of the mode will cause the microwave field received by the receiving horn to be actually different from the radiation field of the high-power microwave source to be measured. Therefore, strictly speaking, this calibration method is also very difficult, and it must use a special mode exciter or a complex mode converter to ensure that the excited mode in the waveguide in front of the radiation horn is the same as the output mode of the high-power microwave source to be measured. At the same time, the output waveguide of the network analyzer needs to be matched to the waveguide in front of the radiation horn, otherwise it is difficult to ensure that the radiation mode can be the same as the mode of the microwave source to be measured in terms of field distribution and relative amplitude. Obviously, the complex mode conversion process will bring additional errors to the calibration of A'.

It can be seen that although the network analyzer calibration method is simple to operate, it can generally only guarantee that the frequency of the calibration source is identical with that of the microwave source to be tested, and it is difficult to realize that the radiation modes of the two are exactly identical. Although the transition waveguide from the output waveguide of the network analyzer to the overmoded waveguide in front of the radiation horn is generally not difficult to obtain good matching performance, the obstacle is that only when the problem of getting a mode generator capable of producing the same output mode as the high-power microwave source is resolved simultaneously, can mode consistency be achieved. Although the contemporary mode excitation or conversion technology can produce the required mode, it is difficult to achieve the ideal requirements due to the performance of this mode exciter or converter, so overall this calibration method is not very convenient and accurate.

3.3.2 Calibration of the Measurement System

1. **Oscilloscope scale calibration**

When the total attenuation measurement method is used, in addition to the need to calibrate the total attenuation A (or A') of the entire system, the corresponding relationship between the voltage scale and the input power of the oscilloscope should also be calibrated. This is because, during the total attenuation calibration, we only get the detection voltage amplitude on the oscilloscope rather than the power value. The relationship between the oscilloscope scale and power, in other words, the relationship between the detection voltage amplitude and power must be calibrated to obtain the calibration curve. Figure 3.3 gives an example of calibration curve for a crystal detector, where the ordinate is the output voltage V_0 of the detector, expressed in mV, the horizontal coordinate is the input power of the detector, expressed in dBm. Using the calibration curve, we can obtain the input power of the detector according to the output voltage of the detector:

$$P_0 = 10^{(V_0/K)/10} \tag{3.13}$$

where P_0 is the input power of the detector (mW) and K is the sensitivity of the detector (mV/dB), which represents the output voltage magnitude measured in mV output by the oscilloscope when the unit power input measured in dB (generally in dBm) is applied. The calibration of K can be performed separately with a known

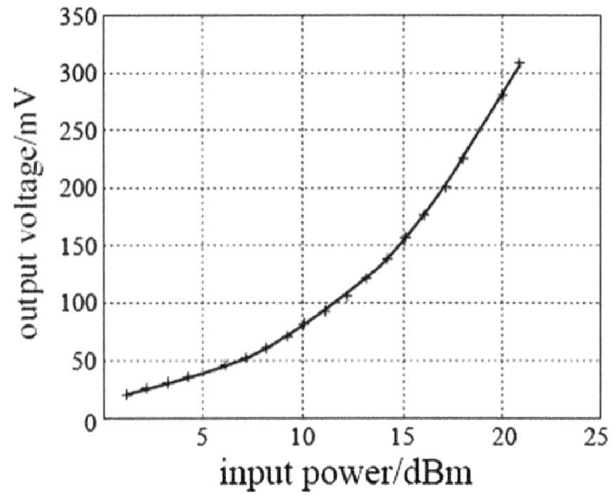

Fig. 3.3 Calibration curve of a certain crystal detector

low-power microwave source. It should be noted that when the microwave signal duty cycle is different, the efficiency of the detector will be different. Similarly, the pulse width also affects the amplitude of the output signal of the detector. Therefore, in addition to the operating frequency of this low-power microwave source being the same as that of the high-power microwave source to be tested, it is best to make the pulse operation way also be the same as the high-power microwave source. But the output mode of this low-power microwave source can be the TE_{10} mode of rectangular waveguide directly, since in the actual system of radiation field measurement, the transmission line after the receiving horn is already the fundamental mode rectangular waveguide. The received microwave signal has been converted to TE_{10} mode, so the low-power microwave source for calibration can output TE_{10} mode.

After the calibration of the detector, the expression (3.10) of the total radiation power of the high-power microwave source can be modified to:

$$P = P_0 \times 10^{-9+A/10} = 10^{-9+(A+V_0/K)/10} \ (\text{MW}) \quad (3.14)$$

where P is the total radiation power of the high-power microwave source radiated from the horn. Since 9 was subtracted from the exponent, the unit of P has been changed from mW to MW (megawatt), and P_0 has been converted to the detection output voltage V_0 via sensitivity (Eq. (3.13)).

2. **Calibration of the attenuation of the receiving system**

Since the output power of the high-power microwave source is very high, the total attenuation A is generally large when the total attenuation measurement method is used to measure the power, which brings great difficulty to directly calibrate the total attenuation A and is prone to large errors. For this reason, in actual measurement, the total attenuation A is generally divided into a number of attenuations for separate calibration and then added up, including attenuation of radiation, attenuation of attenuator (or attenuation of coupling component), attenuation of transmission cable, and sensitivity of the detector, etc. Among them, attenuation of radiation can be calibrated using the experimental calibration method or network analyzer calibration method we introduced above. We have also introduced the calibration of oscilloscope sensitivity above, and the attenuation of other component in receiving systems can be calibrated with the network analyzer, then all attenuations are added up to get the total attenuation A.

From another perspective, for a high-power microwave source with a very high pulse power, the power received by the receiving antenna in the radiation far field is still too large to be directly input to the detector for detection. In order to protect the crystal detector from being broken down by extremely high receiving power, some attenuation components are often added after the receiving horn, such as attenuators, directional couplers, coupling probe, etc. (generally attenuators are used). In addition, for the safety of the measurer, many test instruments such as oscilloscopes, displays of frequency measuring device, displays of low-power meter, etc. are placed in the shielding room, which can also reduce the influence of external interference signals. There will be a certain distance between the measurement site and the shielding room, which requires the use of cables to connect the receiving system and various display instruments, and the cable will also have attenuation. In this way, when calculating the total radiation power, the attenuation caused by these components must be considered separately, that is, these components must also be calibrated.

Among these attenuations, except that the calibration of radiation attenuation requires the radiation mode to be the same as that of the microwave source to be measured, the other attenuations are generated in the microwave system after the receiving horn receives the microwave power, and the microwave mode is already a coaxial line TEM mode or rectangular waveguide TE_{10} mode, so it can be directly calibrated with a network analyzer.

Assuming that the radiation attenuation of the measurement system (that is, the coupling coefficient between the radiation horn and the receiving horn) is L (dB), there is an attenuator with an attenuation of α_1 (dB) and a coupling component (directional coupler or coupling probe) with coupling coefficient of C (dB) between the receiving horn and the detector, and the attenuation of the video cable between the detector and the oscilloscope is α_2 (dB), then Eq. (3.14) should be written in detail as

$$P = 10^{-9+(L+C+\alpha_1+\alpha_2+V_0/K)/10}\ (\text{MW}) \qquad (3.15)$$

Chinese researchers used the total attenuation measurement method to test the output power of the Cerenkov oscillator—conical amplifier they developed. Since they have determined that the output mode of the Cerenkov oscillator—conical amplifier is TM_{01} mode, when calibrating the coupling coefficient from the radiation horn to the receiving horn, a calibration source with known power and TM_{01} output mode was used. The measured attenuation values expressed in decibels are $V_0/K = 12.7$ dB, $L = 44.93$ dB, $\alpha_1 + \alpha_2 = 55.98$ dB, $C = 0$, the power obtained according to Eq. (3.15) is 230 MW [27].

Radiation measurement method requires more calibration. $A = L + C + \alpha_1 + \alpha_2 + A_0/K$ is often very large, there will inevitably be some errors when calibrating them, and a small deviation of A will bring a large error in the calculation of P, which reduces the overall measurement reliability. In particular, there may be different opinions on the accuracy of the calibration methods, so the use of total attenuation measurement often causes scholars to question the results obtained. Even so, due to the simplicity of this method, it has been widely used in the early days of high-power microwave power measurement.

After the total radiation power is obtained, if the output power of the high-power microwave source needs to be calculated, the loss of waveguide transmission line from the output port of the microwave source to the radiation horn, including the transition waveguide and mode converter, should also be considered. Then, the output power of the microwave source can be derived from P.

3.4 Space Integral Measurement Method

If measurements are directly carried out in the radiation field of high-power microwave, and the total radiation power is then obtained through methods such as proportional amplificationor power summation of the annular radiation area, etc., the requirement of a special power source to calibrate each attenuation in Eq. (3.15) can be completely avoided. At the same time, it avoids the need for mode transformation, and there is no longer any error caused by the difference in radiation mode between the calibration and the actual measurement. Thus, the space-integral measurement method of radiation field has become the most widely used method in the power measurement of high power microwave. In this section, we will introduce its various specific methods [28–31].

3.4.1 Single-Point Measurement Method

For completely determined mode (single mode or multimode), the field distribution of the radiation far field is also completely determined in theory. So, when the amplitude coefficient is taken as 1, the spatial distribution of the normalized power of the far field can also be obtained. By measuring the actual power of the radiation field of the microwave source to be tested received by the receiving horn at a certain position (R_0, θ_0, φ_0) in the far field space, and comparing it with the theoretical normalized power at that position, we can obtain the ratio K of the actual power to theoretical power. Then by integrating the theoretical normalized power distribution of the radiation far field in the entire space and multiplying by the ratio K, we can get the total radiation power of the high-power microwave source.

We can think that the radiation field has a spherical distribution, and then the theoretical normalized total power of the radiation field on the sphere of $R = R_0$ is

$$\overline{P}_t = \frac{1}{2}\mathrm{Re}\int\limits_0^{\pi/2}\int\limits_0^{2\pi} R_0^2 \sin\theta\left[\boldsymbol{e}(R_0,\theta,\phi)\times\boldsymbol{h}^*(R_0,\theta,\phi)\right]\mathrm{d}\theta\,\mathrm{d}\phi \tag{3.16}$$

where ***e***, ***h*** are the normalized electric field and magnetic field of the radiation far field.

If the effective cross section of the receiving horn is S_{eff} (we will introduce the calibration of S_{eff} in Sect. 3.6), when the center of the receiving horn is at a certain position (R_0, θ_0, φ_0) in the radiation space, the measured value of received power of the receiving horn is $P(R_0,\theta_0,\varphi_0)$, and the normalized theoretical power $\overline{P}(R_0, \theta_0, \phi_0)$ of the same cross-section at that point should be

$$\overline{P}(R_0,\theta_0,\phi_0) = \frac{1}{2}\mathrm{Re}\iint\limits_{S_{eff}} R_0^2 \sin\theta\left[\boldsymbol{e}(R_0,\theta,\phi)\times\boldsymbol{h}^*(R_0,\theta,\phi)\right]\mathrm{d}\theta\,\mathrm{d}\phi \tag{3.17}$$

Then the ratio K of the actual measured power to the theoretical power should be

$$K = \frac{P(R_0,\theta_0,\phi_0)}{\overline{P}(R_0,\theta_0,\phi_0)} \tag{3.18}$$

We believe that the actual radiation field power of high-power microwave source is amplified by K times in the entire space relative to the normalized theoretical power, therefore, as long as the theoretical normalized power $\overline{P}_t$ (Eq. (3.16)) in the whole radiation space is also multiplied by K, the total power of the actual radiation field can be obtained

$$P_t = K\overline{P}_t \tag{3.19}$$

If K is expressed in decibels, then

$$K = 10\,\mathrm{lg}\frac{P(R_0,\theta_0,\phi_0)}{\overline{P}(R_0,\theta_0,\phi_0)} \tag{3.20}$$

At this time, the total power of the radiation field should be calculated according to Eq. (3.21)

$$P_t = \overline{P}_t \times 10^{K/10} \tag{3.21}$$

The basis of the single-point measurement method is that the actual distribution of the radiation far field is completely consistent with the theoretical distribution, therefore, the ratio K can be considered to be consistent at every point in the space. In fact, there will always be some differences between the actual distribution and the theoretical distribution of the radiation field, so this measurement method will have obvious errorsand is less promoted and applied.

3.4.2 Multi-point Measurement Method

For an axisymmetric radiation field, the distribution of the field in the direction of φ is the same, so we only need to take the distribution at any φ value to represent the entire field distribution, for example, we can take its longitudinal section at the position of $\varphi = 0$ (called the main plane). Figure 3.4 shows the main plane pattern (lobe pattern) of the ideal radiation field of the radar transmitting antenna in polar coordinates. The three-dimensional radiation pattern of the whole space can be obtained by rotating the pattern about 180° around the z axis.

In fact, in the analysis of Sect. 3.2, we already know that the common output modes of high-power microwave sources are TE_{0n} and TM_{0n}, their radiation fields also have the characteristics of rotationally symmetric distribution around the central axis of the radiation horn (but the specific shape of the lobe pattern will be different from the pattern shown in Fig. 3.4), that is, the radiation field is only a function of θ, and it is symmetric to φ. This characteristic is the basis of multi-point measurement in the radiation far field.

1. **Power superposition method of annular area**

In the radiation far field region of the high-power microwave sources to be tested, its wavefront has been basically consistent with the spherical surface, so we can draw a spherical surface at a certain radial position R_0 in the radiation far field to represent the radiating wavefront. With the intersection of the central axis of the radiation horn and the spherical surface as the vertex, the spherical surface is divided into several annular surfaces, and a receiving horn is set at the midpoint or boundary point of each ring (Fig. 3.5). Due to the symmetry of the field, it is considered that the radiation power density is equal everywhere on each annular surface. In this way, the power received by the receiving system that has been calibrated in advance can be used to calculate the power density according to the effective aperture of the receiving horn. The power density is multiplied by the annular area to obtain the radiation power on the entire annular area. The radiation power on all annular surfaces is superimposed to obtain the total radiation power [30, 31]. This is the basic process of the power superposition method of annular area. The calibration of the receiving system can

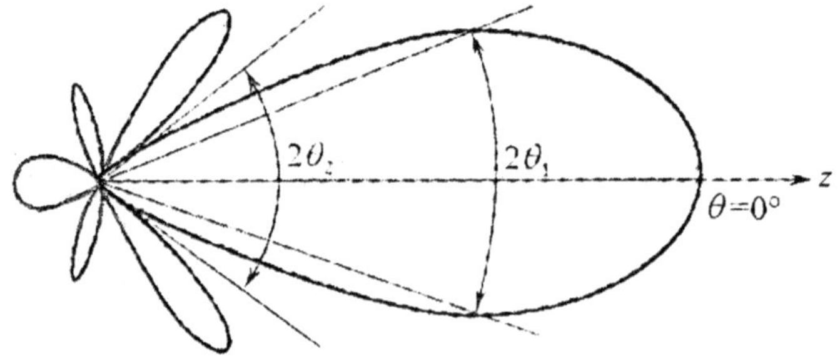

Fig. 3.4 Two-dimensional field lobe pattern in polar coordinates

also be performed by the method described in Sect. 3.2, using the power source whose output power has been measured. At this time, only C, α_1, α_2, and A_0 / K need to be calibrated in Eq. (3.15). Since the measurement of power density is directly carried out using the high-power microwave source to be measured, these quantities that need to be calibrated can be directly calibrated by using the network analyzer or using standard signal source, so the mode conversion is no longer required, and it is no longer necessary to calibrate L.

(1) Computational method 1 of power superposition of the spherical annular area

In this method, the receiving horn is placed at the midpoint of each annular area, and the power density of each annular area obtained according to the effective cross-section of the horn is directly used as the power density of the entire annular surface. Then, the power of the entire annular surface can be calculated, and thus the power of all annular surfaces is superimposed to obtain the total power of the entire radiation space. The measurement coordinate system is shown in Fig. 3.5.

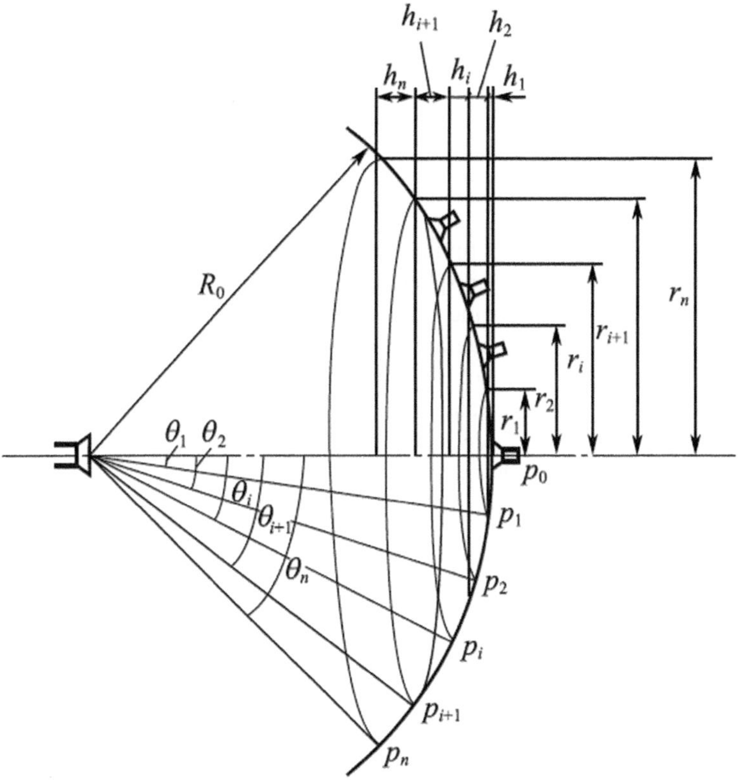

Fig. 3.5 Schematic diagram of the coordinate system for power superposition method of annular area

In the upper half plane of Fig. 3.5, assuming that the radiation power received by the receiving horn at the midpoint $\overline{r}_i = (r_i + r_{i+1})/2$ of the ith ring from r_i to r_{i+1} is p_i, then its power density is

$$\overline{p}_i = \frac{p_i}{S_{eff}} \tag{3.22}$$

where S_{eff} is the effective aperture of the receiving horn, also known as the effective cross section. We believe that the radiation power density on the entire annular surface is equal to $\overline{p}_i$, then the total radiation power on this annular surface is

$$P_i = 2\pi a_i h_i \overline{p}_i \tag{3.23}$$

where $2\pi a_i h_i$ is the area of the ith ring, h_i can be easily obtained according to Fig. 3.5, and

$$a_i^2 = r_{i+1}^2 + \left(\frac{r_{i+1}^2 - r_i^2 - h_{i+1}^2}{2h_{i+1}} \right)^2 \quad (i = 0,\ 1,\ 2,\ 3, \cdots, n.\ r_0 = 0 \text{ when } i = 0) \tag{3.24}$$

Then the total radiation power of the entire radiation space is

$$P = \sum_{i=0}^{n} P_i \tag{3.25}$$

where n is the number of rings for measuring the received power. Theoretically, the ring should cover the entire hemisphere surface, and r_n should be equal R_0. In fact, when i is comparatively large, the radiation field is relatively weak or even close to 0 (see Fig. 3.4), so a certain number of rings can be taken to obtain a sufficiently correct P value, and r_n does not need to equal to R_0. Under a certain value of r_n, the value of n depends on the value of $\Delta r = r_{i+1} - r_i$, the smaller Δr, the larger the number n of rings, and the more accurate the obtained p. But obviously, Δr cannot be smaller than the size of the receiving horn aperture.

(2) **Computational method 2 of power superposition of the spherical annular area**

If the receiving horn is placed on the boundary of each ring surface, the received power is respectively p_0, p_1, p_2, ...p_i, p_{i+1}, ...p_n (as shown in the lower half plane of Fig. 3.5), then the power density received by the receiving horn at the two end boundaries of the ith ring will be

$$\overline{p}_i = \frac{p_i}{S_{eff}} \tag{3.26}$$

$$\overline{p}_{i+1} = \frac{p_{i+1}}{S_{eff}} \tag{3.27}$$

Take the average power of the two points as the power density of the entire ring surface

$$\overline{p}_{i\to i+1} = (\overline{p}_i + \overline{p}_{i+1})/2 \tag{3.28}$$

And the area of the ring can be calculated like this

$$S = 2\pi R_0^2(\cos\theta_i - \cos\theta_{i+1}) \tag{3.29}$$

In this way, the radiation power of the entire ring surface is

$$P_{i\to i+1} = \pi R_0^2(\overline{p}_i + \overline{p}_{i+1})(\cos\theta_i - \cos\theta_{i+1}) \tag{3.30}$$

Of course, the area of the ring can also be calculated according to Eq. (3.23). Since we are measuring at the boundary of each ring, the actual number of rings is one less than the number of the measurement point. If the measurement point is n, the number of rings is $n - 1$. Add up the radiation power calculated on all ring surfaces to get the total radiation power.

$$P_t = \sum_{i=0}^{n-1} P_{i\to i+1} \tag{3.31}$$

Similar to method 1, n does not need to be very large, because when θ is large enough, the power density of the microwave is already close to 0.

In China this calculation method is adopted to measure the output power of the bifrequency magnetically insulated transmission line oscillator. The calculation results showed that the power at the two frequencies was 398 MW and 222 MW, respectively, and the total output power of the magnetically insulated transmission line oscillator was 620 MW [30].

(3) **Power superposition calculation method of planar annular area**

When R_0 is large enough, the spherical wavefront of the radiation field can also be approximately regarded as a plane wavefront, and the spherical annular surface is degraded into a plane annular surface, which brings greater convenience to measurement work and calculation, and the measurement system is also simplified, as shown in Fig. 3.6.

In this case, the calculation of $P_{\rm t}$ is simplified to

$$P_{\rm t} = \sum_{i=0}^{n} \pi\left(r_{i+1}^2 - r_i^2\right)\overline{p}_i \; (i = 0,\ 1,\ 2,\ 3, \cdots, n.\ r_0 = 0 \text{ when } i = 0) \tag{3.32}$$

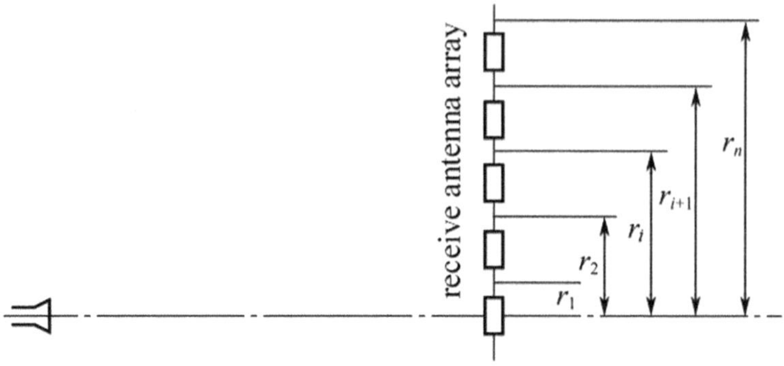

Fig. 3.6 Schematic diagram of the measurement system for the plane annular area power superposition method

where $\overline{p}_i$ is given by Eq. (3.22) or Eq. (3.28), and p_i in Eq. (3.22) or Eqs. (3.26) and (3.27) is corresponding to the measured value at the midpoint of the plane ring or the average value of the measurement at two boundary points, respectively. When measured at the boundary point, the number of plane rings is 1 less than the number of measuring points.

2. **Power normalization measurement method**

Generally speaking, the repeatability of the output pulse of a high-power microwave source is not very good. Even if the radiation mode of the high-power microwave source is stable, the power output of different triggers (called different shots) will be different, especially the pulse waveform is more difficult to maintain consistent. Therefore, when using the multi-point measurement method to measure power, we cannot use the same set of receiving devices, if every time the high-power microwave source fires a shot, the receiving system is moved in advance once, and placed at different ring surfaces where power needs to be measured, and all measurement points are measured one by one, it is obvious that because the radiation power of each shot itself is already different, so the measurement cannot get accurate results. Of course, it is even more impossible to move the receiving horn under the same shot and measure the power of each measuring point. If the receiving horn are is placed at all measuring points at the same time and connected to the receiving system and indicating instrument (generally a memory oscilloscope), all measurement points can get power measurement values at the same shot of the high-power microwave source, which is a reliable method. But this will greatly increase the cost of measurement and the cost and time of calibration of all receiving systems and indicating instruments. It is also difficult to achieve in practical work.

In view of this, the above-mentioned deficiencies can be avoided by normalizing the measured power density. For this purpose, only two measuring horns and the connected receiving system are needed. One of them is always fixed on the measured spherical surface which is the maximum radiation direction of the radiation field (if the radiation mode is unchanged, its maximum radiation direction will not change either). and the receiving horn placed at this position is used to actually measure the maximum radiated power under different shots and calculate its power density. Among them, the power density value of one shot that is considered to be the best representative of the output power level of the measured high-power microwave source is selected as the reference value. While measuring the maximum radiation power of each of the above different shots, another measurement system changes the position of the measuring point in the radiation field once for each shot, and simultaneously measures the radiation power density of different measuring points under different shots, Then, divide the power density value obtained from each measurement by the power density measured in the maximum radiation direction of the same shot to obtain the proportional coefficient K between them, and multiply K by the reference value, the power density of all measurement points can be obtained, which is the actual power density that best represents the level of the microwave source. Then, the power of each ring is calculated based on the area, where the measuring point is located, and the total radiation power is obtained either by power superposition of annular area method or radiation directivity coefficient method (see Sect. 4.3). In essence, this measurement method converts the measured values of different shots to the value of the shot that obtained the reference value, so the total power obtained also represents the total radiation power of the shot that got the reference value.

3.4.3 Radiation Directivity or Antenna Gain Measurement Method

1. **Determination of total radiation power by radiation directivity coefficient**

In order to quantitatively describe the directivity of antenna radiation, the ratio of the power density of the maximum radiation direction of an antenna at a point in the far field region to the power density of an isotropic antenna with the same radiation power at the same point is defined as the antenna directivity coefficient D. Different antennas are compared using an isotropic antenna as a standard, so the relative intensity of the maximum radiation of different antennas can be compared, i.e., the directivity coefficient can compare the strength of the directionality of radiation of different antennas. In general, the directivity coefficient defined in engineering is slightly different from this, which refers to the ratio of the radiation intensity in the maximum radiation direction to the average radiation intensity.

The physical significance of the directivity coefficient is that, with the same radiation power, the field strength of a directional antenna in the maximum direction is $\sqrt{D}$ times that of an omnidirectional antenna ($D = 1$). In other words, for the maximum radiation direction, this is equivalent to an increase in radiation power by a factor of D. The physical nature is that the antenna concentrates some of the power radiated in other directions in this direction. The narrower the main lobe, the stronger the intensity, and the greater the directionality coefficient of the antenna.

Using the directivity coefficient, under far-field measurement conditions, the total radiation power P of a high-power microwave source can be obtained by dividing the power density measured in the maximum radiation direction by the directivity coefficient to obtain the average power density in the entire spherical radiation field, and then multiplying it by the entire spherical surface area [26, 27, 30]:

$$P = \frac{4\pi R_0^2}{D_E}\frac{P_{0m}}{S_{eff}} \tag{3.33}$$

where,

D_{E} is the directivity coefficient of the microwave source in E-plane radiation direction,

P_{0m} is the power measured by the receiving horn in the maximum radiation direction of the microwave source,

S_{eff} is the effective cross-section of the receiving horn; R_0 is the distance from the receiving horn to the microwave radiation port.

In principle, since the radiation field can be theoretically determined, D_{E} can also be derived theoretically. However, due to the fact that the radiation mode of high-power microwave devices is generally not pure, the actual radiation field will have a certain degree of difference from the theoretical radiation field. Therefore, the spatial distribution of the actual radiation power density based on actual measurement should be used to calculate D_{E}.

The definition of the directivity coefficient tells us:

$$D_E = \frac{P_{0m}}{\overline{P}} \tag{3.34}$$

where $\overline{P}$ is the average radiation power over the entire spherical surface, which should be equal to the integral of the power distribution in the radiation field over the entire radiation spherical surface, and then the result is averaged over the surface area of the sphere:

$$\overline{P} = \frac{\int_0^{2\pi}\int_0^{\pi} R_0^2 P_0(\theta)\sin\theta\, d\theta\, d\phi}{4\pi R_0^2} = \frac{\int_0^{\pi} P_0(\theta)\sin\theta\, d\theta}{2} \tag{3.35}$$

From this, we can obtain:

$$D_E = \frac{2}{\int_0^{\pi}\left[P_0(\theta)/P_{0m}\right]\sin\theta\, d\theta} \tag{3.36}$$

where $P_0\ (\theta)$ is the power distribution in the θ direction measured by the receiving horn at the distance of R_0, i.e., the power distribution in space.

In principle, as long as P_{0m} is measured in the maximum radiation direction of the radiation field, and at the same time, the power received by the receiving horn on the

E plane of the microwave source radiation is measured in different θ directions, $P_0(\theta)$ in any θ value can be obtained by interpolation method when the measurement points are sufficient. Thus, D_E can be calculated according to Eq. (3.35). After calibration of S_{eff}, the total radiation power P can be obtained from Eq. (3.33).

2. **Determination of radiation total power by antenna gain**

The relationship between the gain G_t of the transmitting antenna and the directivity coefficient D is as follows:

$$G_t = \eta D \tag{3.37}$$

where η represents the antenna efficiency, which characterizes the ratio of the total power radiated by the antenna to the net power received by the antenna from the microwave source. Obviously, η is mainly determined by the loss of the antenna, including thermal loss, dielectric loss, and induction losses. If the loss can be ignored, or is not considered, then

$$G_t = D \tag{3.38}$$

In this way, Eq. (3.33) can also be expressed as

$$P = \frac{4\pi R_0^2}{G_t}\frac{P_{0m}}{S_{eff}} \tag{3.39}$$

If G_t is represented in decibels, the above equation should be amended to

$$P = \frac{4\pi R_0^2}{10^{G_t/10}}\frac{P_{0m}}{S_{eff}} \tag{3.40}$$

3.5 Numerical Integration Method and Fitting Method

3.5.1 Numerical Integration Method

Obviously, when the total radiation power is obtained by multi-point measurement, the more measurement points, the higher the accuracy of the measurement results, but the measurement points are always limited. A more accurate method than the annular area superposition method is to use numerical integration method based on finite measurement points, such as the variable step Simpson's rule to calculate the total power. The power value at each node required for numerical calculation can be obtained through interpolation using the obtained measurement values. As long as enough nodes are calculated, it is equivalent to increasing the number of divided annular areas, which significantly improves the accuracy of the results.

When using the annular area superposition method, due to the limitations of the receiving horn size and measurement workload, the number of annuli cannot be divided very much, that is, the value of n cannot be very large, which means that there cannot be many measurement points for measuring power. In this way, using the power density measured by receiving horn at the center position of the annular area to represent the power density of the entire annular area will inevitably introduce some error. Further improvement can be achieved by interpolating the obtained power measurement values to obtain power values at more locations, which is equivalent to increasing the measurement points. There are many mathematical interpolation methods available, such as the Lagrange interpolation method and the Newton interpolation method. The Newton interpolation method is suitable for cases of equidistant distribution of r, but for a spherical surface, if r is evenly distributed, as r_i increases with the increase of i, the width of the annular area on the spherical surface will rapidly increase (as h_i shown in Fig. 3.5), which is obviously not conducive to the accurate calculation of $\overline{p}$. However, the Lagrange interpolation method does not have this limitation and is suitable for unequal spacing distribution of r, so we choose the Lagrange interpolation method for convenience.

Assuming that we have measured the power p_i at n measurement points r_i ($i =$ 0,1, ..., n), the change of p_i with r can be represented by an nth-degree polynomial $p(r)$, and the polynomial can be expressed as

$$p(r) = \sum_{i=0}^{n} p_i \frac{\prod\limits_{\substack{k=0 \\ k \neq i}}^{n} (r - r_k)}{\prod\limits_{\substack{k=0 \\ k \neq i}}^{n} (r_i - r_k)} \tag{3.41}$$

Let's take k = 0,1,2,3 as an example, and write out the expression for $p(r)$ [32, 33]

$$\begin{aligned} p(r) = {} & p_0 \frac{(r - r_1)(r - r_2)(r - r_3)}{(r_0 - r_1)(r_0 - r_2)(r_0 - r_3)} + p_1 \frac{(r - r_0)(r - r_2)(r - r_3)}{(r_1 - r_0)(r_1 - r_2)(r_1 - r_3)} \\ & + p_2 \frac{(r - r_0)(r - r_1)(r - r_3)}{(r_2 - r_0)(r_2 - r_1)(r_2 - r_3)} + p_3 \frac{(r - r_0)(r - r_1)(r - r_2)}{(r_3 - r_0)(r_3 - r_1)(r_3 - r_2)} \end{aligned} \tag{3.42}$$

Using the same method, the expression of p(r) can be written-out for any value of n, in this way, with the known r_i and P_i, the P_i value at any other point r can be calculated. This means that we can obtain far more P_i values than the P_i values of original n, then, an annular area is divided for each p_i value. Since the number of P_i is increased, the annular area that can be divided also increases, in other words, each annulus can be divided into smaller parts. Since the new p_i values are calculated rather than directly measured, the further division of the annular is no longer limited by the size of the receiving horn. Using Eq. (3.22), the power density on each divided annulus can be calculated according to $\overline{p}_i$, and consequently, the corresponding radiation power P_i on each annulus can be obtained from Eq. (3.23). Then, the total power P can be calculated according to Eq. (3.25), and due to the increase in P_i points, the obtained P values are more accurate. It should be noted that when the power density $\overline{p}_i$ is calculated by Eq. (3.22), the effective cross-sectional area S_{eff} of the receiving horn is still used, but in this case, p_i is obtained by interpolation and not directly measured by the receiving horn. This is not contradictory because the original measurement data p_i used in the interpolation method are directly measured by the receiving horn. Therefore, the new p_i values obtained using them will represent the p_i obtained by the same receiving horn. Therefore, the power density should obviously still be calculated with the effective cross-section S_{eff} of the horn.

Furthermore, when Simpson numerical integration method is used to replace the superposition calculation of Eq. (3.25) [34], the accuracy of the total power obtained will be further improved. When the Simpson numerical integration method is used, the interval must be equally divided, i.e., it should be

$$r_0\,,\ r_0+h\,,\ r_0+2h\,,\ r_0+3h\,,\ \cdots,\ r_0+nh\,,$$

where h is the step length. First, we use the Lagrange polynomial to obtain $p(r)$, and then use Eq. (3.30) to calculate the power P_i on each annulus area, and express them as

$$P(r_0)=y_0,\quad P(r_0+h)=y_1,\quad P(r_0+2h)=y_2,\ldots P(r_0+nh)=y_n$$

It should be noted that these values are different from the P_i values given in computational method 1 of power superposition of the spherical annular area. y is the power calculated at the boundary point or the power calculated by Eq. (3.30) after measuring the power density, which is the same as the measurement method in computational method 2 of power superposition of the spherical annular area.

If the number of intervals is n, there will be a total of $(n + 1)$ boundary points, and there will be $(n + 1)$ y values (including y_0), the Simpson numerical integration yields

$$P=\int_{r_0}^{r_n}P(r)\mathrm{d}r=\frac{h}{3}(y_0+4y_1+2y_2+4y_3+2y_4+\cdots+4y_{n-1}+y_n)+R \tag{3.43}$$

where R is the error term, which can be expressed as

$$R<\frac{n}{180}h^5M^{(4)} \tag{3.44}$$

where $M^{(4)}$ is the upper limit of the absolute value of the fourth-order derivative of $P(r)$ in the interval (r_0, r_n), generally, R is so small that it can be neglected

3.5.2 Fitting Method for Radiation Pattern

1. **Theoretical basis**

For circular horn antennas, the various multipoint measurement methods described above are quite reliable for measuring the radiation field of rotationally symmetric modes such as TE_{0n} mode or TM_{0n} mode. The total power calculated in this way is theoretically more accurate. However, with the development of high-power microwave technology and the successful development of various mode converters and mode-transformed antennas, many high-power microwave systems have been able to radiate TE_{11} mode or HE_{11} mode with higher directivity through circular horn antennas. In this case, their radiation fields no longer have strict rotational symmetry, and therefore, the total radiation power cannot be simply obtained by annular integration of the power density measured on the main plane. Reference [34] proposes a method to obtain the total radiation power by measuring the radiation power density distribution in two main planes.

For circular horn antennas, no matter its input mode is TE_{11} mode or HE_{11} mode, the radiation field of the horn can be considered as the superposition of TE_{1n} and TM_{1n} fields. If the main polarization direction of the TE_{11} mode is also defined in the direction of the x-axis ($\varphi = 0°$) direction, then according to Eqs. (3.8) and (3.9), the radiation field on the E-plane ($\varphi = 0°$) is

$$E_\theta = \left[-\frac{\omega\mu}{2R} F_{1n}(\theta) + \frac{ka}{2R} Q_{1n}(\theta) \right] \mathrm{e}^{-\mathrm{j}kR} \tag{3.45}$$

And the radiation field on the H-plane ($\varphi = 90°$) is

$$E_\phi = \frac{ka\omega\mu}{2R} f_{1n}(\theta) \mathrm{e}^{-\mathrm{j}kR} \tag{3.46}$$

If the total power of the radiation field is considered to be the average of the spherical crown integral of the radiation power on the E-plane and the H-plane, and the amplitude coefficient is ignored, the normalized total radiation power on the sphere crown cap in the range of θ value from 0 to θ_n (Fig. 3.5) at $R = R_0$ can be expressed as

$$\begin{aligned} P_t &= \frac{1}{2\eta} \int_0^{2\pi} \int_0^{\theta_n} |\boldsymbol{E}(R_0, \theta, \phi)|^2 R_0^2 \sin\theta \, \mathrm{d}\phi \, \mathrm{d}\theta \\ &= \pi R_0^2 \left[\int_0^{\theta_n} \overline{p}_e(\theta) \sin\theta \, \mathrm{d}\theta + \int_0^{\theta_n} \overline{p}_h(\theta) \sin\theta \, \mathrm{d}\theta \right] \end{aligned} \tag{3.47}$$

where the electric field amplitude coefficient is taken as 1 (Normalized with respect to amplitude), $\overline{p}_e(\theta)$ and $\overline{p}_h(\theta)$ are the power density distributions on the E-plane ($\varphi = 0°$) and H-plane ($\varphi = 90°$) on the spherical surface of $r = R_0$, respectively. The first and second terms in the square brackets in Eq. (3.47) represent the integral results of the radiation power density of the E-plane and H-plane on the spherical crown in the range of θ from 0 to θ_n at $r = R_0$.When power measurement is carried out, the total radiation power of the antenna can be obtained by measuring their power density distributions along the E-plane and H-plane respectively, and then calculating according to Eq. (3.47).

The issue lies in the fact that even with the multipoint measurement method, we can only obtain the values of the power density at finite points along the θ direction, which prevents us from integrating them along the θ direction and only becomes an approximate superposition formula. If we take measurements at intervals of every other angle starting from $\theta = 0$ within the range of $\theta = 0$ to $\theta = \theta_n$, we can obtain a total of $n + 1$ measured values of $\overline{p}$, but in reality, we are just dividing it into n angle intervals, so the superposition equation becomes

$$\int_0^{\theta_n} \overline{p}(\theta) \sin\theta \, d\theta \approx \frac{1}{2} \sum_{i=1}^{n} (\overline{p}_{i-1} + \overline{p}_i)(\cos\theta_{i-1} - \cos\theta_i) \tag{3.48}$$

The power of both the E-plane and H-plane is calculated according to Eq. (3.48), and then the total power of the radiation field can be obtained by substituting them into Eq. (3.47).

2. **Fitting function method**

In fact, the method discussed above is essentially the same as computational method 2 of power superposition of the spherical annular area introduced earlier, with the difference being that we measure the power density separately in two directions of the E-plane and the H-plane, then integrate and average the results. Theoretically speaking, the more the number of measuring points of power density and the smaller the angular interval, the smaller the error between the calculated total power and the actual radiated power. However, increasing the number of measurement points may be limited by the size of the receiving horn, and it may also increase the experimental workload or the number of receiving systems. In order to maximize the precision of the final total power result with limited measurement data, Chinese scholars have proposed that the radiation pattern can be fitted using specific functions based on measured data [35]. By comparing the fitted pattern with the theoretical pattern, when there is a good consistency between the two, it is possible to directly use the fitted pattern functions $\overline{p}_e(\theta)$ and $\overline{p}_h(\theta)$ for power integration, instead of the approximate calculation by the superposition method in Eq. (3.48), so as to improve the calculation accuracy.

Taking the circular horn as an example, after repeated comparisons of different fitting functions, Reference [35] proposed that the direction diagram of the radiation power of a circular horn can be fitted using the following form of function

$$\hat{\overline{p}}(\theta) = \sum_{k=0}^{N} A_k J_k(B_k \sin\theta) \tag{3.49}$$

where,

$\hat{\overline{p}}(\theta)$ is the fitted power distribution function,

$J_k(x)$ is the kth order Bessel function of the first kind,

N is generally taken as 3 to 7,

A_k and B_k are the unsolved coefficients, which can be determined by the least square method according to the position and the measured power density of the measuring point.

The principle of least squares method indicates that to obtain the most likely value of a quantity $\overline{p}(\theta)$ from $n + 1$ equally reliable measured values $\overline{p}(\theta_i)$, the sum of the squares of the errors between the obtained $\hat{\overline{p}}(\theta_i)$ at each point i by means of fitting $\hat{\overline{p}}(\theta)$ and the measured values $\overline{p}(\theta_i)$ should be minimized, that is,

$$Q = \sum_{i=1}^{n} \left[\overline{p}(\theta_i) - \hat{\overline{p}}(\theta_i)\right]^2 = \text{minimum} \tag{3.50}$$

where $\overline{p}(\theta_i)$ represents the measured value, and

$$\hat{\overline{p}}(\theta_i) = \sum_{k=0}^{N} A_k J_k(B_k \sin\theta_i) \tag{3.51}$$

represents the most probable value given by the fitting function at that point. Equation (3.50) indicates that the coefficients A_k and B_k should satisfy the following equation

$$Q = \sum_{i=0}^{n} \left[\overline{p}(\theta_i) - \sum_{k=0}^{N} A_k J_k(B_k \sin\theta_i)\right]^2 = \text{minimum} \tag{3.52}$$

According to the method of differential calculus for finding extreme values, it can be known that when A_k and B_k satisfy

$$\frac{\partial Q}{\partial A_k} = 0 \quad \frac{\partial Q}{\partial B_k} = 0 \quad (k = 0, 1, 2, \ldots, N) \tag{3.53}$$

Equation (3.52) can thus hold. By solving the system of Eq. (3.53) using programming, the coefficients A_k and B_k of the fitting function can be obtained. As for the measurement range, in order to reduce the workload of measurements and ensure sufficient accuracy of the fitting function, it is reasonable that we take 15 measurement points symmetrically for each main plane within a 15 dB or 18 dB beam width for TE_{11} mode (or HE_{11} mode) of circular waveguide, and take the average of the measured values at left and right symmetric two points as the power density corresponding to each measurement angle, in this way, the radiation pattern obtained by fitting has a good agreement with the theoretical pattern. Chinese scholars used 11 measurement points for fitting the radiation field of a coaxial conical horn (Fig. 3.7),

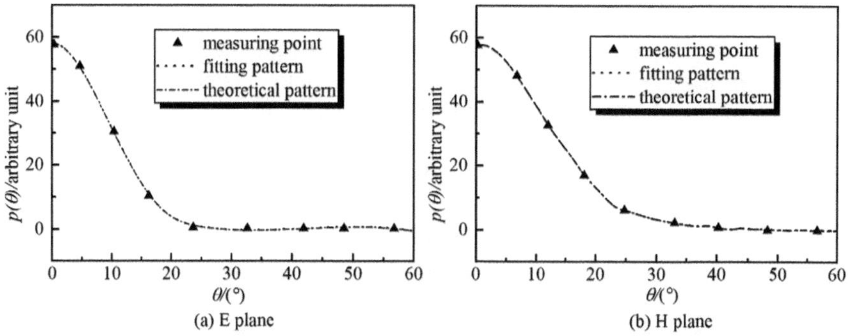

Fig. 3.7 Comparison of the fitting pattern of the radiation field of the coaxial conical horn with the actual pattern

and compared the fitted radiation pattern and the actual radiation pattern. The two were in good agreement, ensuring the accuracy of the power obtained by integrating based on the fitted radiation pattern.

By substituting the expression for the fitted $\overline{p}(\theta)$ into Eq. (3.47), the total radiated power can be calculated. The power obtained in this way will be more accurate than that calculated directly from the superposition Eq. (3.48) [35].

3.6 Calibration of Effective Horn Aperture and Characteristics of Spatial Integral Measurement Method

3.6.1 Calibration of Effective Receiving Horn Aperture

1. **Placement direction of the receiving horn**

In the radiation power measurement of high-power microwave, the antenna receiving low power is typically a rectangular waveguide receiving antenna, which can be either a rectangular waveguide with an open end or a pyramidal horn. This involves the problem of how to place the open-ended rectangular waveguide or pyramidal horn in the radiation field, namely the problem of which direction to choose for the wide wall (or narrow wall) of the waveguide.

According to the coupling principle, the most favorable condition for coupling between two transmission systems (or two microwave components) is that they must have the field component with the same direction at their coupling interface. The electric field of the fundamental TE_{10} mode in the rectangular waveguide as the receiving antenna is perpendicular to the wide side of the waveguide and the magnetic field is parallel to the wide side of the waveguide, therefore, when the rectangular horn antenna is placed in the radiation field, if the electric field of the radiation field is in the direction of φ, the wide side of the horn should be parallel to the θ direction, so that the electric field in the radiation field is perpendicular to the wide side of the horn, ensuring both the radiation field and the field excited in the rectangular horn have an electric field component perpendicular to the wide edge at the mouth of the horn. If the magnetic field in the radiation field is in the direction of φ, the wide side of the horn antenna should also be parallel to the direction of φ, so that the radiation field and the field excited in the rectangular horn both have magnetic field components parallel to the wide side of the horn at the mouth of horn.

Of course, the spatial distribution of the radiation field is complex, and the electric field may not necessarily be exactly perpendicular or parallel to the direction of φ. Distortions of field will also occur at the mouth of the horn due to the metal boundary. In any case, the wide side of the receiving horn should be placed as perpendicular as possible to the electric field of the radiation field, and as parallel as possible to the magnetic field of the radiation field for the best reception. Of course, the above coupling principle is only strictly valid to small hole coupling. The complexity of the radiation field distribution and the disturbance of the field by effects of the receiving antenna boundary will make the coupling more complicated, but this does not affect the coupling principle to determine the placement direction of the rectangular waveguide or horn, as the main coupling will still occur between the field components that satisfy the above coupling principle.

2. **Calibration of the effective aperture of the receiving horn**

In the various power measurement methods discussed above, in order to obtain the power density of the measuring point, it is necessary to consider the effective cross-section of the receiving horn. In other words, we must first determine the effective cross section S_{eff} of the receiving antenna in order to calculate the power density. Therefore, we should calibrate S_{eff}.

The effective cross-section is caused by the boundary conditions of the horn antenna. When a rectangular horn antenna is placed in a uniform plane electromagnetic wave, its response to the incident wave is not uniform. This is because the electric field parallel to the metal wall on the boundary of the horn must be zero, while the radiation field at the same position is generally not zero, causing a mismatch between the horn and the radiation field, resulting in reflection, which weakens the actual field received by the horn. In other words, if the electromagnetic field is still considered as a plane wave, it is as if the effective receiving aperture of the horn reduces, and the reduced aperture size is called the effective cross-section. The effective cross-section can also be called the effective aperture. It can be seen that the effective cross-section reflects the ability of the antenna to receive electromagnetic waves, which is a function of the direction and frequency of the received electromagnetic waves, representing the ability of the receiving antenna to absorb radiation from any specific direction at this frequency and deliver power to the output port. For a lossless antenna, if it can absorb all incident electromagnetic waves without reflection, the actual aperture area is its effective area. The ratio of the maximum effective area of the antenna to the actual aperture area is called the surface utilization factor, or the aperture efficiency of the antenna.

(1) Empirical equation

The effective area (cross-section) of the horn can be expressed by the following equation:

$$S_{eff} = \frac{G\lambda^2}{4\pi} \tag{3.54}$$

where G is the gain of the openend waveguide or pyramidal horn antenna, and λ is the wavelength of the electromagnetic wave.

By determining the gain of the antenna, the effective aperture of the antenna can be calculated. The U.S. National Bureau of Standards has proposed empirical equations for calculating the gain of an end open waveguide or pyramidal horn antenna.

(2) Gain of open waveguide

The U.S. National Bureau of Standards has proposed the following equation for calculating the gain of a rectangular end open waveguide with a wide-to-narrow side ratio of 2:1 [24],

$$G = 21.6fa \tag{3.55}$$

where f is the frequency of the electromagnetic wave (GHz), and a is the width of the rectangular waveguide (m). When the gain is to be expressed in decibels, Eq. (3.55) should be rewritten as:

$$G_{dB} = 10\,\lg(fa) + 13.34 \tag{3.56}$$

As long as the distance from the opening of the waveguide to the field point is greater than $2a$, the deviation of Eq. (3.56) is $\pm$ 0.5 dB.

It should be noted that the U.S. National Bureau of Standards obtained this equation in the frequency range of 200–500 GHz. Experimental results from Chinese scientific research workers have shown that the calculation results of this equation for the S, C, X band calculation results are also basically accurate.

(3) Gain of pyramidal horn

The U.S. National Bureau of Standards has proposed the following equation for calculating the gain of a pyramidal horn:

$$\begin{aligned} G &= (113.3whf^2) \times 10^{-(R_H+R_E)} \\ G_{\mathrm{dB}} &= 20.54 + 10\lg(wh) + 20\lg f - 10(R_H + R_E) \end{aligned} \tag{3.57}$$

where R_{H} and R_{E} are "near-field gain simplification factors" introduced to simplify the expression of the near-field gain.

$$\begin{aligned} R_H &= (0.01p)(1 + 10.19p + 0.51p^2 - 0.097p^3) \\ R_E &= (0.1q^2)(2.31 + 0.053q) \end{aligned} \tag{3.58}$$

where:

$$p = \left(\frac{w^2 f}{0.3}\right)\left(\frac{1}{l_H} + \frac{1}{R_0}\right)$$
$$q = \left(\frac{h^2 f}{0.3}\right)\left(\frac{1}{l_E} + \frac{1}{R_0}\right) \tag{3.59}$$

where,

w, h, l_H, l_E are the dimensions of the horn (m), as shown in Fig. 3.8,

f is the frequency (GHz),

R_0 is the distance from the receiving horn port to the radiant electromagnetic field point.

As long as the distance from the horn mouth to the field point is greater than 0.5 m, the difference between the calculated value of Eq. (3.57) and the measured value is within ± 0.5 dB.

(4) **Determining the effective cross-section by the directivity coefficient**

In the case of no loss, in other words, without considering the loss, the gain of the horn antenna is equal to the directivity coefficient. Therefore, we can use the directivity coefficient D instead of gain G to calculate the antenna's effective cross-section. The value of D can be calculated using Eq. (3.36).

It is proposed in Ref. [24] that when the physical aperture size of the antenna is at least 1λ, the ratio of the effective cross-section to the physical area can be approximated as 0.6, i.e.,

$$S_{eff} = 0.6S \tag{3.60}$$

where S is the actual aperture area of the antenna.

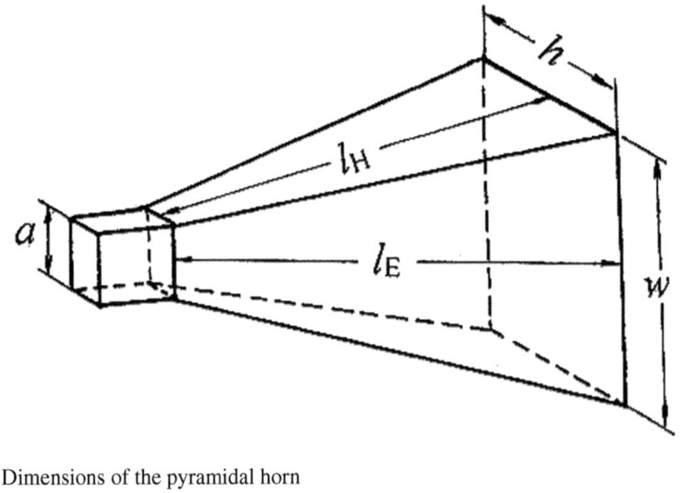

Fig. 3.8 Dimensions of the pyramidal horn

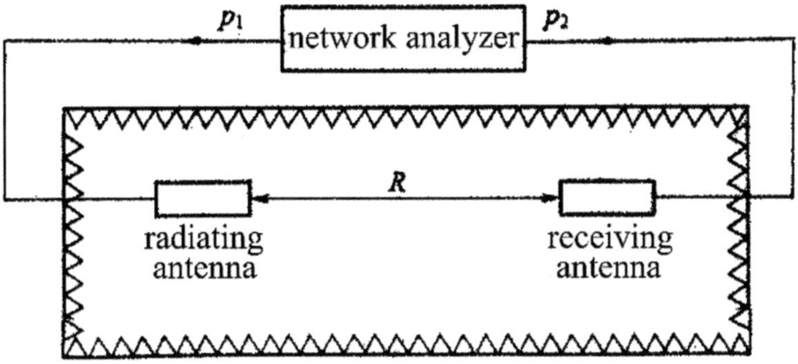

Fig. 3.9 Schematic diagram of the experimental system for calibrating the effective cross-section of the antenna

3. **Experimental calibration**

The system diagram used for experimental calibration of the effective area of the receiving antenna is shown in Fig. 3.9.

When the polarization of the transmitting antenna and the receiving antenna is matched, the directions of the main lobe are aligned with each other, and when the loss is not considered and the distance R between the two antennas meets the far field condition, according to Eqs. (3.38) and (3.54), it is not difficult to obtain the power received by the receiving antenna.

$$P_{0m} = \left(\frac{\lambda}{4\pi R}\right)^2 PG_tG_r \tag{3.61}$$

where,

P is the radiated power of the transmitting antenna,

P_{0m} is the microwave power measured by the measuring horn in the maximum radiation direction of the microwave source,

G_t is the gain of the transmitting antenna,

G_r is the gain of the receiving antenna,

R is the center distance between the transmitting antenna and the receiving antenna.

If the transmitting antenna is exactly the same as the receiving antenna, then

$$G_t = G_r = G \tag{3.62}$$

Substituting Eqs. (3.62) to (3.61), we can obtain

$$G = \left(\frac{P_{0m}}{P}\right)^{1/2} \frac{4\pi R}{\lambda} \tag{3.63}$$

Because

$$\frac{P_{0m}}{P} = [S_{21}]^2 \tag{3.64}$$

so

$$G = S_{21}\frac{4\pi R}{\lambda} \tag{3.65}$$

Substituting Eqs. (3.65) to (3.54), we get the final expression of the effective cross-section of the antenna

$$S_{eff} = S_{21}R\lambda \tag{3.66}$$

In this way, by using a network analyzer to measure S_{21}, we can calculate the effective cross-section S_{eff} using Eq. (3.66).

4. **Using the cutoff waveguide to avoid breakdown of the receiving system**

When using the radiation field method to measure the power of high-power microwave, if the output power of the high-power microwave source is high enough to cause the received power by the receiving horn to be too large, it will cause breakdown of the fundamental mode rectangular waveguide in the receiving system or the probe in the waveguide-coaxial converter connected with the waveguide. In this case, directly connecting an attenuator to the output port of the receiving horn to reduce the power is a feasible method, but the power capacity of the attenuator is often limited. When it cannot withstand the power level received by the horn, this method must be abandoned. Therefore, scholars home and abroad have proposed to use a cutoff waveguide to reduce the transmission power level in the fundamental mode rectangular waveguide to avoid such a breakdown [36].

Their approach is to connect a section of rectangular waveguide with the receiving horn, and this waveguide is cut-off for the frequency of the measured high-power microwave, so that the microwave forms an attenuated wave inside the waveguide. The amplitude of the wave decreases exponentially with the increase of the waveguide length, reducing the output microwave power at the end of the waveguide. The microwave then passes through the waveguide-coaxial converter and is input to the detector for pulse waveform and power detection, and indicated by the oscilloscope. Adjusting the length of the cutoff waveguide can control the amount of attenuation of the microwave.

Since the microwave power attenuated by the cutoff waveguide is not absorbed by the waveguide loss but is reflected back to the original radiation space through the receiving horn, this situation is equivalent to the reduction of microwave power received by the horn. Therefore, the reduction of microwave power caused by the cutoff waveguide can be reflected by the gain of the horn. As long as the experimental calibration of the receiving horn gain is carried out together with the cutoff waveguide and the waveguide-coaxial converter, it is not necessary to calibrate the attenuation of the cut-off waveguide separately.

Using two completely the same horns, along with waveguide and waveguide-coaxial converters, according to the method given in Fig. 3.9, the gain of the antenna (Eq. (3.65)) and the effective cross-section (Eq. (3.66)) can be calibrated.

3.6.2 Merits and Faults of the Spatial Integration Measurement

1. **Advantages of spatial integration measurement method**

(1) **A single-mode generator is no longer required**

The biggest advantage of the spatial integration measurement method compared to the total attenuation measurement method is that it is not necessary to calibrate all factors causing power attenuation in the entire measurement system from the radiation antenna to the final indicating device, or to directly calibrate the total attenuation. In the total attenuation measurement method, whether using experimental calibration method or using network analyzer method, a specially—designed calibration source with known power and at least the same frequency and output mode as the microwave source to be tested must be used, but generally, different pulse formats or even continuous wave sources can be allowed. However, such a calibrated source must have sufficient output power so that the receiving system can receive enough power for the indicator to show the received power level. More importantly, achieving consistency between the radiation mode of the calibration source and the output mode of the microwave source to be tested will bring a lot of work. And the different modes will lead to different radiation field, causing the radiation field distribution at the location of the receiving horn to be inconsistent with the actual radiation field distribution of the high-power microwave source at that location, which makes the calibration completely worthless. The spatial integration measurement method, as a direct measurement of the high-power microwave radiation field under test, does not require separate calibration of the total attenuation, thus avoiding the need for a single-mode generator that can produce the same mode as the high-power microwave radiation mode. This fundamentally overcomes the defect of the total attenuation measurement method.

(2) **High measurement accuracy**

In addition, in the total attenuation measurement method, due to the total attenuation is generally large, especially the attenuation (or coupling coefficient) of the radiation space between the radiating horn and the receiving horn, which typically accounts for a relatively large proportion of the total attenuation and under the same relative error, the absolute error is also relatively large, which significantly affects the accuracy of the final calculated total power. The spatial integration measurement method, by measuring the power density at selected measuring points and then integrating (superposition) to obtain the total power, just avoids the calibration of this part of the attenuation, thereby avoiding the errors caused by the calibration of such attenuation.

Using spatial numerical integration or radiation pattern fitting methods to calculate the total radiation power, the calculated total power is closer to the actual radiation power, and the calculation accuracy is further improved.

(3) **Simple and convenient**

The spatial integration measurement method is easy to achieve in the microwave darkroom and generally does not require special dedicated equipment. Conventional instruments such as horn antenna, network analyzer, crystal detector, microwave attenuator, and memory oscilloscope and so on can meet the measurement requirements.

The spatial integration measurement method is widely used in high-power microwave power measurement due to its convenience, accuracy, and reliability. Many papers on far-field measurement of high-power microwave radiation have been published at home and abroad. The specific measurement methods, especially the methods to improve the measurement accuracy, and the calibration of each segment of the measurement system such as receiver horn and detector are discussed in detail. Here, we only introduce the basic principles and main methods of this measurement method. Due to the different actual measurement systems and measurement devices used by each organization, it is not possible for us to discuss each specific detail in depth. Readers can refer to these references when applying the far-field measurement method in practice.

2. **Shortcomings of the spatial integration measurement method**

The basis of the spatial integration measurement method lies in the complete consistency between the radiation field of the high-power microwave source and the theoretical radiation field. When the actual radiation field differs from the theoretical radiation field for various reasons, this measurement basis is destroyed, which leads to a decrease in measurement accuracy. The greater the difference, the lower the reliability of the measurement results, which is the biggest shortcoming of the spatial integration measurement method. The radiation pattern fitting method partially overcomes this drawback to a certain extent. The measurement accuracy can be improved by integrating the fitted power density of the measured range instead of the superposition of the annular area power. However, the fitting process of this method is slightly complicated, and it is not easy to achieve ideal consistency between the fitting results and the actual radiation field distribution.

Additionally, for the multi-point measurement method, it is only suitable for measuring the radiation field of TM_{0n} or TE_{0n} modes with rotational symmetry of the central axis, which limits its application. If the radiation field is not ideally rotationally symmetric, this measurement method is no longer applicable. However, with the development of mode conversion technology, various mode converters and mode conversion antennas have been successfully developed, and the output of many high-power microwave sources after mode conversion can radiate the TE_{11} mode or quasi-HE_{11} mode by circular horns. For these modes, under the condition of a sufficient number of measurement points, using the E surface and H surface measurement, and then the total power can be calculated by using numerical integration or radiation pattern fitting methods. It can be seen that this method is essentially a multi-point measurement method.

Chapter 4
Coupling Field Measurement of High-Power Microwave Power

4.1 Introduction

In Chap. 3, we mainly introduced the radiation field measurement method of high-power microwave power. This chapter will mainly discuss the coupling field measurement of high-power microwave power.

The basic idea of the coupling field measurement of high-power microwave power is to separate a very small part of the extremely high power through some coupling method, input it into a calibrated detector-oscilloscope indicating system (or directly input the power into a peak power meter) for power measurement. If the coupled power still exceeds the measurement range of the indicating device, an attenuator should be used to further reduce the power. The total power of high-power microwaves can be obtained by converting the measured power through the coupling coefficient of the coupling method, the attenuation of the attenuator, and the loss of the entire measurement system. The key to this measurement method lies in the calibration of the coupling degree, attenuation, and system loss. Their uncertainties will cause large errors in the calculated high power. The accuracy (calibration precision) of the power indicating device will also have a certain impact on the measurement error.

The biggest difference between coupling field measurement and radiation field measurement is the method of reducing the power level of high-power microwaves to a small power level that the measurement indicating system can accept. Radiation field measurement relies on the receiving horn to receive only a small part of the power in the entire radiation field to achieve the purpose of reducing the power level, while coupling field measurement uses the coupling device to directly couple a very small part of the power from the high-power microwave transmission system for measurement, thereby reducing the power level.

There are many forms of coupling methods for coupling power measurement, such as probe coupling, directional coupler coupling, etc., but probe coupling is currently more widely used in practical measurements. In addition, the wavenumber spectrum

W. Wang et al., *High-Power Microwave Measurement*,
https://doi.org/10.1007/978-981-96-7763-4_4

coupling method and the small hole mode selective coupling method can also be used for power measurement. Regarding these two measurement methods, because they are mainly used for mode measurement, we will discuss them in Chap. 7. The power indication of the coupling field measurement can use a peak power meter, but more often it uses a combination of a detector and an oscilloscope.

The biggest advantage of coupling field power measurement is that it can be performed online and dynamically. On the one hand, the coupling device is directly connected to the output transmission system of high-power microwaves instead of its terminal, so its connection does not affect the connection between the system terminal and the load, such as the connection with the radiation antenna, that is, power measurement can be performed while the high-power microwave system maintains normal working status; on the other hand, the power coupled out of the main system by the coupling device only accounts for a very small part of the total power, so its connection does not affect the normal working status of the high-power microwave system; third, after the coupling device is connected to the system, there is no need to install or move parts during the measurement process, that is, the output power of the microwave source can be detected throughout the experimental process of the high-power microwave source. Therefore, it is not only an online measurement method, but also a dynamic and real-time measurement method.

Obviously, the radiation field measurement of high-power microwaves cannot be an online measurement method. The radiation field is formed specifically for measurement. At this time, the high-power microwave system cannot be connected to the working load for normal experiment. In this measurement method, because the receiving horn needs to be moved to perform measurements at multiple points, it is also impossible to achieve dynamic and real-time measurements.

4.2 Probe Coupling Method

4.2.1 Overview

The probe is a widely used coupling method in microwave technology. Generally, it uses the inner conductor of the coaxial line to extend into various waveguide components or cavities that transmit or store microwave energy, and couples out a part of the microwave energy from them to perform various signal or power monitoring. The outer conductor of the coaxial line is connected to the metal outer wall of the waveguide component or cavity. The probe is usually divided into two forms. If the inner conductor of the coaxial line is linearly inserted into the waveguide component or cavity in the direction parallel to the electric field in the microwave system, it will mainly couple with the microwave electric field. When any mode field in the waveguide component or cavity has an alternating electric field along the direction of the probe at the probe position, an alternating current will be excited on the probe, and microwave power will be output from the probe. This probe is called an electric probe. If the inner conductor of the coaxial line is inserted into the waveguide component or cavity and then bent into a ring shape, its end is connected to the metal outer wall of the waveguide component or cavity, that is, closed with the outer conductor of the coaxial line to form a coupling ring, so that the ring surface is perpendicular to the microwave magnetic field direction in the system, mainly coupling with the microwave magnetic field. When any mode in the waveguide has an alternating magnetic field passing through the coupling ring, an induced current will be generated in the closed loop, and microwave power will be output. This probe is called a magnetic probe. The structure of the two probes is shown in Fig. 4.1.

In high-power microwave power measurement, because the power level transmitted in the waveguide is very high, so it is hoped that the coupling of the probe should be very small (the positive coupling coefficient is very large) to ensure that the coupled power falls within the measurement range of the detector or small-power meter, or the power can be indicated after an external attenuator is used. Therefore, electric probes are generally used in power measurement, and magnetic probes are rarely used. This is because the end face area of the electric probe is much smaller than the annular area of the magnetic probe, so the coupling with the electric field is much weaker than the coupling of the magnetic probe with the magnetic field. In order to further reduce the coupling of the electric probe, the front end of the probe can also be made into a conical shape to make its end face area smaller; at the same time, the electric probe can be completely parallel to the waveguide wall without extending into the waveguide, so as to reduce the coupling coefficient to a greater extent, and also reduce the possibility of the probe causing field breakdown.

The main disadvantage of probe coupling is that it is prone to high-frequency breakdown. Due to the effect of tip discharge, its breakdown power is far less than the power capacity of the waveguide itself. In addition, whether it is an electric probe or a magnetic probe, the calibration of the coupling coefficient can only be based on a single mode. When there are parasitic modes in the high-power overmoded transmission system, the accuracy of its calibration will be reduced, resulting in inaccuracy of the calculated resulting microwave power transmitted in the waveguide.

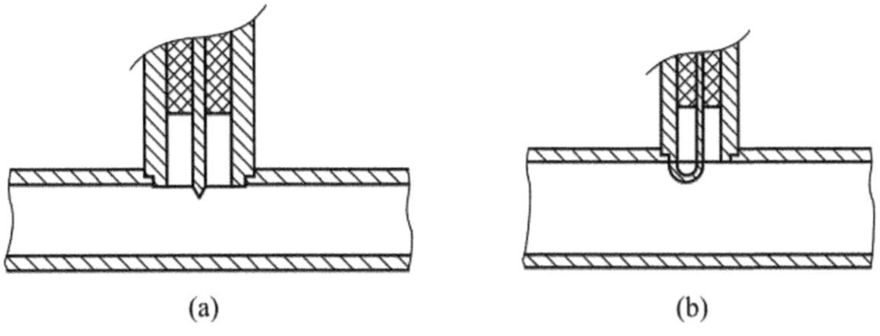

Fig. 4.1 Structure diagram of coupling probes, **a** electric probe, **b** magnetic probe

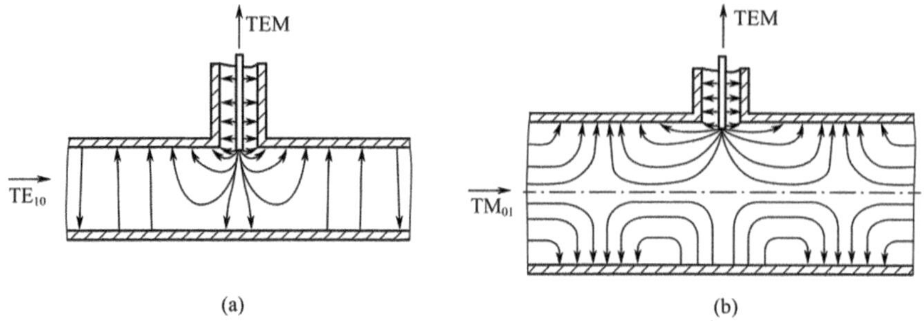

Fig. 4.2 Electric probe coupling with the electric field in the rectangular waveguide and in the circular waveguide, **a** coupling with the TE_{10} mode in the rectangular waveguide, **b** coupling with the TM_{01} mode in the circular waveguide

In Fig. 4.2, we give a schematic diagram of the electric probe coupling with the TE_{10} mode in the rectangular waveguide and the TM_{01} mode in the circular waveguide. Since the electric field of the TE_{10} mode in the rectangular waveguide is distributed along the wide side of the waveguide with a sine function, perpendicular to the wide side, the coupling will be strongest when the electric probe is inserted into the waveguide vertically on the centerline of the wide side, and it will decrease as the probe position deviates from the centerline, and the coupling is weakest near the narrow side. As for the TM_{01} mode in the circular waveguide, due to its circular symmetry, the coupling caused by the probe when inserted at any position on the circumference along the normal direction of the cylindrical surface is the same.

Magnetic robes are relatively less commonly used, but for the sake of completeness, we have included a necessary discussion on them here.

4.2.2 Coupling Power of Probe [37]

1. **Coupling power of electric probe**

The charge excited on the surface of the electric probe by the electromagnetic field in the waveguide is

$$Q(t) = S_{eff} D(t) \tag{4.1}$$

where, S_{eff} is the effective cross-section of the probe, $D(t)$ is the electrical flux density at the location of the probe. For a simple harmonic electric field (a field with a time $e^{\pm j\omega t}$ factor), Eq. (4.1) can be used to calculate the current on the electric probe and the power coupled to it

$$D(t) = \varepsilon E_n e^{-j\omega t}$$

$$I(t) = \frac{\mathrm{d}Q(t)}{\mathrm{d}t} = -\mathrm{j}S_{eff}\,\omega\varepsilon E_n \mathrm{e}^{-\mathrm{j}\omega t}$$

$$P_{ce} = \frac{1}{2}|I|^2 Z_0 = \frac{Z_0}{2}\omega^2\varepsilon^2 S_{eff}^2 |E_n|^2 \tag{4.2}$$

where, Z_0 is the impedance of the electric probe, which is generally set to 50 Ω for coaxial lines, E_n is the electric field parallel to the probe at the location of the probe. Since the probe is always inserted into the waveguide in a direction perpendicular to the waveguide wall, the electric field E_n that couples with it must be the normal component of the electric field perpendicular to the waveguide wall.

2. **Coupling power of magnetic probe**

If a wire is made into a closed loop, when the magnetic flux passing through the area enclosed by the closed loop changes, that is, when the magnetic field changes, an induced current will be generated in the loop. The fact that a current can be generated in the loop implies that there is a potential difference and an electromotive force, at the two ends of the circuit. Faraday provided a formula for calculating this electromotive force u.

$$u = -\frac{\mathrm{d}B(t)}{\mathrm{d}t} S_{eff}|\cos\theta| \tag{4.3}$$

The formula is called Faraday's law of electromagnetic induction. Where, $\mathrm{d}B/\mathrm{d}t$ is the rate of change of the magnetic field B with time, θ is the angle between the normal direction of the coupling ring plane and the direction of the magnetic lines of force passing through the coupling ring, S_{eff} is the effective area of the magnetic lines of force enclosed by the closed loop when the closed loop is perpendicular to the magnetic field lines (that is, θ is 0° or 180°, $|\cos\theta|=1$). Since the probe itself is inserted into the waveguide in the direction perpendicular to the waveguide wall, the magnetic field component that can pass through the coupling ring must be the tangential component of the magnetic field parallel to the waveguide wall. Similarly, for a simple harmonic magnetic field, we can obtain the coupling power of the magnetic probe as follows:

$$B(t) = \mu H_t \mathrm{e}^{-\mathrm{j}\omega t}$$

$$P_{cm} = \frac{1}{2}\frac{|u|^2}{Z_0} = \frac{1}{2Z_0}\omega^2\mu^2 S_{eff}^2 |H_t|^2 \tag{4.4}$$

where, Z_0 is the impedance of the magnetic probe, which can also be set to 50 Ω like the electric probe; H_t is the tangential magnetic field component parallel to the waveguide wall at the location of the probe. It should be noted that, unlike the electric field where the normal component is unique, there are two mutually perpendicular directions for the magnetic field that can be considered as tangential. By adjusting the orientation of the coupling ring plane, it is possible for the plane to be perpendicular to each of the two tangential components of the magnetic field. Therefore, there are two possibilities when calculating the coupling power of the magnetic probe.

4.2.3 Fields and Power in Waveguides [38, 39]

To calculate the coupling power of the probe, according to the formula given above, we first need to know the expression of the field components in the waveguide, and then calculate the coupling coefficient of the probe. According to the definition Eq. (4.22) of the coupling coefficient, we should also know the power transmitted in the waveguide.

1. **Field and power flow in circular waveguide**

The circular waveguide is represented by a cylindrical coordinate system, as shown in Fig. 4.3.

In the expression for the field components in a circular waveguide, the change of the field in the φ direction should have a starting angle of φ_0, unless there are special non-uniformities in the waveguide, such as small holes, probes, and other excitation mechanisms. In general, for an ideally uniform circular waveguide, this starting angle is uncertain, or the change of the field in the φ direction can start from any angular position. Therefore, when using a probe for power coupling from a circular waveguide, the probe can be inserted into the waveguide at any position on the circular waveguide circumference, and it will not affect the size of the coupling coefficient, or in other words, the coupling coefficient of the probe is independent of φ and only related to the depth of the probe insertion into the waveguide. However, it should be noted that if two or more probes (or coupling holes) are inserted at different angular positions on the circular waveguide circumference at the same

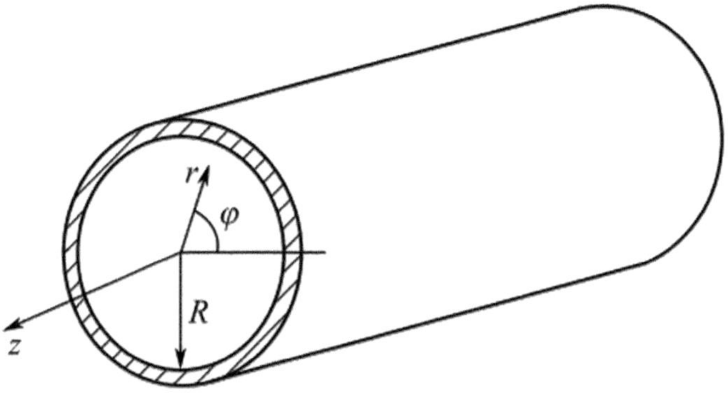

Fig. 4.3 The cylindrical coordinate system of the circular waveguide

time, their coupling will be related to the φ direction. We are currently discussing only the coupling situation of a single probe. We can assume that the probe will couple with the maximum value of the field distribution in the angular direction. Therefore, the angular distribution of the field can be omitted in the expression of the field component, and the factor containing the φ direction will no longer appear in the following field component expression.

(1) **TM$_{mn}$ mode**

In a circular waveguide with a radius of R, the field expression of the TM$_{mn}$ mode is

$$\begin{aligned}
E_z &= E_{zm}J_m(k_c r)\mathrm{e}^{-\mathrm{j}(\omega t-\beta z)}\\
E_r &= -\mathrm{j}\frac{\beta}{k_c}E_{zm}J_m'(k_c r)\mathrm{e}^{-\mathrm{j}(\omega t-\beta z)}\\
E_\phi &= -\mathrm{j}\frac{\beta m}{k_c^2 r}E_{zm}J_m(k_c r)\mathrm{e}^{-\mathrm{j}(\omega t-\beta z)}\\
H_z &= 0\\
H_r &= \mathrm{j}\frac{\omega\varepsilon m}{k_c^2 r}E_{zm}J_m(k_c r)\mathrm{e}^{-\mathrm{j}(\omega t-\beta z)}\\
H_\phi &= -\mathrm{j}\frac{\omega\varepsilon}{k_c}E_{zm}J_m'(k_c r)\mathrm{e}^{-\mathrm{j}(\omega t-\beta z)}
\end{aligned} \tag{4.5}$$

where, $m = 0, 1, 2, \ldots, n = 1, 2, 3, \ldots, n$ cannot be 0, indicating the change of the field in the radial direction in the circular waveguide cannot be 0; $J_m'(k_c r)$ is derivative of the Bessel function $J_m(k_c r)$; k_c is the cutoff wave number of the TM$_{mn}$ mode.

$$k_c = \frac{\mu_{mn}}{R} \tag{4.6}$$

μ_{mn} is the nth root of the mth order Bessel function of the first kind $J_m(k_c r)$.

The power transmitted by the TM$_{mn}$ mode in the circular waveguide is.

$$P_t = \frac{1}{2}\int_S (\boldsymbol{E}_{mn} \times \boldsymbol{H}_m^*) \cdot \mathrm{d}\boldsymbol{S} = \frac{\pi R^2}{2N_m}\left(\frac{\beta}{k_c}\right)^2 \frac{E_{zm}^2}{\eta_{TM}} J_m'^2(k_c R) \tag{4.7}$$

where, β is the phase constant of the TM$_{mn}$ mode, η_{TM} is the wave impedance of the TM$_{mn}$ mode, which is expressed as

$$\eta_{TM} = \frac{\beta}{\omega\,\varepsilon} \tag{4.8}$$

N_m is called the Neumann coefficient.

$$N_m = \begin{cases} 1 \ (\mathrm{m} = 0) \\ 2 \ (\mathrm{m} \neq 0) \end{cases} \tag{4.9}$$

(2) **TE$_{mn}$ mode**

In a circular waveguide, the field component expression of the TE$_{mn}$ mode is

$$
\begin{aligned}
H_z &= H_{zm} J_m(k_c r) \mathrm{e}^{-\mathrm{j}(\omega t - \beta z)} \\
H_r &= -\mathrm{j}\frac{\beta}{k_c} H_{zm} J'_m(k_c r) \mathrm{e}^{-\mathrm{j}(\omega t - \beta z)} \\
H_\phi &= -\mathrm{j}\frac{\beta m}{k_c^2 r} H_{zm} J_m(k_c r) \mathrm{e}^{-\mathrm{j}(\omega t - \beta z)} \\
E_z &= 0 \\
E_r &= -\mathrm{j}\frac{\omega \mu m}{k_c^2 r} H_{zm} J_m(k_c r) \mathrm{e}^{-\mathrm{j}(\omega t - \beta z)} \\
E_\phi &= \mathrm{j}\frac{\omega \mu}{k_c} H_{zm} J'_m(k_c r) \mathrm{e}^{-\mathrm{j}(\omega t - \beta z)}
\end{aligned}
\tag{4.10}
$$

where, $m = 0, 1, 2, \ldots, n = 1, 2, 3, \ldots$, similarly, n cannot be 0, k_c is the cutoff wave number of the TE$_{mn}$ mode.

$$k_c = \frac{\mu'_{mn}}{R} \tag{4.11}$$

μ'_{mn} is the nth root of the derivative $J'_m(k_c r)$ of the mth order Bessel function of the first kind $J_m(k_c r)$.

The power transmitted by the TE$_{mn}$ mode in the circular waveguide is

$$P_t = \frac{1}{2}\int_S \left(\boldsymbol{E}_{mn} \times \boldsymbol{H}^*_m\right) \cdot \mathrm{d}\boldsymbol{S} = \frac{\pi R^2}{2N_m}\left(\frac{\beta}{k_c}\right)^2 \eta_{TE} H^2_{zm}\left(1 - \frac{m^2}{k_c^2 R^2}\right) J^2_m(k_c R) \tag{4.12}$$

where, β is the phase constant of the TE$_{mn}$ mode, η_{TE} is the wave impedance of the TE$_{mn}$ mode, which can be expressed as

$$\eta_{TE} = \frac{\omega \mu}{\beta} \tag{4.13}$$

The Neumann coefficient N_m is shown in Eq. (4.9).

2. **Field and power flow in rectangular waveguide**

The rectangular waveguides are usually represented by a rectangular coordinate system (Fig. 4.4).

(1) **TM$_{mn}$ mode**

In a rectangular waveguide with dimensions $a \times b$ (a is the wide side, b is the narrow side), the field expression of the TM$_{mn}$ mode is:

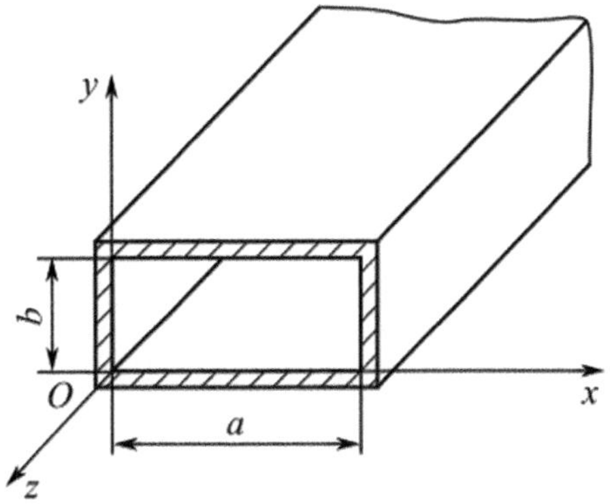

Fig. 4.4 The rectangular coordinate system of the rectangular waveguide

$$
\begin{aligned}
E_z &= E_{zm}\sin(k_x x)\sin(k_y y)e^{-j(\omega t-\beta z)}\\
E_x &= -j\frac{\beta}{k_c^2}k_x E_{zm}\cos(k_x x)\sin(k_y y)e^{-j(\omega t-\beta z)}\\
E_y &= -j\frac{\beta}{k_c^2}k_y E_{zm}\sin(k_x x)\cos(k_y y)e^{-j(\omega t-\beta z)}\\
H_z &= 0\\
H_x &= j\frac{\omega\varepsilon}{k_c^2}k_y E_{zm}\sin(k_x x)\cos(k_y y)e^{-j(\omega t-\beta z)}\\
H_y &= -j\frac{\omega\varepsilon}{k_c^2}k_x E_{zm}\cos(k_x x)\sin(k_y y)e^{-j(\omega t-\beta z)}
\end{aligned}
\tag{4.14}
$$

where, $m = 1, 2, 3, \ldots$, and $n = 1, 2, 3, \ldots$, that is, both m and n cannot be 0. k_c is the cutoff wave number of the TM_{mn} mode in the rectangular waveguide, which can be calculated according to Eq. (4.15):

$$
\begin{aligned}
k_c^2 &= \left(\frac{m\pi}{a}\right)^2 + \left(\frac{n\pi}{b}\right)^2\\
k_x^2 &= \left(\frac{m\pi}{a}\right)^2,\quad k_y^2 = \left(\frac{n\pi}{b}\right)^2
\end{aligned}
\tag{4.15}
$$

The power transmitted by the TM_{mn} mode in the rectangular waveguide is:

$$
P_t = \frac{ab}{8}\frac{\omega\varepsilon\beta}{k_c^2}E_{zm}^2 = \frac{ab}{8}\left(\frac{\beta}{k_c}\right)^2\frac{1}{\eta_{TM}}E_{zm}^2 \tag{4.16}
$$

where, η_{TM} is the wave impedance of the TM_{mn} mode in the rectangular waveguide, which can be expressed as:

$$\eta_{TM} = \frac{\beta}{\omega\varepsilon} \tag{4.17}$$

where β is the phase constant of the TM_{mn} mode.

(2) **TE_{mn} mode**

The field expression of the TE_{mn} mode in the rectangular waveguide is:

$$\begin{aligned}
H_z &= H_{zm}\cos(k_x x)\cos(k_y y)\mathrm{e}^{-\mathrm{j}(\omega t-\beta z)} \\
H_x &= \mathrm{j}\frac{\beta}{k_c^2}k_x H_{zm}\sin(k_x x)\cos(k_y y)\mathrm{e}^{-\mathrm{j}(\omega t-\beta z)} \\
H_y &= \mathrm{j}\frac{\beta}{k_c^2}k_y H_{zm}\cos(k_x x)\sin(k_y y)\mathrm{e}^{-\mathrm{j}(\omega t-\beta z)} \\
E_z &= 0 \\
E_x &= \mathrm{j}\frac{\omega\mu}{k_c^2}k_y H_{zm}\cos(k_x x)\sin(k_y y)\mathrm{e}^{-\mathrm{j}(\omega t-\beta z)} \\
E_y &= -\mathrm{j}\frac{\omega\mu}{k_c^2}k_x H_{zm}\sin(k_x x)\cos(k_y y)\mathrm{e}^{-\mathrm{j}(\omega t-\beta z)}
\end{aligned} \tag{4.18}$$

where, $m = 0, 1, 2, \ldots$, and $n = 1, 2, 3, \ldots$, but m and n cannot be 0 at the same time. k_c is the cutoff wave number of the TE_{mn} mode in the rectangular waveguide, which can also be calculated according to Eq. (4.15). The power transmitted by the TE_{mn} mode in the rectangular waveguide is:

$$P_t = \frac{ab}{2N_m N_n}\frac{\omega\mu\beta}{k_c^2}H_{zm}^2 = \frac{ab\eta_{TE}}{2N_m N_n}\left(\frac{\beta}{k_c}\right)^2 H_{zm}^2 \tag{4.19}$$

where, η_{TE} is the wave impedance of the TE_{mn} mode in the rectangular waveguide, which can be expressed as:

$$\eta_{TE} = \frac{\omega\mu}{\beta} \tag{4.20}$$

N_m and N_n are the Neumann coefficients.

$$N_m = \begin{cases} 1 \ (\mathrm{m}=0) \\ 2 \ (\mathrm{m}\neq 0) \end{cases}, \quad \mathrm{N_n} = \begin{cases} 1 \ (\mathrm{n}=0) \\ 2 \ (\mathrm{n}\neq 0) \end{cases} \tag{4.21}$$

4.3 Coupling Coefficient and Effective Section of Probe

4.3.1 *Definition of Coupling Coefficient and Coupling Rules*

1. **Definition of coupling coefficient and total power**

(1) **Definition of coupling coefficient**

When the coupling coefficient takes a positive value, its definition is

$$C = 10\lg\frac{P_t}{P_c}\ (\text{dB}) \tag{4.22}$$

where P_t is the power of the mode wave transmitted in the waveguide and P_c is the power coupled by the electric probe or magnetic probe. We have already given the power flow P_t of the TM_{mn} mode and TE_{mn} mode in the circular waveguide and rectangular waveguide. Now, we only need to calculate the coupling power of the electric probe or magnetic probe according to Eqs. (4.2) and (4.4) to get their coupling coefficient.

(2) **Calculating the total power of the system from the coupling coefficient**

On the contrary, knowing the coupling coefficient C and measuring the power P_c coupled by the probe, we can calculate the total power P_t transmitted in the waveguide.

$$P_t = P_c \times 10^{C/10} \tag{4.23}$$

This is the principle of the coupling method for measuring high-power microwave power. As can be seen, determining the coupling coefficient of the probe is the basis of this measurement.

In theory, the coupling coefficient of the probe can be directly measured using a network analyzer, but in practical measurements, serious difficulties may be encountered. This is mainly due to the problem of the mode—field excitation source in the system, which we have encountered more than once before. Because the coupling coefficient of the probe is closely related to the microwave modes being coupled, the microwave modes generated by high-power microwave sources are often diverse. When calibrating the coupling coefficient using a network analyzer, it is difficult to excite the required modes for calibration in the test waveguide that has the probe installed. Therefore, we generally can only determine the coupling coefficient of the probe through theoretical calculations (experimental methods can also be used in relatively simple mode cases), as all factors that affect the coupling coefficient, such as frequency, modes, coupling position, etc., are taken into consideration in the theoretical formulas.

The Chinese researchers have conducted research on measuring the output power of relativistic backward wave tubes using electric probes and obtained a power ranging from 400 to 800 MW. Compared with the measurement results of the power probe manufactured in Russia, the relative error of the two is less than 15%.

2. **Basic Rules of Probe Coupling** [38, 40]

(1) **The field coupled by the probe is generally the field at the waveguide wall**

In the power measurement of high-power microwaves, the power level of high-power microwaves is generally quite large, but it is desired that the power coupled by the probe is as small as possible, and at the same time, it is also required to reduce the probability of breaking through in the waveguide caused by the introduction of the probe. Therefore, in general, the front end of the electric probe is only flush with the waveguide wall, and rarely extends into the waveguide; the coupling ring of the magnetic probe inevitably has to extend into the waveguide, otherwise it is difficult for the other end of the ring to be welded with the waveguide wall to form a closed ring, but one will also try to minimize the degree of extension into the waveguide.

(2) **The field component coupled by the electric probe**

For the electric probe, the electric field component that can couple with it is the normal component of the electric field at the waveguide wall, and the normal direction of the waveguide wall is unique. Therefore, each waveguide wall with an electric probe inserted can only correspond to a unique electric field component that can be coupled with the electric probe. If a certain mode does not have this electric field component, it is impossible to couple with the electric probe.

In the circular waveguide, the normal of the waveguide wall is the r direction, which means that only the E_r component of the electric field can be coupled with the electric probe. For the TE_{0n} mode, since there is only the E_φ component and no E_r component, the electric probe cannot couple with the TE_{0n} mode and couple out microwave power from the waveguide. Of course, this is only an ideal situation in theory, in fact, the introduction of the probe itself will also cause field distortion, thereby generating the E_r component and forming coupling, but this coupling is generally quite weak.

In the rectangular waveguide, if a Cartesian coordinate system is taken, and by convention, the wide side is in the x direction, and the narrow side is in the y direction, then for the wide side of the waveguide, its normal is the y direction; and for the narrow side, its normal is the x direction. In the rectangular waveguide, no matter what mode, it is impossible that the electric field has neither the E_y component nor the E_x component (note: this state may exist in the resonant cavity), so the electric probe can always couple with the electric field in the rectangular waveguide, but for the TE_{m0} mode, because the electric field only has the E_y component, the probe must be inserted from the wide side of the waveguide; and for the TE_{0n} mode, because the electric field only has the E_x component, the probe can only be inserted from the narrow side.

(3) The field component coupled with the magnetic probe

In the circular waveguide, the tangential component of the magnetic field at the waveguide wall that can be coupled with the magnetic probe is H_φ and H_z, and this component has only one in the TM mode, and the TE mode may have two. The coupling ring plane of the magnetic probe should be perpendicular to the magnetic field component that is expected to be coupled with it, allowing the magnetic field lines to pass through the coupling ring.

The TM_{0n} mode only has one tangential component H_φ, so the magnetic probe can only couple with the H_φ of the TM_{0n} mode, and the plane of the coupling ring should be perpendicular to the φ direction. And the TE_{0n} mode, although it has both H_r and H_z components, but only H_z is tangential, so when coupling with the TE_{0n} mode, it should couple with the H_z component, and the plane of the coupling ring should be perpendicular to the z direction. Other modes that have both H_φ and H_z tangential components at the waveguide wall, if the magnetic probe hopes to couple with H_φ, the plane of the coupling ring is perpendicular to the φ direction, if it hopes to couple with H_z, the plane of the coupling ring is perpendicular to the z direction, if the magnetic probe is inserted in any other direction, it will simultaneously couple with the components of H_φ and H_z in the vertical direction of the coupling ring.

We know that for the magnetic field, the magnetic lines of force must form a closure by themselves, so, in the rectangular waveguide, the magnetic lines of force at the waveguide wall want to form a closed loop. For the TE_{mn} mode, it may be composed of H_x and H_z components, or it may be composed of H_y and H_z components; for the TM_{mn} mode, it will be composed of H_x and H_y components to form a closed loop of force lines. Therefore, when placing a magnetic probe in a rectangular waveguide, if it is coupled with the H_z component, the coupling ring can be inserted from the wide side or the narrow side, and the plane of the coupling ring should be perpendicular to the z direction; if it is coupled with the H_x component, the coupling ring must be inserted from the wide side, and the plane of the coupling ring should be perpendicular to the x direction; if it is coupled with the H_y component, the coupling ring must be inserted from the narrow side, and the plane of the coupling ring should be perpendicular to the y direction. But it should be noted that the position where the coupling ring is placed, the magnetic field component cannot be exactly at the 0 value.

(4) Adjustment of probe coupling coefficient

For the rectangular waveguide, because the field distribution of the given mode in the waveguide is determined, in general, whether it is an electric field or a magnetic field, their distribution along the inner wall of the waveguide is uneven in the majority of cases therefore, we can conveniently adjust the strength of the coupling between the probe and the field by changing the insertion position of the probe on the waveguide wall. For example, for the most common TE_{10} mode, E_y is distributed along the wide side according to the sine function, so the field is strongest in the middle of the wide side, weakens on both sides, and becomes 0 at the end point (the connection point with the narrow side). Correspondingly, the electric probe is inserted at the center position of the wide side and the coupling is strongest, when it deviates to both sides, the coupling gradually weakens. The coupling of the magnetic probe will also have similar situations. For example, for the same TE_{10} mode, when the magnetic probe is inserted on the wide side of the waveguide and coupled with H_z, the probe is closer to the edge of the wide side and the coupling is stronger, and the coupling is 0 on the center line of the wide side, because here H_z equals 0. In addition, the magnetic probe can also adjust the strength of the coupling by adjusting the angle θ of the plane of the coupling ring relative to the direction of the magnetic field. If the magnetic field is evenly distributed along a certain side of the rectangular waveguide without change, the probe can either adjust the angle of the plane of the coupling ring relative to the direction of the magnetic field, or adjust the insertion depth of the probe to change the coupling strength.

For the uniform circular waveguide, although the electromagnetic field generally also has a sine or cosine function distribution along the circumference (except for the circularly symmetric mode TE_{0n} mode and TM_{0n} mode), the position of this distribution on the circumference is not fixed, so, the insertion of the probe from any angular position of the circumference does not affect the probe coupling. For a given mode, to adjust the coupling coefficient of the probe, for the electric probe, it can only be achieved by adjusting the insertion depth of the probe; for the magnetic probe, it can be achieved by changing the angle θ of the plane of the coupling ring and the magnetic line of force. For the circularly symmetric mode TE_{0n} mode, because it does not have the E_r component, it is impossible to couple with the electric probe, and can only use the magnetic probe to couple with the H_z component; and for the circularly symmetric mode TM_{0n} mode, at the position of the waveguide inner wall, it has both the E_r components and the H_φ components the component, so both the electric probe and the magnetic probe can couple with it.

3. **Basic assumptions of probe coupling** [41, 42]

Before deriving the coupling coefficient of the probe, we first give some basic assumptions to simplify the derivation process.

(a) For the electric probe, we assume that its end face is flush with the inner wall of the waveguide, which means that the electric field it couples is the field value at the inner wall of the waveguide, that is, the field value when $r = R$ (for circular waveguide) or $x = a$, or $y = b$ (for rectangular waveguide). Obviously, the unevenness of the probe and the inner wall of the waveguide will affect the coupling coefficient.
(b) For the magnetic probe, we assume that its coupling ring has just extended into the inner wall of the waveguide, which couples with the field at the inner wall of the waveguide as well, that is, the field value when $r = R$ (for circular waveguide) or $x = a$, or $y = b$ (for rectangular waveguide). Strictly speaking, the coupling ring has a certain area, and the magnetic lines of force passing through it will have a certain distribution, more accurate calculations should seek the integral or average value of the magnetic field distribution passing through the coupling ring. We think that the size of the magnetic probe coupling ring is

much smaller than the size of the waveguide, it can be approximated that the magnetic field passing through the coupling ring is uniform, and it is equal to the magnetic field near the waveguide wall corresponding to the center position of the coupling ring. Of course, this will bring some errors to the calculation of the coupling coefficient.

(c) In the derivation, we assume that the waveguide is filled with air or vacuum, that is, $\varepsilon = \varepsilon_0$, $\mu = \mu_0$, so there is the following relationship between the speed of light c and the dielectric constant and magnetic permeability:

$$c = \frac{1}{\sqrt{\varepsilon\mu}} = \frac{1}{\sqrt{\varepsilon_0\mu_0}} \tag{4.24}$$

where, $\varepsilon_0 = 8.854 \times 10^{-12}\,\mathrm{F/m}$; $\mu_0 = 1.2566 \times 10^{-6}\,\mathrm{H/m}$.

Similarly, we can also get:

$$\eta_0 = \sqrt{\frac{\mu}{\varepsilon}} = \sqrt{\frac{\mu_0}{\varepsilon_0}} \approx 377\,\Omega \tag{4.25}$$

η_0 is called the free space wave impedance.

(d) Assume that the characteristic impedance Z_0 of the coaxial line part of the probe is 50 Ω.

4.3.2 Coupling Coefficient of Probe

1. **Coupling coefficient of electric probe**

(1) **The coupling coefficient of electric probe in circular waveguide** [43]

As pointed out earlier, the electric field component coupled with the electric probe in the circular waveguide is E_r. We should calculate the power coupled by the electric probe based on the value of E_r at $r = R$.

(a) TM_{mn} mode.

From Eq. (4.5), when $r = R$, we have:

$$|E_r|_{r=R} = \frac{\beta}{k_c}\left|E_{zm}J'_m(k_cR)\right| \tag{4.26}$$

The power of the probe coupling can be obtained by substituting Eq. (4.26) into Eq. (4.2).

$$P_{ce} = \frac{Z_0}{2}\omega^2\varepsilon^2 S_{eff}^2 \frac{\beta^2}{k_c^2} E_{zm}^2 J_m'^2(k_cR) = \frac{Z_0}{2}\left(\frac{\beta}{k_c}\right)^2 (\omega\varepsilon)^2 S_{eff}^2 E_{zm}^2 J_m'^2(k_cR) \tag{4.27}$$

According to the size of P_t given by Eq. (4.7), the coupling coefficient of the electric probe with the TM_{mn} mode in the circular waveguide can be obtained

$$C = 10\lg\frac{P_t}{P_{ce}} = 10\lg\frac{R^2\eta_0}{4\pi N_m S_{eff}^2 Z_0}\frac{\lambda^2}{\sqrt{1-(\lambda/\lambda_c)^2}} \tag{4.28}$$

where,

λ_c is the cutoff wavelength corresponding to k_c
η_0 is the free space wave impedance
Z_0 is the coaxial line impedance, taken as 50 Ω
N_m see Eq. (4.9).

(b) TE_{mn} mode.

From Eq. (4.10), when $r = R$, we have

$$|E_r|_{r=R} = \frac{\omega\mu\, m}{k_c^2 R}H_{zm}|J_m(k_c R)| \quad (\mathrm{m} \neq 0) \tag{4.29}$$

The power of the probe coupling can be obtained by substituting Eq. (4.29) into Eq. (4.2).

$$P_{ce} = \frac{Z_0}{2}\omega^2\varepsilon^2 S_{eff}^2\frac{\omega^2\mu^2 m^2}{k_c^4 R^2}H_{zm}^2 J_m^2(k_c R) = \frac{Z_0}{2}\frac{\omega^4\varepsilon^2\mu^2 m^2}{k_c^4 R^2}S_{eff}^2 H_{zm}^2 J_m^2(k_c R) \quad (\mathrm{m} \neq 0) \tag{4.30}$$

From Eq. (4.29), it can be seen that if $m = 0$, that is, for the TE_{0n} mode, the coupling power of the electric probe is 0, and the electric probe cannot couple with the circular electric mode in the circular waveguide. Obviously, this is because there is no E_r field component in the TE_{0n} mode.

According to the size of P_t given by Eq. (4.12), the coupling coefficient of the electric probe with the TE_{mn} mode in the circular waveguide can be obtained, but at this time in Eq. (4.12), because $m \neq 0$, $N_m = 2$.

$$C = 10\lg\frac{R^2\lambda^2\eta_0}{8\pi Z_0 S_{eff}^2}\left[\left(\frac{2\pi R}{m\lambda_c}\right)^2 - 1\right]\sqrt{1-(\lambda/\lambda_c)^2} \quad (\mathrm{m} \neq 0) \tag{4.31}$$

(2) **The coupling coefficient of electric probe in rectangular waveguide**

The electric probe in the rectangular waveguide can be coupled with the E_y field component of the waveguide, and can also be coupled with the E_x component. When the probe is inserted from the wide side of the waveguide, it is coupled with E_y, and when it is inserted from the narrow side, it is coupled with E_x. Here we only give the coupling coefficient of the probe inserted from the wide side and coupled with E_y. The situation where the electric probe in the rectangular waveguide is inserted from the narrow side of the waveguide and coupled with the E_x component is relatively less used in actual applications. When needed, readers can use a completely similar method to derive its coupling coefficient. We will not list it here.

(a) TM_{mn} mode.

When the probe is inserted from the wide side, at $y = b$, the E_y field component can be written according to Eq. (4.14).

$$\left|E_y\right|_{y=b} = \left|E_{yb}\right| = \frac{\beta}{k_c^2} k_y E_{zm} |\sin(k_x x)| = \frac{\beta n\pi}{k_c^2 b} E_{zm} \left|\sin\left(\frac{m\pi}{a}x\right)\right| \quad (m \neq 0,\ n \neq 0) \tag{4.32}$$

Then the power coupled by the electric probe is

$$P_{ce} = \frac{Z_0}{2}\left|\omega\varepsilon S_{eff} E_{yb}\right|^2 = \frac{Z_0}{2}\frac{\beta^2}{k_c^4}\omega^2\varepsilon^2 S_{eff}^2 \left(\frac{n\pi}{b}\right)^2 E_{zm}^2 \sin^2\left(\frac{m\pi}{a}x\right) \quad (m \neq 0,\ n \neq 0) \tag{4.33}$$

It can be seen immediately that when $m = 0$ or $n = 0$, $P_{ce} = 0$, which is easy to understand, because there is no TM_{0n} mode and TM_{m0} mode in the rectangular waveguide, and its simplest mode is TM_{11} mode.

The power of the TM_{mn} mode in the rectangular waveguide has been given by Eq. (4.16), so the coupling coefficient of the electric probe to it can be obtained

$$C = 10\lg \frac{ab^3\eta_0}{4n^2\pi^2 Z_0 S_{eff}^2}\left(\frac{\lambda}{\lambda_c}\right)^2 \frac{1}{\sqrt{1-(\lambda/\lambda_c)^2\sin^2(m\pi x/a)}} \quad (\mathrm{n} \neq 0,\ \mathrm{m} \neq 0) \tag{4.34}$$

It can be seen from Eq. (4.34) that adjusting the probe's position along the x-axis can adjust the size of the coupling coefficient.

(b) TE_{mn} mode.

The probe is still inserted from the wide side of the waveguide. At $y = b$, the E_y field of the TE_{mn} mode can be written according to Eq. (4.18)

$$\left|E_y\right|_{y=b} = \left|E_{yb}\right| = \frac{\omega\mu}{k_c^2}\frac{m\pi}{a} H_{zm}\left|\sin\left(\frac{m\pi}{a}x\right)\right| \quad (m \neq 0) \tag{4.35}$$

Then the power coupled by the electric probe is

$$P_{ce} = \frac{Z_0}{2}\left|\omega\varepsilon S_{eff} E_{yb}\right|^2 = \frac{Z_0}{2}\frac{m^2\pi^2}{a^2}\left(\frac{k}{k_c}\right)^4 S_{eff}^2 H_{zm}^2 \sin^2\left(\frac{m\pi}{a}x\right) \quad (\mathrm{m} \neq 0) \tag{4.36}$$

From Eq. (4.36), it can be seen that m cannot be equal to 0, otherwise the coupled power P_{ce} will be 0. This conclusion can be easily drawn from the field structure of the TE_{0n} mode. When $m = 0$, its electric field only has the E_x component, no the E_y component, and the E_x component cannot be coupled with the electric probe inserted from the wide side of the waveguide, and can only be coupled with the electric probe inserted from the narrow side of the waveguide.

The power of the TE_{mn} mode in the rectangular waveguide has been given by Eq. (4.19), but at this time $m \neq 0$ in the equation, so $N_m = 2$, the coupling coefficient of the electric probe to it should be:

$$C = 10\lg \frac{a^3 b\eta_0}{2N_n m^2 \pi^2 Z_0 S_{eff}^2}\left(\frac{\lambda}{\lambda_c}\right)^2 \frac{\sqrt{1-(\lambda/\lambda_c)^2}}{\sin^2(m\pi x/a)} \quad (m \neq 0) \tag{4.37}$$

2. **Coupling coefficient of magnetic probe**
(1) **The coupling coefficient of magnetic probe in circular waveguide**

In a circular waveguide, the tangential component of the magnetic field coupled with the magnetic probe can be either the H_φ or H_z component. We should calculate the power coupled with the magnetic probe based on the value of the magnetic field component at $r = R$.

Here, we only discuss the coupling of the magnetic probe with the H_φ component. The coupling with the H_z component can be similarly derived, and we will not go into detail about its process.

(a) TM_{mn} mode.

The field component expression of TM_{mn} in a circular waveguide has been given by Eq. (4.5), and when $r = R$, it is

$$\left|H_\phi\right|_{r=R} = \frac{\omega\varepsilon}{k_c} E_{zm}\left|J'_m(k_c R)\right| \tag{4.38}$$

The power coupled with the magnetic probe should be calculated according to Eq. (4.4)

$$P_{cm} = \frac{1}{2Z_0}\omega^2\mu^2 S_{eff}^2 \frac{\omega^2\varepsilon^2}{k_c^2} E_{zm}^2 J_m'^2(k_c R) = \frac{\omega^4\mu^2\varepsilon^2 S_{eff}^2}{2Z_0 k_c^2} E_{zm}^2 J_m'^2(k_c R) \tag{4.39}$$

The coupling coefficient of the magnetic probe with the TM_{mn} mode in the circular waveguide can be obtained by the size of P_t of the TM_{mn} mode given by Eq. (4.7).

$$C = 10\lg \frac{R^2 Z_0}{4\pi N_m S_{eff}^2}\frac{\lambda^2}{\eta_0}\sqrt{1-(\lambda/\lambda_c)^2} \tag{4.40}$$

where, N_m is given by Eqs. (4.9) or (4.21).

(b) TE_{mn} mode.

When $r = R$, it can be obtained from Eq. (4.10) that.

$$\left|H_\phi\right|_{r=R} = \frac{\beta m}{k_c^2 R} H_{zm} |J_m(k_c R)| \quad (\mathrm{m} \neq 0) \tag{4.41}$$

From this, it can be obtained that

$$P_{cm} = \frac{1}{2Z_0}\omega^2\mu^2 S_{eff}^2 \frac{\beta^2 m^2}{k_c^4 R^2} H_{zm}^2 J_m^2(k_c R) \quad (\mathrm{m} \neq 0) \tag{4.42}$$

From Eq. (4.42), it can be seen that m cannot be equal to 0, which is very obvious because the TE_{0n} mode is a circular electric mode, and its magnetic field only has the H_r component and H_z component, and there is no H_φ component. Therefore, when coupled with the H_φ component, the coupling power is 0. At this time, it can be changed to be coupled with the H_z component, as long as the magnetic probe is rotated 90° to make its coupling ring plane change from perpendicular to the angular direction to perpendicular to the z direction.

Combining Eq. (4.42) with Eq. (4.12), the coupling coefficient of the magnetic probe with the H_φ component of the TE_{mn} mode can be obtained (in this equation N_m is already equal to 2).

$$C = 10\lg \frac{R^2 Z_0}{8\pi S_{eff}^2}\left[\left(\frac{2\pi R}{m\lambda_c}\right)^2 - 1\right]\frac{\lambda^2}{\eta_0\sqrt{1-(\lambda/\lambda_c)^2}} \quad (m \neq 0) \tag{4.43}$$

(2) **The coupling coefficient of magnetic probe in rectangular waveguide**

We still assume that the magnetic probe is inserted from the wide side of the waveguide. At this time, it can be coupled with the H_x or H_z, depending on the direction of the coupling ring of the magnetic probe. Here, we only consider the case where it is coupled with the H_x magnetic field component. As for the case where it is coupled with the H_z component, readers can make similar deductions.

(a) TM_{mn} mode.

When $y = b$, it can be obtained from Eq. (4.14) that

$$|H_x|_{y=b} = \frac{\omega\varepsilon}{k_c^2}\frac{n\pi}{b}E_{zm}\left|\sin\left(\frac{m\pi}{a}x\right)\right| \quad (\mathrm{m} \neq 0,\, n \neq 0) \tag{4.44}$$

We know that m and n of the TM_{mn} mode in the rectangular waveguide cannot be 0, that is, there is no TM_{0n} mode and TM_{m0} mode.

The power coupled with the magnetic probe is

$$P_{cm} = \frac{1}{2Z_0}\left(\frac{k}{k_c}\right)^4 \frac{n^2\pi^2}{b^2} S_{eff}^2 E_{zm}^2 \sin^2\left(\frac{m\pi}{a}x\right) \quad (m \neq 0,\ n \neq 0) \tag{4.45}$$

Compared with the total power of the TM_{mn} mode transmitted in the waveguide (Eq. (4.16)), the coupling coefficient of the magnetic probe is obtained.

$$C = 10\lg \frac{ab^3 Z_0}{4n^2\pi^2 S_{eff}^2}\left(\frac{\lambda}{\lambda_c}\right)^2 \frac{\sqrt{1-(\lambda/\lambda_c)^2}}{\eta_0 \sin^2(m\pi x/a)} \quad (\mathrm{m} \neq 0,\ \mathrm{n} \neq 0) \tag{4.46}$$

(b) TE_{mn} mode.

When $y = b$, it can be obtained from Eq. (4.18) that

$$|H_x|_{y=b} = \frac{\beta}{k_c^2}\frac{m\pi}{a} H_{zm}\left|\sin\left(\frac{m\pi}{a}x\right)\right| \quad (\mathrm{m} \neq 0) \tag{4.47}$$

The reason why m cannot be 0 is easy to understand in the rectangular waveguide, because if $m = 0$, the magnetic field only has H_y and H_z components, and there is no H_x component.

When $m \neq 0$, the power coupled with the magnetic probe when TE_{mn} mode is coupled with the magnetic probe is

$$P_{cm} = \frac{1}{2Z_0}\frac{\omega^2\mu^2\beta^2}{k_c^4} S_{eff}^2 \left(\frac{m\pi}{a}\right)^2 H_{zm}^2 \sin^2\left(\frac{m\pi}{a}x\right) \quad (\mathrm{m} \neq 0) \tag{4.48}$$

Compared with the total power of the TE_{mn} mode transmitted in the rectangular waveguide (Eq. (4.19)), the coupling coefficient of the probe is obtained. Note that in Eq. (4.19), because $m \neq 0$, N_m should be directly equal to 2.

$$C = 10\lg \frac{a^3 b Z_0}{2N_n m^2\pi^2 S_{eff}^2}\left(\frac{\lambda}{\lambda_c}\right)^2 \frac{1}{\eta_0 \sin^2(m\pi x/a)\sqrt{1-(\lambda/\lambda_c)^2}} \quad (\mathrm{m} \neq 0) \tag{4.49}$$

4.3.3 Calibration of Effective Section of Probe

After obtaining the coupling coefficient of the probe and measuring the power coupled by the probe, the total power transmitted in the waveguide can be calculated using Eq. (4.23) in principle. However, in all the calculation formulas for the coupling coefficient, there is still an effective cross section S_{eff} of the probe that has not been determined. Therefore, we will discuss the calibration of S_{eff} below.

Obviously, the effective section S_{eff} of the probe is a function of the microwave mode and frequency coupled by the probe. Therefore, strictly speaking, the calibration of the effective cross section of the probe must be performed under the same state of the microwave field mode and frequency actually coupled by the probe during the probe operation. However, this will inevitably bring great difficulties to the calibration because it requires that the microwave signal used for calibration must have the same working mode and frequency as the high-power microwave source. The frequency problem is not big, and the signal source is usually frequency adjustable, but the requirement of the mode will make the calibration system quite complicated in general. We must first obtain the required working mode through a single-mode exciter or mode converter.

Here, we will only introduce the method of calibrating the effective cross section of the probe directly using a network analyzer in the most readily available waveguide mode, and the calibration of the effective cross section of the probe under other modes can be carried out using exactly the same method. The simplest and most convenient mode that can be obtained may be the TE_{10} mode in the rectangular waveguide or the TM_{01} mode in the circular waveguide, because these are the two modes that are easiest to excite relative to other modes, especially the TE_{10} mode in the rectangular waveguide, which is the fundamental mode and can achieve single-mode transmission.

1. **Calibration in rectangular waveguide**

The biggest advantage of using a rectangular waveguide to calibrate the effective cross section of the probe is that in a standard rectangular waveguide, the TE_{10} mode can be efficiently and uniquely excited through the waveguide-coaxial converter, and there will be no other modes. However, in practical high-power microwave power measurements using probes, generally, an overmoded waveguide is employed. To calibrate the probe under realistic operating conditions as much as possible, the same overmoded waveguide should be used. At this time, if the calibration is still performed on the TE_{10} mode, the high-quality waveguide-coaxial converter of the overmoded rectangular waveguide becomes a key component in the calibration system and must be carefully designed and manufactured. Alternatively, the standard waveguide can be used to excite the TE_{10} mode first, and then the gradual change waveguide can be used to transition to the overmoded waveguide, which may be a more convenient and reliable solution.

(1) **Calibration of effective section of electric probe**

The calibration system is shown in Fig. 4.5.

Using the above system, the S_{21} expressed in decibels is measured using a network analyzer. This S_{21} is defined as

$$S_{21} = 10\lg\frac{P'_c}{P'_t} = -10\lg\frac{P'_t}{P'_c} = -C \tag{4.50}$$

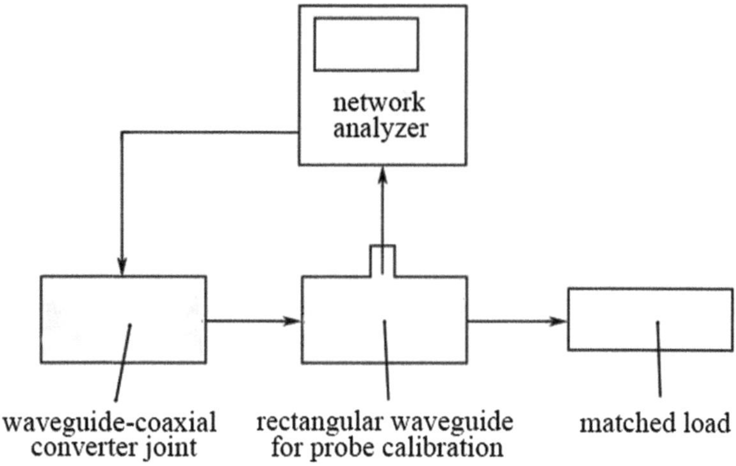

Fig. 4.5 Block diagram of the system calibrating the probe in a rectangular waveguide

where, P_t' is the input power of the measured waveguide, P_c' is the power coupled by the probe from the waveguide, C is the coupling coefficient represented by a positive value. Obviously, due to $P_c' < P_t'$, S_{21} itself is the negative value representing the coupling coefficient of the probe, and it should be equal to the negative value of the coupling coefficient of the rectangular waveguide electric probe given by Eq. (4.37), that is,

$$S_{21} = -10\lg \frac{a^3 b\eta_0}{2N_n m^2 \pi^2 Z_0 S_{eff}^2}\left(\frac{\lambda}{\lambda_c}\right)^2 \frac{\sqrt{1-(\lambda/\lambda_c)^2}}{\sin^2(m\pi x/a)} \tag{4.51}$$

Substituting the relevant parameters of the TE_{10} mode in the rectangular waveguide: $N_n = 1$, $m = 1$, $Z_0 = 50\ \Omega$, $\lambda_c = 2a$, $\eta_0 = 377\ \Omega$ into Eq. (4.51) and assuming that the probe is inserted from the center of the wide side of the waveguide, $x = a/2$, we get

$$S_{21} = -10\lg \frac{0.0955ab\lambda^2\sqrt{1-(\lambda/2a)^2}}{S_{eff}^2} \tag{4.52}$$

The effective section of the electric probe can be obtained from the formula below.

$$S_{eff}^2 = 0.0955ab\lambda^2\sqrt{1-(\lambda/2a)^2} \times 10^{-S_{21}/10} \tag{4.53}$$

(2) Calibration of effective section of magnetic probe

Similar to the electric probe, the effective section of the magnetic probe can also be calibrated using the rectangular waveguide working in the TE_{10} mode. The calibration system remains unchanged, only the electric probe is replaced with the magnetic probe. However, at this time, because the probe is still inserted from the center of the wide side, the H_z of the TE_{10} mode here is 0, so it can only be coupled with the H_x component, and H_x is the largest at $a/2$.

The coupling ring plane of the magnetic probe is perpendicular to the x direction during measurement. According to Eq. (4.49), the S_{21} expressed in decibels is:

$$S_{21} = -10\lg \frac{a^3 b Z_0}{2N_n m^2 \pi^2 S_{eff}^2}\left(\frac{\lambda}{\lambda_c}\right)^2 \frac{1}{\eta_0 \sin^2(m\pi x/a)\sqrt{1-(\lambda/\lambda_c)^2}} = -C \quad (4.54)$$

Substituting the relevant parameters of the TE_{10} mode in the rectangular waveguide: $N_n = 1$, $m = 1$, $Z_0 = 50\ \Omega$, $\lambda_c = 2a$, $\eta_0 = 377\ \Omega$, into the above equation, we get

$$S_{21} = -10\lg \frac{ab\lambda^2}{8\pi^2 S_{eff}^2}\frac{0.1326}{\sqrt{1-(\lambda/2a)^2}} \quad (4.55)$$

$$S_{eff}^2 = \frac{1.68ab\lambda^2 \times 10^{-3}}{\sqrt{1-(\lambda/2a)^2}} \times 10^{-S_{21}/10} \quad (4.56)$$

As can be seen, the effective section of the magnetic probe can be obtained by measuring the S_{21} expressed in decibels using a network analyzer.

2. Calibration in circular waveguide

When calibrating the effective section of the probe using TM_{01} mode in a circular waveguide, it is less convenient than using TE_{10} mode in a rectangular waveguide. Because the TM_{01} mode in the circular waveguide is not a fundamental mode, therefore, it is more difficult to excite the TM_{01} mode with the mode exciter in the circular waveguide than to excite the TE_{10} mode with a waveguide-coaxial converter in the rectangular waveguide. Not only may the reflection be larger, but more importantly, parasitic modes are easily generated. The calculation formula for calibration is written for a single specific mode, so the existence of parasitic modes will inevitably affect the accuracy of calibration.

The TM_{01} mode in the circular waveguide is generally excited by inserting the inner conductor of the coaxial line from a transverse end face of the circular waveguide, utilizing the high-frequency current in the inner conductor to generate a magnetic field around it, thus, exciting the TM_{01} mode. In order to improve the matching performance of the exciter and improve the mode conversion efficiency, the circular waveguide is usually made into a cone shape, as shown in Fig. 4.6. The minimum radius R_{min} of the conical section must satisfy the transmission condition of the TM_{01} mode, and it is best to cut off the TE_{21} mode that is one order higher than the TM_{01} mode, that is, R_{min} is required to be

$$0.486\lambda > R_{min} > 0.383\lambda \tag{4.57}$$

The maximum radius of the conical section should be the same as the radius of the overmoded circular waveguide actually used for the calibrated probe in the high-power microwave measurement, and connected to it. If there is a uniform circular waveguide with R_{min} as the radius in front of the conical section, it will be better, which can make the higher-order spurious modes have sufficient attenuation and disappear, and then connect the conical waveguide to the required overmoded waveguide.

The measurement system for calibrating the effective section of the probe in the circular waveguide is shown in Fig. 4.7.

(1) **Calibration of effective section of electric probe**

The probe is coupled with the E_r component in the circular waveguide, and the S_{21} expressed in decibels is measured in the system where the TM_{01} mode has been excited. According to Eq. (4.28), it can be obtained that

$$S_{21} = -10\lg\frac{P_t}{P_{ce}} = -10\lg\frac{R^2\eta_0}{4\pi N_m S_{eff}^2 Z_0}\frac{\lambda^2}{\sqrt{1-(\lambda/\lambda_c)^2}} \tag{4.58}$$

For the TM_{01} mode, substituting $N_m = 1$, $\lambda_c = 2.62R$, and $Z_0 = 50\ \Omega$, $\eta_0 = 377\ \Omega$, into above equation, it can be obtained that

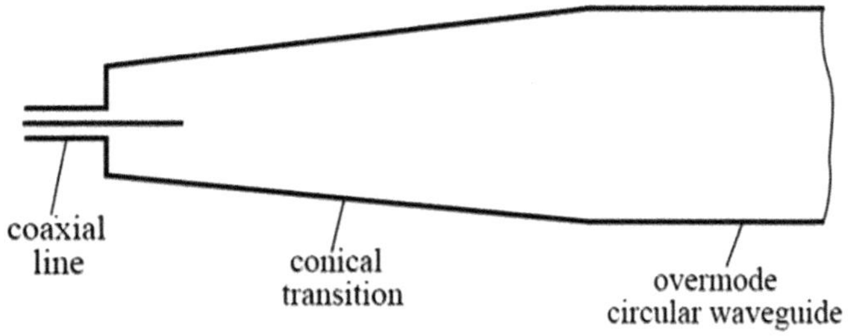

Fig. 4.6 The exciter of the TM_{01} mode in a circular waveguide

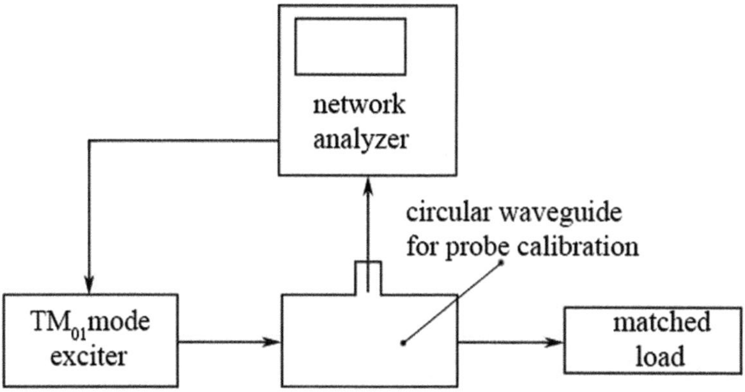

Fig. 4.7 Block diagram of the system for calibrating the probe in a circular waveguide

$$S_{21} = -10\lg \frac{0.6R^2}{S_{eff}^2} \frac{\lambda^2}{\sqrt{1-(\lambda/2.62R)^2}} \tag{4.59}$$

From this, the effective section of the electric probe can be obtained.

$$S_{eff}^2 = \frac{0.6R^2\lambda^2}{\sqrt{1-(\lambda/2.62R)^2}} \times 10^{-S_{21}/10} \tag{4.60}$$

(2) **Calibration of effective section of magnetic probe**

The magnetic probe is coupled with the H_φ component in the circular waveguide. After measuring $S_{21,}$ according to Eq. (4.40), it should be equal to following formula

$$S_{21} = -10\lg \frac{R^2 Z_0}{4\pi N_m S_{eff}^2} \frac{\lambda^2}{\eta_0} \sqrt{1-(\lambda/\lambda_c)^2} \tag{4.61}$$

For the TM_{01} mode, $N_m = 1$, $\lambda_c = 2.62\text{R}$, $Z_0 = 50\ \Omega$, and $\eta_0 = 377\ \Omega$. In this way, S_{21} is obtained.

$$S_{21} = -10\lg \frac{1.0544 \times 10^{-2} R^2 \lambda^2}{S_{eff}^2} \sqrt{1-(\lambda/2.62R)^2} \tag{4.62}$$

Therefore, the effective section of the magnetic probe can be obtained.

$$S_{eff}^2 = 1.0544 \times 10^{-2} R^2 \lambda^2 \sqrt{1-(\lambda/2.62R)^2} \times 10^{-S_{21}/10} \tag{4.63}$$

All the expressions of the effective cross section S_{eff} of the probe obtained above can be seen to contain both λ and λ_c, that is, frequency and mode. It can be seen that

S_{eff} is a function of the frequency and mode of high-power microwaves. Therefore, strictly speaking, for high-power microwaves of different frequencies and modes, a single-mode generator or mode converter needs to be used to obtain the required mode to calibrate S_{eff}.

4.4 Small-Hole Coupling Method

The small hole coupling method is also a widely used method for measuring the coupling field power of high-power microwaves. In practical measurements, directional couplers are used for small hole coupling, and the design and calculation technology for the couplers is already quite mature, which brings great convenience to the application of the small hole coupling method.

Small holes generally refer to coupling holes whose radius r_0 is much smaller than the microwave wavelength λ.

4.4.1 *Technical Specifications of Directional Coupler*

The directional coupler is a widely used waveguide component, which can be regarded as a directional power divider. As shown in Fig. 4.8, the directional coupler is composed of two microwave transmission lines coupled to each other. The transmission line that inputs microwaves is called the main line or main waveguide, and the transmission line that couples a portion of microwave power from the main line through small holes is called the secondary line or secondary waveguide. For the directional coupler, the microwave power P_1 input from the main line is coupled to the secondary waveguide through the small holes, and most of the power in the secondary line only propagates in one direction (P_2^+), and only a small amount of power propagates in the reverse direction ($P_2^- \ll P_2^+$), which is the directivity of the directional coupler. The remaining power P_1' in the main line is output from the other end of the main line. The input port of the microwave power P_1 in the main line is called the input port, and most of the power will still be output from the main line except for a small part of the power coupled to the secondary line. This output port is called the through port. In the secondary line, the port where the coupled power enters and propagates in the same direction as P_1 (forward) is called the output port, and the port where the power is transmitted in the reverse direction (reverse) and output is called the isolation port. For a single-hole directional coupler, the output port and the isolation port are exactly opposite, that is, the output port is in the reverse direction and the isolation port is in the forward direction.

Due to the reciprocity principle of linear elements, the left end of the main line in Fig. 4.8 can be used as the microwave power input port, or the right end can also be used as the input port, and the effect is the same. Note that the output port and the isolation port in the secondary line are interchanged simultaneously. At this time,

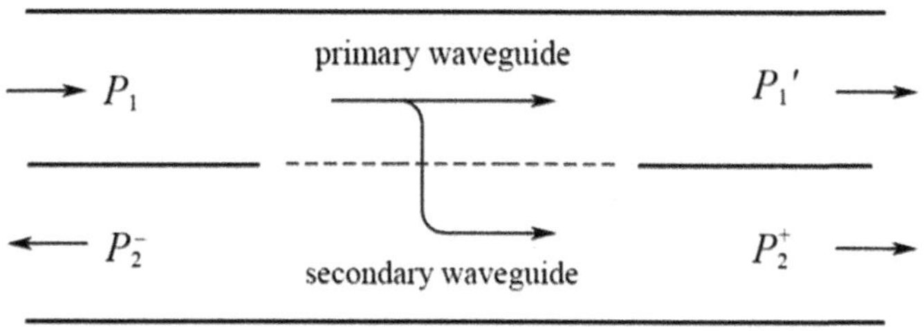

Fig. 4.8 The schematic diagram of the directional coupler

not only does the coupling coefficient remain unchanged, but the directivity between them also remains unchanged. This characteristic of the directional coupler allows us to sample the forward and reverse waves in the main line separately and perform measurements, monitoring, etc. It can be seen that this is a very useful waveguide component.

The main technical indicators of directional couplers include coupling coefficient, directivity and isolation, input standing wave ratio, and operating bandwidth.

1. **Coupling coefficient**

The decibel value of the ratio of the power P_2^+ in the forward propagation coupled from the main line to the secondary line by the directional coupler to the power P_1 of the incident wave in the main line is called the coupling coefficient, expressed as C

$$C = 10\lg\frac{P_2^+}{P_1}\ (\text{dB}) \tag{4.64}$$

Obviously, since P_2^+ is always less than P_1, the coupling coefficient is a negative value. It is customary for people to directly call its absolute value as the coupling coefficient, i.e.

$$C = \left|10\lg\frac{P_2^+}{P_1}\right| = 10\lg\frac{P_1}{P_2^+}\quad (\text{dB}) \tag{4.65}$$

In this way, the value of the coupling coefficient becomes a positive value. In this book, the coupling coefficient mentioned refers to the positive value of the coupling coefficient represented by Eq. (4.65), unless otherwise specified.

2. **Directivity and isolation**

Directivity is an important index of directional couplers. Its definition is the distribution ratio of microwave power transferred from the main line coupling to the secondary line in forward and backward propagation, expressed in decibels, that is,

$$d = 10\lg \frac{P_2^+}{P_2^-} \text{ (dB)} \tag{4.66}$$

Directivity d is a quality index that reflects the ability of the directional coupler to propagate power in a directional manner from the main line coupling to the secondary line. The larger d is, the better the directivity of the directional coupler.

Based on the coupling coefficient and directivity, the isolation between the isolation port and the input port of the directional coupler can be reflected as follows,

$$I = 10\lg \frac{P_1}{P_2^-} = 10\lg \frac{P_1}{P_2^+}\frac{P_2^+}{P_2^-} = C + d \text{ (dB)} \tag{4.67}$$

3. **Input standing wave ratio**

The input standing wave ratio is defined as the standing wave coefficient measured at the input port of the main line when the direct-through port (P_1' output port), the forward coupling port (P_2^+ output port), and the reverse isolation port (P_2^- output port) of the directional coupler are all connected to a matching load. The input standing wave ratio reflects the size of the reflection observed at the input end.

4. **Operating bandwidth**

The operating bandwidth refers to the operating frequency range of the directional coupler when the coupling coefficient, directivity, and input standing wave ratio all meet the given requirements.

The theoretical basis of the directional coupler is the small hole diffraction theory and the principle of phase superposition. We will introduce their basic theories of these two aspects separately.

4.4.2 Fundamental Theory of Small-Hole Coupling

1. **Small hole diffraction theory**

(1) **Coupling strength** [44, 45]

The electromagnetic wave uses the diffraction characteristics of the small hole to realize energy coupling. When the size of the small hole is much smaller than the electromagnetic wave wavelength, the small hole can be equivalent to a combination of electric dipole and magnetic dipole (Fig. 4.9). The dipole moment of the electric dipole is proportional to the normal electric field $\boldsymbol{E}_{1\mathrm{n}}$ of the incident wave at the small hole, and the dipole moment of the magnetic dipole is proportional to the tangential magnetic field $\boldsymbol{H}_{1\mathrm{t}}$ of the incident wave at the small hole. The tangential magnetic field can be divided into two components in orthogonal directions of u and v. These dipole moments, which are related to the shape and size of the small hole, are expressed as

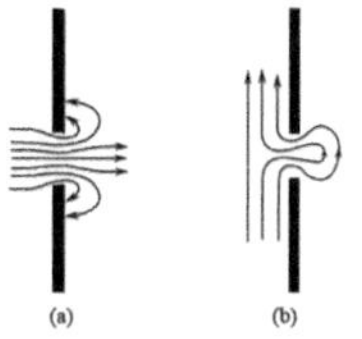

Fig. 4.9 The coupling of small holes, **a** equivalent electric dipole, **b** equivalent magnetic dipole

electric dipole

$$\boldsymbol{P} = -\varepsilon_0 p_{\rm n} \boldsymbol{E}_{1\rm n} \tag{4.68}$$

magnetic dipole

$$\boldsymbol{M} = m_{\rm u} \boldsymbol{H}_{1\rm u} + m_{\rm v} \boldsymbol{H}_{1\rm v} \tag{4.69}$$

where, $p_{\rm n}$ is the normal electric polarization rate, $m_{\rm u}$ and $m_{\rm v}$ are the two orthogonal tangential magnetic polarization rates. They depend on the shape and size of the small hole. The most commonly used small hole is a circular hole, and in rare cases, elliptical holes and rectangular holes are used. Their polarization rates are as follows:

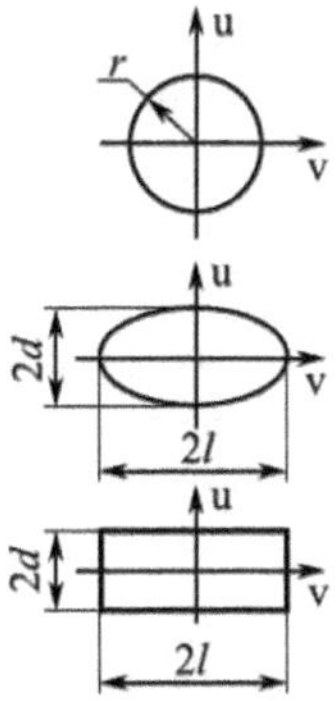

circular hole	$m_{\rm u} = \frac{4}{3} r^3$	$m_{\rm v} = \frac{4}{3} r^3$	$p_{\rm n} = \frac{2}{3} r^3$
elliptical hole	$m_{\rm u} = \frac{\pi}{3} \frac{l^3}{\ln(4l/d) - 1}$	$m_{\rm v} = \frac{\pi}{3} l d^2$	$p_{\rm n} = \frac{\pi}{3} l d^2$
reclanoular hole	$m_{\rm u} = 0$	$m_{\rm v} = \frac{\pi}{2} l d^2$	$p_{\rm n} = \frac{\pi}{2} l d^2$

Thus, the relative amplitude—coupling strength of the wave in the secondary waveguide excited by the mode field in the main waveguide through the small hole can be calculated. Assuming that the field value of the incident wave mode field in the main waveguide at the position of the small hole is represented by the subscript 1, and the field value of the excited mode wave in the secondary waveguide at the position of the small hole is represented by the subscript 2, then according to the small hole coupling theory, the small hole coupling strength between the main and secondary waveguides can be expressed as

$$\begin{aligned} a^{\pm} &= \frac{A_2^{\pm}}{A_1} = -j\frac{\omega}{2}\left(\mu_0 \boldsymbol{M} \cdot \boldsymbol{H}_{2\rm t}^{\mp} - \boldsymbol{P} \cdot \boldsymbol{E}_{2\rm n}^{\mp}\right) \\ &= -j\frac{\omega}{2}\left(\mu_0 m_{\rm u} H_{1\rm u} H_{2\rm u}^{\mp} + \mu_0 m_{\rm v} H_{1\rm v} H_{2\rm v}^{\mp} + \varepsilon_0 p_{\rm n} E_{1\rm n} E_{2\rm n}^{\mp}\right) \end{aligned} \tag{4.70}$$

where,

A_1 is the amplitude coefficient of the incident wave in the main waveguide,

A_2 is the amplitude coefficient of the excited wave in the secondary waveguide,

a is relative ratio of A_2 and A_1, called the coupling strength, the superscript "+" represents the forward wave, and the "-" represents the reverse wave. In the main waveguide, only the forward incident wave is present, so the "+" sign is omitted,

H_{1v},$H_{2v}^{\pm}$ are the normalized magnetic field tangential components in the v direction in the main and secondary waveguides at the position of the small hole, respectively,

H_{1u},$H_{2u}^{\pm}$ are the normalized magnetic field tangential components in the u direction,

E_{1n}, $E_{2n}^{\pm}$ are the normalized electric field normal components at the position of the small hole in the main and secondary waveguides, respectively,

Since the incident wave in the main waveguide and the excited wave in the secondary waveguide may have different modes, and even the waveguide types of the main and secondary waveguides themselves may be different, in order to make the field components of the two comparable, all field components in the formula should be normalized values.

Equation (4.70) shows that the condition for the existence of coupling between the main and secondary waveguides through small holes is: at the position of the small hole, there exists either a normal component of the electric field or a tangential component of the magnetic field in the same direction in both the main and secondary waveguides, and this component is non-zero at the location of the small hole. We can use Eq. (4.70) to determine whether there is small hole coupling between two waveguides or to determine what kind of mutual coupling position should be taken between two waveguides to cause small hole coupling. For example, when the narrow side of a single-mode (TE_{10} mode) rectangular waveguide as the secondary waveguide is coupled with the main waveguide, that is, the coupling hole is located on the common wall between the narrow side of the secondary rectangular waveguide and the main waveguide, then regardless of whether the main waveguide is a rectangular waveguide or a circular waveguide, the TM mode cannot couple with the TE_{10} mode in the secondary waveguide. This means that the TM mode in the main waveguide cannot excite the TE_{10} mode in the secondary waveguide (According to the reciprocity principle, the reverse is also true. When a rectangular waveguide is used as the main waveguide to input microwaves and couples with a small hole on its narrow side, the TE_{10} mode cannot excite the TM mode in the secondary waveguide, regardless of whether it is a rectangular waveguide or a circular waveguide.). This is because the TE_{10} mode does not have a normal electric field perpendicular to the narrow side at the position of the small hole, and the tangential magnetic field parallel to the narrow side at the position of the small hole has only a H_z component, but for the TM mode, it is just the opposite. Although there is a normal electric field component E_n perpendicular to the common wall, there is no magnetic field tangential component H_z, therefore there is no common same direction field component between the two at the coupling hole position, so coupling cannot occur.

The above coupling principle also applies to the coupling of fields between small holes in other microwave components, such as between resonant cavities, between resonant cavities and waveguides, and between different types of transmission lines.

It should be emphasized that Eq. (4.70) is only applicable to the case where the small hole size is much smaller than the electromagnetic wave wavelength, that is, $r_0 \ll \lambda$. r_0 is the small hole radius, and λ is the working wavelength.

(2) **The correction of coupling strength of small hole**

Equation (4.70) is derived under the assumption of $r_0 \ll \lambda$ and neglecting the thickness of the waveguide wall, so the field components therein only reflect the distribution and magnitude of the fields at the center of the small hole, without considering the influence of the radial distribution of the field components along the small hole when the coupling hole does not meet the small hole condition, nor considering the attenuation of the electromagnetic waves passing through the small hole with a certain wall thickness, thereby causing a certain error in the calculation of $a^{\pm}$.

F. Spolaore and H. G. Engler pointed out that the radius and wall thickness of the coupling hole have an impact on the coupling strength, and can be corrected by multiplying the electric dipole moment and magnetic dipole moment by the aperture correction factor and the wall thickness correction factor, respectively [46].

(a) Correction factor of wall thickness

When considering the influence of the wall thickness on the coupling, it can be divided into two cases: when the modes in the main waveguide and the secondary waveguide are the same and when they are different. They respectively provide expressions for the wall thickness correction factors K_c and K_m for the electric dipole moment and the magnetic dipole moment under these two situations.

When the modes in the main and secondary waveguides are the same, thickness correction factors are:

$$\begin{aligned} K_e &= 1 - 0.13\mathrm{e}^{-2\gamma_e t} \\ K_m &= 1 - \left[0.35 + 0.15(kr_0)^2\right](1 - \mathrm{e}^{-2\gamma_m t}) \end{aligned} \tag{4.71}$$

When the modes in the main and secondary waveguides are different, then:

$$\begin{aligned} K_e &= \left[1 - 0.14(1 - \mathrm{e}^{-2\gamma_e t})\right]\mathrm{e}^{-\gamma_e t} \\ K_m &= \left[1 - \left\{0.1 + \left[0.3 + 0.15(kr_0)^2\right]^2\right\}(1 - \mathrm{e}^{-2\gamma_m t})\right]\mathrm{e}^{-\gamma_m t} \end{aligned} \tag{4.72}$$

In Eqs. (4.71) and (4.72): k is the free-space wave number; t is the wall thickness of the coupling hole, γ_e and γ_m are the propagation constants of the circular waveguide TM_{01} and TE_{11} modes, respectively.

$$\gamma_e = \left[\left(\frac{2.405}{r_0}\right)^2 - k^2\right]$$
$$\gamma_m = \left[\left(\frac{1.84}{r_0}\right)^2 - k^2\right] \quad (4.73)$$

From this, it can be seen that in the correction of the coupling strength for the wall thickness, γ_e and γ_m are in physical essence to regard the small hole as a section of cutoff circular waveguides transmitting the TM_{01} and TE_{11} modes, respectively.

(b) Correction factor of large aperture

When the radius of the coupling hole does not satisfy the small hole condition $r_0 \ll \lambda$, it is necessary to consider the correction for the large hole of the electric dipole moment and the magnetic dipole moment F. Spolaore and H. G. Engler proposed the calculation formula for the large hole correction factor R_e, R_m.

$$R_e = \frac{1}{1 + 0.4(kr_0)^2}$$
$$R_m = \frac{1}{1 - 0.4(kr_0)^2} \quad (4.74)$$

The factors K_e, K_m, R_e, and R_m have quite high accuracy under the condition of $kr_0 \leq 1$, that is, $r_0 \ll \lambda$. Obviously, under the conditions of thin walls and small apertures, $K_e \approx 1$, $K_m \approx 1$, $R_e \approx 1$, and $R_m \approx 1$, and thus Eq. (4.70) is obtained. After introducing the correction factors K_e, K_m, R_e, and R_m, Eq. (4.70) should be modified as follows:

$$a^{\pm} = -j\frac{\omega}{2}\left[(\mu_0 m_u H_{1u} H_{2u}^{\mp} + \mu_0 m_v H_{1v} H_{2v}^{\mp}) K_m R_m + \varepsilon_0 p_n E_{1n} E_{2n}^{\mp} K_e R_e\right] \quad (4.75)$$

In high-power microwave systems, the main waveguide is generally an overmoded waveguide, with a relatively large size and a correspondingly thick waveguide wall. Therefore, the influence of the wall thickness on the coupling cannot be completely ignored. Relatively speaking, the coupling required for high-power measurements is very small (the positive coupling coefficient is very large), which means that the coupling hole is very small and generally satisfies the small hole condition $r_0 \ll \lambda$, so the influence of the aperture size on the coupling can often be ignored.

2. **Phase superposition principle**

In general, directional couplers are often composed of multiple coupling holes, and multi-hole directional couplers can significantly expand the operating bandwidth and improve directionality. When the coupling is not done through a single hole but through multiple holes, the total coupling depends not only on the coupling strength of each hole determined by the small hole diffraction theory, but also on the phase relationship between the waves excited by each hole.

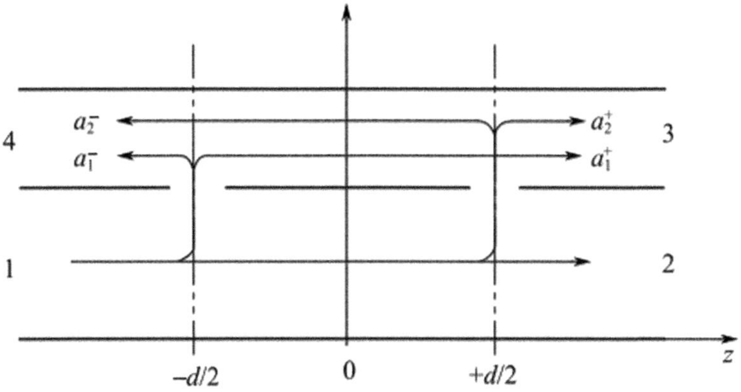

Fig. 4.10 Schematic diagram of two-hole coupling

(1) **Dual-hole coupling**

Let's first analyze the situation of two coupling holes (Fig. 4.10). Ports 1 to 2 represent the main waveguide, while ports 3 to 4 represent the secondary waveguide. In our analysis, we consider that the coupling of the small holes is very weak and therefore will not affect the amplitude of the incident wave in the main waveguide, that is, the amplitude of the incident wave in the main waveguide remains unchanged. In this way, if the two holes have the same shape and size, their relative coupling strength is also the same, assumed to be $a^{\pm}$, $a^{\pm}$ determined by Eqs. (4.70) or (4.75). At the same time, let's assume that the phase constant of the incident wave in the main waveguide is β_1, and the phase constant of the wave excited in the secondary waveguide is β_2, with the two holes located at $z = \pm d/2$.

We can consider that for the situation of the composite wave resulting from the excitation of two forward waves and two backward waves in the secondary waveguide by two holes, without considering waveguide losses, we only need to consider their superposition results at the positions $z = + d/2$ and $z = - d/2$ in the secondary waveguide. This is because, for the two backward waves excited by the two holes, the composite wave after superposition at the position $z = - d/2$ in the secondary waveguide, when it continues to propagate in the waveguide at $z < - d/2$, as x changes, the composite wave will remain in its original form for transmission. Therefore, even if it is output at any position of $z < - d/2$, the amplitude will be exactly the same as at the position $z = - d/2$. So, thus it can be seen that we only need to consider the superposition results of the two backward waves at the position $z = - d/2$. Similarly, for the two forward waves excited by the two holes in the secondary waveguide, we only need to consider their composite wave resulting after superposition at the position $z = + d/2$.

If the incident wave is input at $z = - d/2$ in the main waveguide, with an amplitude of 1 and a phase of 0. The relative amplitude of the forward and backward waves excited by the first hole at $- d/2$ in the secondary waveguide will be $a_1{}^+$ and $a_1{}^-$, respectively. We examine the case of superposition of two backward waves at z

$= -\ d/2$ and the superposition of two forward waves at $z = +\ d/2$. Where a_1^- remains unchanged at $-\ d/2$, and a_1^+ must propagate a distance of d in the secondary waveguide to reach $z = +\ d/2$, where its relative amplitude has become $a_1^+ \mathrm{e}^{-\mathrm{j}\beta_2 d}$. The incident wave will first propagate a distance of d in the main waveguide before reaching the second hole at $+\ d/2$, at which point the relative amplitude of the incident wave will become $\mathrm{e}^{-\mathrm{j}\beta_1 d}$. Therefore, the relative amplitude of the forward wave excited in the secondary waveguide by the second hole at $+\ d/2$ is $a_2^+ \mathrm{e}^{-\mathrm{j}\beta_1 d}$, and the relative amplitude of the backward wave excited by the second hole at $+\ d/2$ is $a_2^- \mathrm{e}^{-\mathrm{j}\beta_1 d}$, but the backward wave will propagate another distance of d in the secondary waveguide before returning to the first hole at $-\ d/2$, at which point its relative amplitude will be $a_2^- \mathrm{e}^{-\mathrm{j}(\beta_1+\beta_2)d}$. As mentioned earlier, two identical holes have the same coupling strength, that is, $a_1^- = a_2^- = a_1^-$, $a_1^+ = a_2^+ = a_1^+$.

So, at $z = +\ d/2$, the composite waves of the two forward waves excited by two coupling holes in the secondary waveguide will be

$$A^+ = a_1^+ \mathrm{e}^{-\mathrm{j}\beta_2 d} + a_2^+ \mathrm{e}^{-\mathrm{j}\beta_1 d} = 2a^+ \mathrm{e}^{-\mathrm{j}(\beta_1+\beta_2)d/2} \cos(\beta_1 - \beta_2)d/2 \tag{4.76}$$

And at $z = -d/2$, the composite waves of the two backward waves in the secondary waveguide will be

$$A^- = a_1^- + a_2^- \mathrm{e}^{-\mathrm{j}(\beta_1+\beta_2)d} = 2a^- \mathrm{e}^{-\mathrm{j}(\beta_1+\beta_2)d/2} \cos(\beta_1 + \beta_2)d/2 \tag{4.77}$$

Because ultimately all we care about is the size of amplitude of the waves generated by the excitation, we can just consider their overall relative amplitudes, i.e.

$$A^\pm = \left|2a^\pm \cos\theta^\pm\right| \tag{4.78}$$

where,

$$\theta^\pm = \left|(\beta_1 \mp \beta_2)d/2\right| \tag{4.79}$$

Obviously, when, $\theta^+ = i^+\pi(i^+ = 0, 1, 2, \ldots), \left|\cos\theta^+\right| = 1$, the coupling of the two holes will be in-phase superposition, and when $\theta^- = (i^- - 1/2)\pi(i^- = 1, 2, \ldots), \cos\theta^- = 0$, the coupling of the two holes will be out-of-phase cancellation, meaning there is no output in the backward direction, thus achieving ideal directional coupling.

(2) **Multi-hole coupling**

If there are a total of $N = 2n$ symmetrically distributed coupling holes relative to the centerline (Fig. 4.11a), each pair of symmetric holes not only has symmetrically distributed positions, but also symmetric shapes and sizes. If the single-hole coupling strengths of each pair of holes are $a_1^\pm$, $a_2^\pm$, ..., $a_k^\pm$, ..., $a_n^\pm$, then similar to the derivation for double-hole coupling, we can obtain the total relative amplitude of the waves coupled into the auxiliary waveguide as

$$A^{\pm} = \left|2a_1^{\pm}\cos\theta_1^{\pm} + 2a_2^{\pm}\cos\theta_2^{\pm} + \cdots + 2a_k^{\pm}\cos\theta_k^{\pm} + \cdots + 2a_n^{\pm}\cos\theta_n^{\pm}\right|$$
$$= 2\left|\sum_{k=1}^{n} a_k^{\pm}\cos\theta_k^{\pm}\right| \quad (\text{when } N = 2n) \tag{4.80}$$

where,

$$\theta_k^{\pm} = (\beta_1 \mp \beta_2)\frac{d_k}{2}$$
$$d_k = 2\sum_{k=1}^{k} S_k - S_1 \quad (\text{k} = 1, 2)\ldots k, \ldots n) \tag{4.81}$$

The definition of S_k, d_k are shown in Fig. 4.11a.

If there is a single coupling hole with a coupling strength of $a_0{}^{\pm}$ at the center of symmetry (Fig. 4.11b), the total number of coupling holes becomes $N = 2n + 1$, and the total relative amplitude of the wave excited in the secondary waveguide becomes

$$A^{\pm} = a_0^{\pm} + 2\left|\sum_{k=1}^{n} a_k^{\pm}\cos\theta_k^{\pm}\right| \quad (\text{when } N = 2n + 1) \tag{4.82}$$

where,

$$\theta_k^{\pm} = (\beta_1 \mp \beta_2)\frac{d_k}{2}$$
$$d_k = 2\sum_{k=1}^{k} S_k \quad (\text{k} = 1, 2, \ldots k, \ldots n) \tag{4.83}$$

Regardless of whether $N = 2n$ or $N = 2n + 1$, only when the values of θ_k corresponding to d_k for different values of k satisfy

$$\theta_k^{+} = 2i_k^{+}\pi \left(i_k^{+} = 0, 1, 2, \ldots\right) \tag{4.84}$$

or satisfy

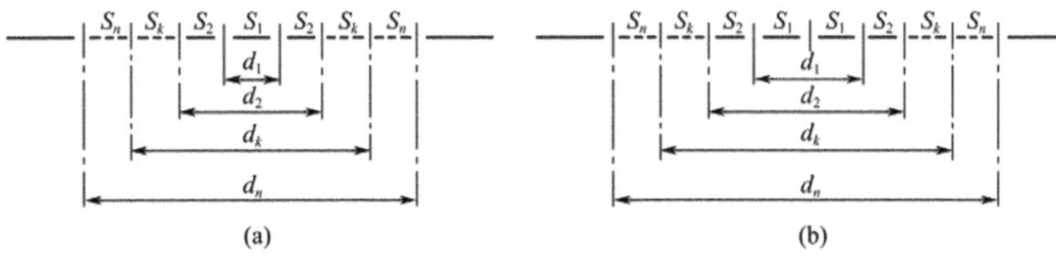

Fig. 4.11 Holes distribution for multi-hole coupling, **a** the number of holes $N = 2n$, **b** the number of holes $N = 2n + 1$

$$\theta_k^+ = \left(2i_k^+ - 1\right)\pi \ \left(i_k^+ = 1, 2, \ldots\right) \tag{4.85}$$

all forward waves can achieve in-phase superposition. It should be noted here: the values of θ_k must either all satisfy Eq. (4.84) or all satisfy Eq. (4.85); there cannot be some θ_k that satisfy Eq. (4.84) and some θ_k that satisfy Eq. (4.85), because their cosines one is + 1 and another is -1, and they will cancel each other out when superimposed.

And when

$$\theta_k^- = \left(i_k^- - 1/2\right)\pi \ \left(i_k^- = 1, 2, \ldots\right) \tag{4.86}$$

all backward waves (except a_0^-) can be out-of-phase cancellation

4.4.3 Practical Directional Coupler

In high-power microwave power measurements, practical directional couplers often consider mainly large power capacity and high coupling coefficient (low coupling power), and the operating bandwidth of directional couplers is generally not the main consideration. After strictly calibrating the coupling coefficient of the directional coupler, as well as the attenuators and power meters (or detector-oscilloscope combinations), the directional coupler can be used for high-power measurements and power sampling and monitoring of high-power microwave sources.

1. **Waveguide-coaxial single-hole directional coupler for high-power microwave power monitoring** [47–49]

Single-hole directional couplers are very simple. Because there is only one coupling hole, the coupling is weak, the coupling coefficient expressed with positive values is relatively high, but their operating frequency band is narrow and its directionality is relatively low. It is often used for power monitoring and sampling, and when necessary, it can also be used for power measurements. The simplest single-hole directional coupler is a rectangular waveguide single-hole directional coupler, where the main waveguide and the secondary waveguide are standard rectangular waveguides with the same dimensions, and a part of their wide or narrow sides overlap to form a common wall, with the coupling hole located at the center of the common wall. This type of directional coupler is commonly used in conventional microwave systems. However, it is obviously not suitable for high-power microwaves, as high-power microwave transmission and measurement require the main waveguide of the directional coupler to be an overmoded waveguide, while the secondary waveguide generally is a standard fundamental mode waveguide. As for the directional coupler specifically used for power and mode measurements in overmoded waveguides (including rectangular waveguides and circular waveguides), we will introduce it in Chap. 7. Here, we introduce a special single-hole directional coupler mainly used for power monitoring and sampling—the waveguide-coaxial directional coupler.

When the single-hole directional coupler is used for microwave signal sampling and monitoring, the signal coupled from the secondary waveguide always enters the low-power system for measurement. Many measurement instruments have input ports that are often coaxial connectors. Therefore, if the output of the secondary waveguide of the directional coupler is directly a coaxial line rather than a waveguide, it will be much more convenient to connect it to the measurement instrument. The waveguide-coaxial single-hole directional coupler is developed to meet this need and is now being used more and more widely. It is not only easy to use, but also has a simple structure, small size, and because the main and secondary waveguides are different types of transmission lines, it is very convenient to use overmoded rectangular waveguides for the main waveguide. However, in the power monitoring and measurement of high-power microwaves, the application of this directional coupler is still relatively rare. In fact, if technicians are preparing to use the small-hole coupling method for high-power microwave monitoring and measurement, the waveguide-coaxial single-hole directional coupler should be one of the options worth considering.

The main waveguide of the waveguide-coaxial single-hole directional coupler is a rectangular waveguide, while the secondary waveguide is a strip line located in a closed cylindrical cavity. The baseboard of the strip line is the bottom of the cylindrical cavity, which forms a common wall with the wide side of the rectangular waveguide, and the coupling small hole is located in the center of the common wall. The two ends of the strip line are connected to the inner conductor of the coaxial line, and the outer conductor of the coaxial line is connected to the upper cover plate of the cylindrical cavity, which is the top plate of the strip line. Figure 4.12 shows a schematic diagram of this type of directional coupler structure.

Waveguide-coaxial single-hole directional couplers can also be designed using small hole diffraction theory. The main difference is that the secondary waveguide is a strip line, while the main waveguide is an overmoded rectangular waveguide. We can easily write down the field components of the high-order modes in the overmoded waveguide, but the problem is how to get the expression of the field component of the strip line in the cylindrical cavity. The authors of this book have

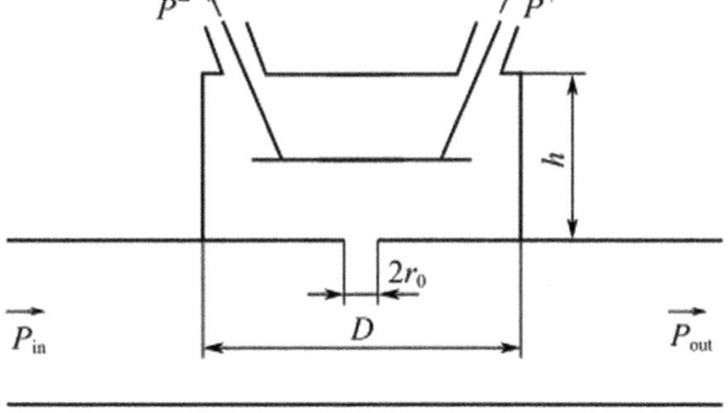

Fig. 4.12 Waveguide-coaxial single-hole directional coupler

proposed three methods to solve this problem. The basic idea is to use the Schwarz-Christoffel conformal transformation method to transform the strip line in the closed cylindrical cavity into a flat capacitor, in order to determine its field distribution. Then, the coupling coefficient of the coupler is calculated based on Eq. (4.70). The authors have progressively presented three transformation methods, from simple to complex, gradually increasing the accuracy, effectively solving the design and calculation problems of this single-hole directional coupler.

(1) **Ideal strip line model I**

When the width of the central conducting strip in the strip line is much smaller than the size of the ground plate, and the thickness is much smaller than the distance between the upper and lower ground plates, and when assuming that the central conducting strip matches the output coaxial line at both ends, as an approximation, we can consider the actual strip line as an ideal strip line with infinitely large ground plates and zero thickness for the central conducting strip. Under this approximation, using the transformation function

$$\begin{aligned} z_1 &= \mathrm{e}^{\frac{\pi z}{h}} \\ \omega &= \operatorname{arc}\ \cos z_1 \end{aligned} \tag{4.87}$$

The ideal stripline on the original $z = x + \mathrm{j}y$ plane can be transformed onto the $z_1 = x_1 + \mathrm{j}y_1$ plane, and then transformed from the z_1 plane to the $\omega = u + \mathrm{j}v$ plane to become a plate capacitor, where the plate spacing is $\pi/2$, the width is $\operatorname{arccosh}(\mathrm{e}^{\pi b/2h})$, b represents the length of the central conductive strip, and h denotes the distance between the upper and lower bottom of the cavity.

The potential distribution $\varphi\ (\omega)$ on a plate capacitor can be expressed as

$$\varphi(\omega) = V_0 - \left(\frac{2V_0}{\pi}\right)u \tag{4.88}$$

From this, the electric field distribution of the stripline in the z plane can be obtained

$$\begin{aligned} E(z) &= -\left(\frac{\mathrm{d}\phi}{\mathrm{d}z}\right)^* = -\left[\left(\frac{\mathrm{d}\phi}{\mathrm{d}\omega}\right)\left(\frac{\mathrm{d}\omega}{\mathrm{d}z_1}\right)\left(\frac{\mathrm{d}z_1}{\mathrm{d}z}\right)\right]^* \\ &= -\frac{2V_0}{h}r^{-1/2}\left(\cos\frac{\phi}{2} + \mathrm{j}\sin\frac{\phi}{2}\right) \end{aligned} \tag{4.89}$$

where, the "*" sign denotes the conjugation, and V_0 is the potential of the central conductive strip, and since the field in the stripline is a TEM wave, V_0 can be understood as the amplitude coefficient of the field. Since on the z plane, x is the real axis and y is the imaginary axis, Eq. (4.89) can be directly written as

$$\boldsymbol{E}^{\pm}(z) = Em^{\pm}r^{-\frac{1}{2}}\left[\boldsymbol{a}_x\cos\frac{\varphi}{2} + \boldsymbol{a}_y\sin\frac{\varphi}{2}\right] \tag{4.90}$$

From

$$\frac{E_x^{\pm}}{H_y^{\pm}} = -\frac{E_y^{\pm}}{H_x^{\pm}} = \pm\eta \tag{4.91}$$

we can also obtain the magnetic field distribution of the striplines in the z-plane

$$\boldsymbol{H}(z) = \frac{\pm Em^{\pm}r^{-\frac{1}{2}}\left[\boldsymbol{a}_x \sin\frac{\varphi}{2} - \boldsymbol{a}_y \cos\frac{\varphi}{2}\right]}{\eta} \tag{4.92}$$

where, $\eta = \sqrt{\mu_0/\varepsilon_0}$, in each of the above equations

$$\begin{aligned} r &= \sqrt{2}\exp\left[\left(-\frac{\pi}{h}\right)x\right]\left[\mathrm{ch}\left(\frac{2\pi}{h}\right)x - \cos\left(\frac{2\pi}{h}\right)y\right]^{1/2} \\ \phi &= \mathrm{arc\ tg}\frac{-\exp\left[\left(-\frac{2\pi}{h}\right)x\right]\sin\left[\left(\frac{2\pi}{h}\right)y\right]}{\exp\left[\left(-\frac{2\pi}{h}\right)x\right]\cos\left[\left(\frac{2\pi}{h}\right)y\right]} \\ &-1\frac{\pi}{2} \le \phi \le \frac{3\pi}{2},\quad \mathrm{x} \ge 0,\quad -\frac{h}{2} \le y \le \frac{h}{2} \end{aligned} \tag{4.93}$$

Thus, using Eqs. (4.90) and (4.92) and the expressions of the field components in the rectangular waveguide of the main waveguide, we can calculate the coupling strength from Eq. (4.70) and thus the coupling coefficient and directionality.

(2) **Ideal stripline model II**

Unlike the ideal stripline model I, we use the Schwarz-Christoffel transform described below

$$\begin{aligned} W &= -\cosh^2\frac{\pi z}{h} \\ W' &= \frac{\mathrm{sn}^{-1}\left[(-W)^{\frac{1}{2}},\mathrm{k}\right] - \mathrm{sn}^{-1}\left(\frac{1}{k},\mathrm{k}\right)}{\mathrm{sn}^{-1}(1,\mathrm{k}) - \mathrm{sn}^{-1}\left(\frac{1}{k},\mathrm{k}\right)} \end{aligned} \tag{4.94}$$

to transform the stripline on the $z = x + \mathrm{j}y$ plane into the $W = u + \mathrm{j}v$ plane, and then we transform it from the W plane into the $W' = u' + \mathrm{j}v'$ plane to become a plate capacitor. Where sn^{-1} is the anti-elliptic sine function and k is its modulus.

$$k = \cosh^{-1}\frac{\pi d}{2h} \tag{4.95}$$

The calculation of the potential of the plate capacitor and the electromagnetic field component of the stripline can be performed using a similar method as in model I. Comparing the calculated coupling coefficient of the directional coupler obtained using this method with the measured results, the consistency between the two is higher than that of model I.

(3) **Stripline model III obtained by considering the length of the ground plate and the thickness of the central conductive strip**

Considering the finite size of the stripline ground plate and approximating the cross-section of the central conductive strip as a very flat elliptical band, the size of the stripline portion of the actual waveguide-coaxial single-hole directional coupler is always limited, and the thickness of the central conductive strip is not zero. We can use the Schwarz-Christoffel transform described below.

$$Z = x + \mathrm{j}y = C_1\left\{\lambda F(\sin^{-1} t|m) - \frac{\mathrm{j}}{(1 - m\alpha^2)^{1/2}} F\left[\sin^{-1}\left(\frac{1 - m\alpha^2}{1 - mt^2}\right)^{1/2} |g\right]\right\} + C_2$$

$$W = u + \mathrm{j}v = C_3 \int_0^{t/\alpha} \frac{\mathrm{d}t}{\left[(1 - t^2)(\alpha^2 - t^2)\right]^{1/2}}$$

$$+ C_4 = C_3 F(\sin^{-1}\frac{t}{\alpha}|\alpha^2) + C_4 \tag{4.96}$$

where,

$$g = \frac{1 - m}{1 - m\alpha^2} \tag{4.97}$$

Transform the strip line in the Z-plane to the t-plane, and then transform it into a finite-sized flat plate capacitor in the W-plane. Here, F represents the elliptic integral of the first kind, λ, α, m, C_1, C_2, C_3, and C_4 are all constants, which can be determined during the transformation process. They are related to the actual dimensions of the strip line, including the cylindrical cavity diameter and the thickness of the conducting strip.

The calculations show that, after considering the finite size of the strip line ground plane and the thickness of the central conducting strip, the calculated coupling coefficient compared with the measured value of the directional coupler is more accurate than that of the ideal strip line model II.

It should be noted that the specific transformation processes for models II and III are quite complex, so we will not further derive them here. Interested readers can refer to the author's relevant papers (Refs. [47–49]). Just like the rectangular waveguide single-hole directional coupler, the waveguide-coaxial single-hole directional coupler also requires adjustment of the angle θ between the center line of the central conducting strip in the strip line and the center line of the wide side of the waveguide to improve the directivity. However, its directivity is not very high, and in this type of

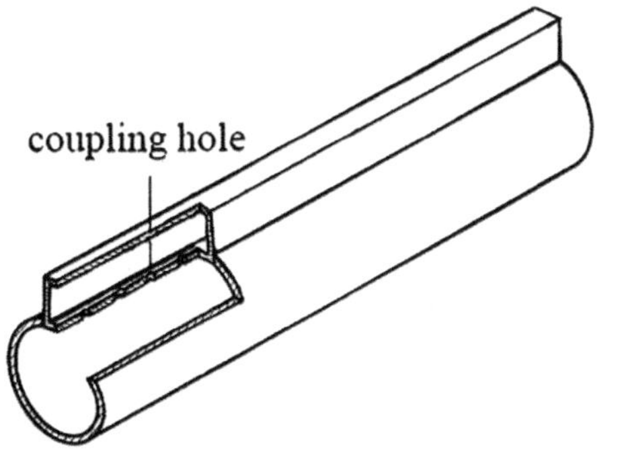

Fig. 4.13 Schematic diagram of a circular waveguide-rectangular waveguide directional coupler

directional coupler, the output is actually at the opposite direction of the secondary waveguide (port $P-$ in Fig. 4.12).

2. **Circular waveguide multi-hole coupler for high-power microwave power measurement**

In high-power microwave systems, overmoded circular waveguides are the most widely used transmission lines. Therefore, for power measurement using coupling methods, circular waveguide directional couplers are the most important coupling components. Figure 4.13 shows the most common type of circular waveguide directional coupler, where the main waveguide is an overmoded circular waveguide, and the secondary waveguide generally uses the fundamental mode rectangular waveguide to facilitate the coupling output signal to connect the measurement system. This is because measurement instruments are generally connected using coaxial lines or fundamental mode rectangular waveguides, with almost no use of circular waveguides. If the measurement instrument requires a coaxial connection, the rectangular waveguide-coaxial converter joint is also a very mature product. Therefore, using the fundamental mode rectangular waveguide output is the most convenient for connecting to the measurement system.

The directional coupler shown in Fig. 4.13 uses a series of small holes on the common wall of the narrow side of the rectangular waveguide and the circular waveguide. According to our coupling principle described in Eq. (4.70), this type of directional coupler is most suitable for coupling the TE_{0n} mode in the circular waveguide, This is because, for the circular waveguide serving as the main waveguide and the secondary rectangular waveguide, the TE0n mode in the former and the TE_{10} mode in the latter both have Hz components at the location of the coupling hole, which can enable magnetic field coupling. If the TM_{0n} mode is transmitted in the main circular waveguide, the wide side of the rectangular waveguide should be used as the common wall with the circular waveguide. In this case, the E_r component and H_φ component of the TM_{0n} mode in the circular waveguide, and the E_y and H_x components in the rectangular waveguide, not only have the same direction at the coupling hole location, but are also non-zero, so coupling can occur, both electrically and magnetically.

If in this type of coupler, all coupling holes not only have the same hole interval, equal to S, but also have the same hole radius r, that is, the single hole coupling strength is the same, assumed to be $a^{\pm}$, then we call this situation equal-interval coupling and equal-intensity coupling. In this case, Eqs. (4.80) and (4.82) can be further simplified to:

$$A^{\pm} = \left| a^{\pm} \frac{\sin N\phi^{\pm}}{\sin \phi^{\pm}} \right| \tag{4.98}$$

The technical indicators of the directional coupler are as follows:

$$\left.\begin{aligned} C &= 20\lg A^{+} = 20\lg\left| a^{+} \frac{\sin N\varphi^{+}}{\sin \varphi^{+}} \right| \\ d &= 20\lg \frac{A^{+}}{A^{-}} = 20\lg\left| \frac{a^{+} \sin N\varphi^{+} \sin \varphi^{-}}{a^{-} \sin N\varphi^{-} \sin \varphi^{+}} \right| \\ \varphi^{\pm} &= \frac{(\beta_1 \mp \beta_2)S}{2} \end{aligned}\right\} \tag{4.99}$$

where, S is the hole interval, $a^{\pm}$ is the single hole coupling strength, and N is the number of coupling holes. It must be pointed out that, since the modes in the main and secondary waveguides are different, $a^{\pm}$ should be calculated under power normalization conditions. We will discuss this situation in Sect. 7.3.2 and provide the formula for calculating $a^{\pm}$.

If the phase constants of the transmission modes in the main and auxiliary waveguides are the same, that is, $\beta_1 = \beta_2 = \beta$, $\varphi^{+} = 0$, $\varphi^{-} = \beta S$, then

$$\lim_{\varphi^{+}\to 0} \left| \frac{\sin N\varphi^{+}}{\sin \varphi^{+}} \right| = N \tag{4.100}$$

In this way, Eq. (4.99) can be further simplified to

$$\left.\begin{aligned} C &= 20\lg\left| Na^{+} \right| \\ d &= 20\lg\left| \frac{Na^{+} \sin \beta S}{a^{-} \sin N\beta S} \right| \end{aligned}\right\} \tag{4.101}$$

In this case, a^{+} can be calculated from Eqs. (4.70) or (4.75) after the given coupling hole radius r_0, or after the given required C value and selected N, the value of a^{+} can be obtained from the first equation in Eq. (4.101), and then r_0 and a^{-} can be calculated.

From Eq. (4.101), it can be seen that as long as N is large enough, the directivity d can also maintain a high value over a much wider frequency range than the dual-hole directional coupler. This can be understood from a physical standpoint as the dual-hole directional coupler only has the two backward waves formed in the secondary waveguide by the coupling of the two holes, the two backward waves will just cancel each other out only at the center frequency, or come close to canceling each other out within a small frequency range. Therefore, overall, it has a relatively narrow operating bandwidth. In contrast, in a multi-hole directional coupler, N coupling holes will excite N backward waves in the secondary waveguide, when these N backward waves are superimposed, it is entirely possible that they will come close to canceling each other out not only at one frequency, but at a range of frequencies. As N increases, the chances of approaching cancellation at different frequencies also increase, meaning that the frequency range where the directivity values can be maintained sufficiently widens. Therefore, multi-hole couplers can achieve broadband operation.

Directional couplers also have various coupling methods, such as coupling through a continuous coupling slot made in a certain function distribution form, or coupling through crisscross-shaped holes, etc. The hole radius or hole interval of the small holes also have various distribution methods, such as Chebyshev distribution, binomial distribution, etc., all of which are specifically designed to further improve the performance of directional couplers.

3. **Mode-selective couplers**

When we analyze the working principle of directional couplers, we already know that the coupling strength of the small holes depends on the modes of the main and secondary waveguides. Therefore, the coupling coefficient of directional couplers changes with different modes. Specifically, at a given frequency, if the secondary waveguide operates in the TE_{10} mode of the rectangular waveguide and the size of the coupling hole is determined, the coupling coefficient depends on the mode of operation in the main waveguide. When there are multiple modes of transmission in the main waveguide, using this characteristic and combining the phase superposition principle (by changing the hole interval), we can design mode-selective couplers, so that the secondary waveguide only couples with a certain mode of the main waveguide, i.e., it has a certain coupling strength with that mode, while coupling with other modes is almost negligible (i.e., the positive coupling coefficient with other modes is very large). In this way, by designing such a coupler for each mode in the main waveguide, we can separate each mode and have each coupler output only one mode.

We will introduce the design method of mode-selective couplers in the description of mode measurement in Chap. 7. Here, we just want to point out that by using mode-selective couplers, power measurement can also be performed. Clearly, measuring the power of each mode and adding them up gives the total power in the main waveguide. The specific power measurement method will also be discussed when introducing the mode measurement using mode-selective couplers.

4.5 Calibration of Directional Coupler

The calibration of the coupling coefficient and directivity of directional coupler can generally be achieved using a network analyzer. However, there is a significant obstacle to this method, which is the manufacturing of the mode exciter. The main and secondary waveguides are both standard rectangular waveguides for conventional directional couplers, and they primarily operate in the TE_{10} mode of the rectangular waveguide. Therefore, it's convenient to excite the TE_{10} mode using a commercial coaxial-waveguide converter joint, without the presence of spurious modes. However, in the field of high-power microwave, the output of high-power microwave devices typically is higher-order modes, and circular waveguide is often used. Even if rectangular waveguide is used, it is also operating in higher-order modes. As a result, the directional couplers need to couple the microwave energy of higher-order modes. During calibration, we should first input the required higher-order modes into the main waveguide (while the secondary waveguide generally operates in the TE_{10} mode of the rectangular waveguide), in order to calibrate the coupling coefficient of that mode. For many higher-order modes, obtaining high-purity mode excitation is quite difficult, making the calibration of these directional couplers quite troublesome. This difficulty is similar to the challenges encountered in calibrating the effective section S_{eff} of probe mentioned earlier.

For some modes with relatively simple structures and low eigenvalues, we can directly use probes or coupling holes for excitation. The most familiar and commonly used example is the coaxial-waveguide joint, which is a mode exciter that uses a probe to excite the TE_{10} mode in the rectangular waveguide by means of the coaxial line's TEM mode. These exciters have high mode conversion efficiency, resulting in high mode purity and low spurious mode content. However, most modes cannot be directly excited using probes or coupling holes, and instead require relatively complex mode converters, also known as mode transformers, to obtain them. Although the terms "mode exciter" and "mode converter" are different, there is no clear distinction. The authors believe that, generally speaking, the element that directly excites the required mode through probes or coupling holes (including large and small holes) using the TE_{10} mode of the rectangular waveguide or the TEM mode of the coaxial line is referred to as mode exciters, while the element that obtain the required mode through relatively complex waveguide changes from an already high-order mode are referred to as mode converters. However, this distinction is not strict, and the terminology is often used interchangeably.

Since the calibration of directional coupler using a network analyzer only requires low power, the power capacity of the mode exciter or mode converter can be disregarded.

Numerous mode exciters and mode converter solutions have been proposed by scholars worldwide, especially a great number of complex and widely applicable mode transformation solutions have been reported. Due to the abundance of literature, it is not possible to discuss them all in detail here. The following briefly introduces some simple mode exciters or mode converters that may be suitable for parameters calibration in high-power microwave measurements.

4.5.1 Mode Exciters

1. **Circular waveguide TE_{11} mode exciter**

The TE_{11} mode in a circular waveguide is the fundamental mode, and its field structure is very similar to the TE_{10} mode in a rectangular waveguide. Therefore, it can be directly transformed from a rectangular waveguide to a circular waveguide, and the electromagnetic field naturally transitions from the TE_{10} mode of the rectangular waveguide to the TE_{11} mode of the circular waveguide (see Fig. 4.14). This type of circular waveguide TE_{11} mode exciter is commonly referred to as a square-to-circular transition.

2. **Circular waveguide TE_{01} mode exciter**

In a gyrotron, especially in conventional gyrotrons, the TE_{01} mode of the circular waveguide is a commonly used operating mode. Therefore, for gyrotron amplifiers, a signal in the TE_{01} mode must be input. However, in general, the input signal from the signal source is the TE_{10} mode of the rectangular waveguide, or the TEM mode of a coaxial line. Even if it is the TEM mode of a coaxial line, it can easily be transformed into the TE_{10} mode of the rectangular waveguide through a coaxial-waveguide transition. Therefore, using the TE_{10} mode of the rectangular waveguide to excite the TE_{01} mode of the circular waveguide is a very important excitation issue in gyrotrons.

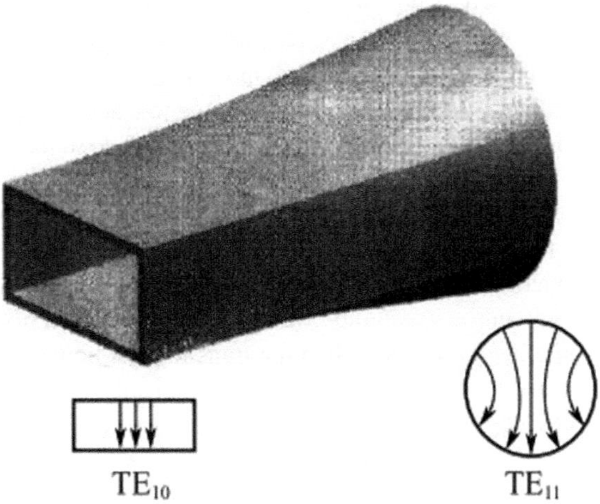

Fig. 4.14 The transition from the rectangular waveguide TE_{10} mode to the circular waveguide TE_{11} mode

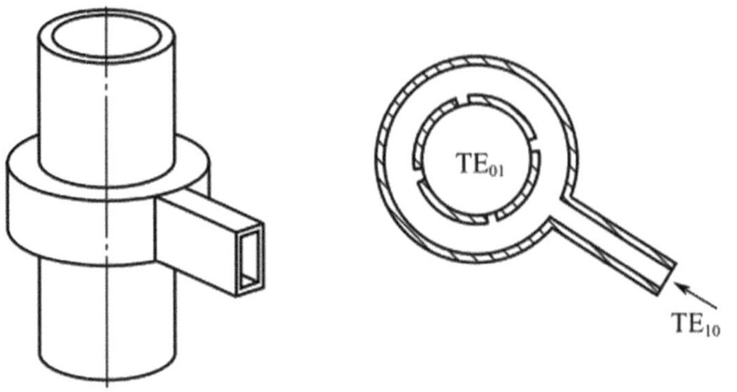

Fig. 4.15 Schematic diagram of exciting TE_{01} mode in a circular waveguide using TE_{10} mode of a rectangular waveguide

Using a rectangular waveguide surrounding a circular waveguide, four rectangular coupling holes are uniformly distributed along the circumference on the common wall between the wide edge of the rectangular waveguide and the circular waveguide. Since the coupling holes are large enough, generally reaching a length close to or equal to the width of the rectangular waveguide, we can use the TE_{10} mode of rectangular waveguides surrounding the circular waveguide to excite the TE_{01} mode in the circular waveguide. Figure 4.15 shows a common exciter of TE_{01} mode in gyrotron amplifiers [50].

Figure 4.16 shows another structure of the circular waveguide TE_{01} mode exciter proposed by scholars in Taiwan. This is a method of excitation that can be applied in a wider range. It uses a two-stage Y-branch power divider to divide the input TE_{10} mode electromagnetic wave into four routes with equal amplitude and phase. The ports of the four routes are directly connected to coupling holes distributed at equal angles on the side wall of the circular waveguide. The coupling holes are the same size as the ports of the rectangular waveguide. The length of the four rectangular waveguides from the first power divider to the connection end with the circular waveguide must be equal to ensure that the microwave reaches the coupling port in the same phase. At the coupling port, the H_x of the TE_{10} mode in the rectangular waveguide couples with the H_z of the TE_{01} mode in the circular waveguide [51].

The TE_{01} mode in the circular waveguide has outstanding characteristics. Because the TE_{01} mode is a circular electric mode, on the waveguide wall at $r = R$, the magnetic field of the TE_{01} mode only has the z component H_z, so its high-frequency current on the waveguide wall only has the φ direction component J_φ and no longitudinal wall current. When the transmission power is determined, as the frequency increases, the waveguide loss monotonically decreases. Therefore, when the operating frequency increases, the attenuation of the TE_{01} wave in the circular waveguide

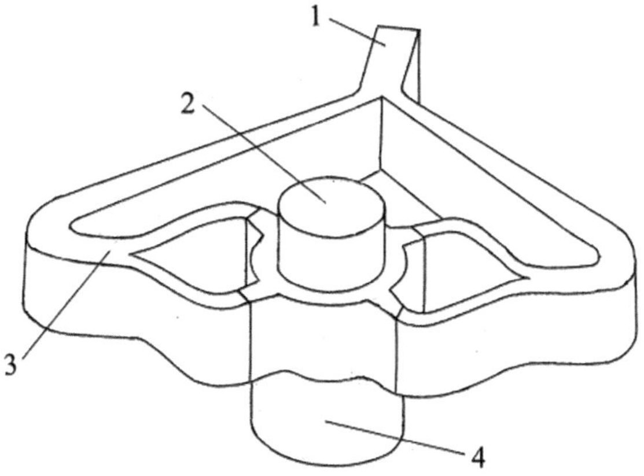

Fig. 4.16 Another structural schematic diagram of rectangular waveguide TE_{10} mode exciting circular waveguide TE_{01} mode. 1—TE_{10} mode input, 2—cut = off waveguide segment, 3—Y-branch power divider, 4—TE_{01} mode output

during transmission can be very small, making it suitable for long-distance transmission of millimeter waves and as a working mode of high-Q cavities. It has important applications in electron cyclotron maser, plasma processing systems, linear colliders, and other fields.

3. **Circular waveguide TM_{01} mode probe exciter**

The circular waveguide TM_{01} mode exciter is relatively simple, and Fig. 4.6 shows its structural diagram.

4. **Rectangular waveguide TE_{20} mode exciter**

The structure of TE_{20} mode exciter is relatively simple, and it can be conveniently excited by changing the waveguide section shape from the TE_{10} mode standard waveguide, as shown in Fig. 4.17. This exciter is actually the first section of the improved Marie-type rectangular waveguide TE_{10} mode to circular waveguide TE_{01} mode converter that we will introduce in Sect. 4.5.2 [52, 53]. When necessary, it can also be further transformed into an overmoded rectangular waveguide that can transmit the TE_{20} mode through a linear transition waveguide. Similarly, there are other schemes for obtaining TE_{20} mode exciters by changing the waveguide cross-sectional shape.

In addition, by directly using a probe, we can also excite the TE_{20} mode. The structure of the exciter is shown in Fig. 4.18. Feed TE_{10} mode microwaves into the E side branch of the E-T branch waveguide. According to the characteristics of the E-T branch, the electromagnetic field will be output in equal amplitude and opposite phase from the two directions of the main waveguide. After propagating through waveguides of the same length, energy is coupled into the excited waveguide using probes located at a distance of $a/4$ from the narrow side on both sides of the excited rectangular waveguide (where a is the width of the waveguide). Because the phases of the two microwaves are opposite, their amplitudes are equal, and the positions where they are excited all satisfy the field structure requirements of the rectangular waveguide TE_{20} mode, the TE_{20} mode is excited.

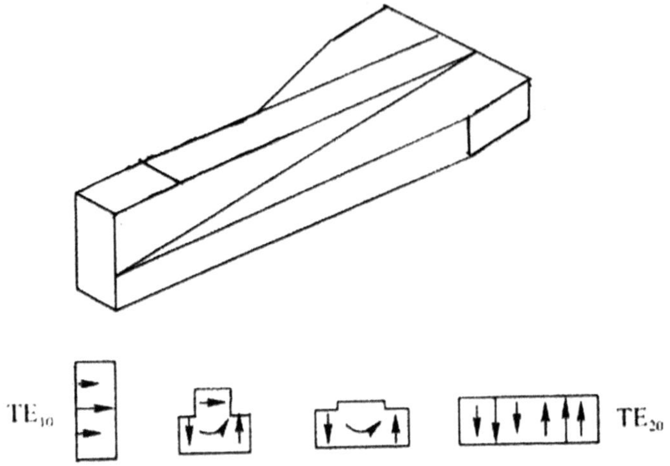

Fig. 4.17 TE20 mode exciter of a rectangular waveguide

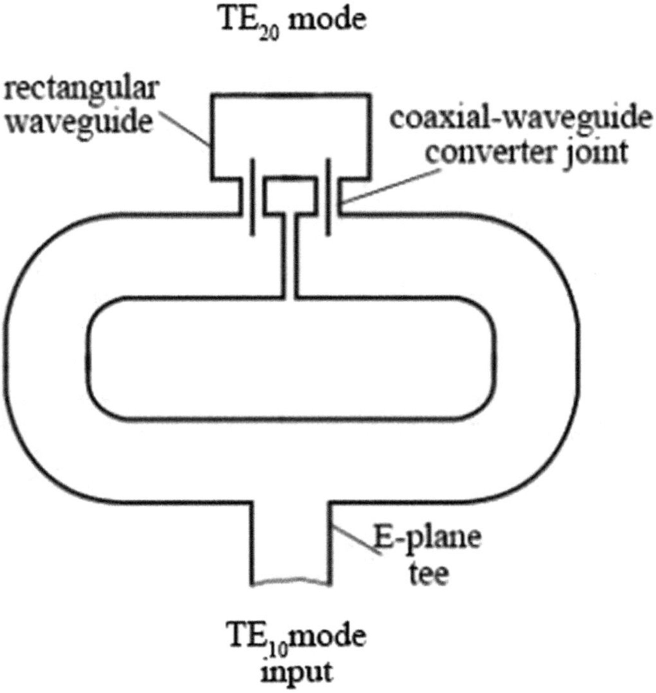

Fig. 4.18 TE_{20}-mode dual-probe exciter of a rectangular waveguide

5. **Rectangular waveguide-circular waveguide exciter of various modes of coupling hole**

In 1960, Japanese scholar Kenichi Noda proposed connecting the wide or narrow side of a rectangular waveguide perpendicularly to a circular waveguide port at the short-circuited end of the rectangular waveguide. By opening coupling holes of different distributions and shapes on the connecting wall of the rectangular waveguide, various TE modes or TM modes can be excited in the circular waveguide. To excite the TM mode in the circular waveguide, holes should be opened on the wide side of the rectangular waveguide and coupled to the circular waveguide port. To excite the TE mode in the circular waveguide, holes should be opened on the narrow side of the rectangular waveguide and coupled to the circular waveguide port, thus exciting various modes in the circular waveguide, including TE_{11}, TM_{01}, TE_{01}, TM_{11}, TE_{12}, TE_{21}, TE_{31}, TE_{41}, and so on. Kenichi Noda not only provided the coupling structure diagram but also presented the measurement results of mode field distribution, which were conducted at a frequency of 24 GHz. This type of exciter structure is very simple, and the report suggests that the mode conversion efficiency is also very high. Readers are advised to conduct further research based on reference [54].

4.5.2 Mode Converters

Mode converters can generally be divided into waveguide converters and quasi-optical converters. The former mainly achieves mode conversion through the change of waveguide dimensions (including changes in section shape, size, and axis bending) and sometimes supplements the waveguide with diaphragms and coupling holes. The latter is primarily composed of a Vlasov antenna and multi-level metal reflective mirror, using the mode conversion function of the Vlasov antenna and the focusing and phase calibration actions of the reflective mirror to transform high-order modes in circular waveguides into Gaussian beam outputs. Mode converters are generally structurally complex, and as we've mentioned before, there are a large number of proposed solutions. Therefore, we'll only introduce a few simple structures relatively suitable for calibrating probes and small hole couplers in high-power microwave measurements. Therefore, we'll only introduce a few relatively simple structures of mode converters suitable for calibrating probes and small hole couplers in high-power microwave measurements. For a large number of mode converters used in various high-power microwave system output devices, such as TE_{0n}–HE_{11} converters and TM_{0n}-HE_{11} converters, each of which includes several conversion sections or is a combination of several converters, more specific details can be found in references [5] and [55], and we won't discuss them here.

1. **Circular waveguide TE_{01} mode converter** [52, 53]

The circular waveguide TE_{01} mode is the most commonly used operating mode in conventional gyrotrons. The converter proposed by G. R. F. Marie, which converts

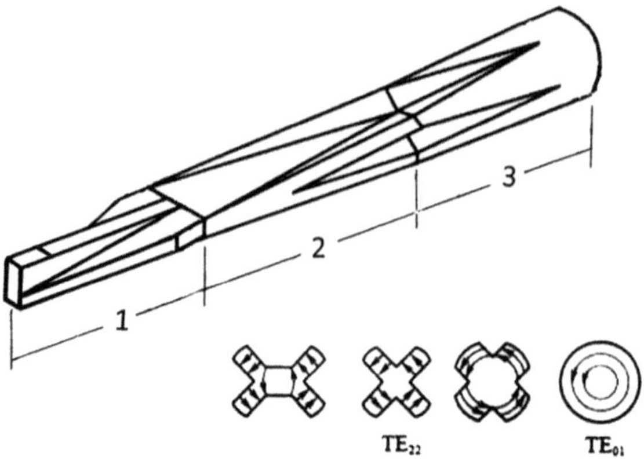

Fig. 4.19 The mode converter of Marie-type rectangular waveguide TE_{10} mode to circular waveguide TE_{01} mode

the rectangular waveguide TE_{10} mode to the circular waveguide TE_{01} mode, has been widely used. It consists of three sections: first, the first section converts the rectangular waveguide TE_{10} mode to the rectangular waveguide TE_{20} mode; the second section transitions the rectangular waveguide to a cross-shaped waveguide, during which the mode changes from the TE_{20} mode of the rectangular waveguide to the TE_{22} mode in the cross-shaped waveguide; and finally, the last section expands the cross-shaped waveguide into a circular waveguide, during which the mode naturally changes from the TE_{22} mode in the cross-shaped waveguide to the TE_{01} mode in the circular waveguide (Fig. 4.19).

The design method of the improved Marie-type mode converter is relatively simple.

The first section starts with a standard rectangular waveguide of dimensions $a_1 \times b_1$, and ends with a rectangular waveguide whose wide side is doubled, having dimensions of $a_2 \times b_2$, where $a_2 = 2a_1$. For convenience, the dimensions of the narrow side can remain unchanged, that is, $b_2 = b_1$. However, a_2 and a_1, as well as b_2 and b_1, are exactly interchanged in spatial position, meaning the wide side a_1 at the starting end becomes the narrow side b_2 at the end (Fig. 4.20a), and the narrow side b_1 at the starting end becomes the wide side a_2 at the end (Fig. 4.20b). The simplest method is to use linear transformation, and the starting positions and ending positions of the transformations $a(z)$ and $b(z)$ are offset from each other by a length of $L_1/4$. L_1 can be taken as 5–8 times the half waveguide wavelength $\lambda_g/2$ of the TE_{10} mode in the standard waveguide $a_1 \times b_1$ at the design frequency, i.e., $L_1 = (5\text{–}8)\lambda_g/2$.

The change law of $a(z)$ and $b(z)$ in the first section is as follows:

$$\begin{cases} a(z) = a_1 & (0 \le z \le 0.25L_1) \\ a(z) = a_1 - \frac{(a_1 - b_2)}{0.75L_1}(z - 0.25L_1) & (0.25L_1 \le z \le L_1) \end{cases} \tag{4.102}$$

$$\begin{cases} b(z) = a_2 & (0.75L_1 \le z \le L_1) \\ b(z) = b_1 + \frac{(a_2 - b_1)}{0.75L_1} z & (0 \le z \le 0.75L_1) \end{cases} \tag{4.103}$$

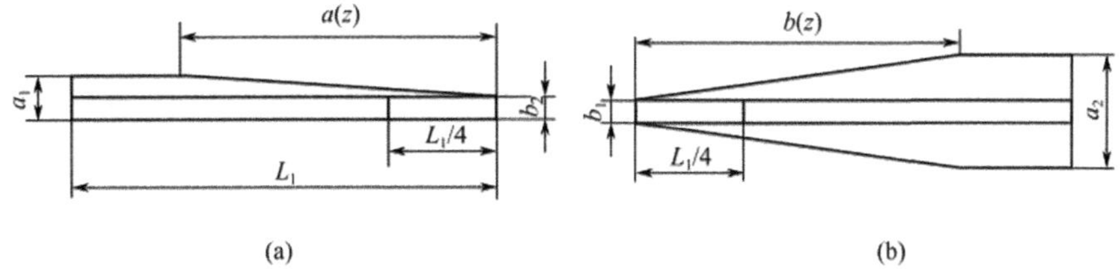

Fig. 4.20 The size change of the first stage of the Marie mode converter

In the second section, the b_2 edge of the $a_2 \times b_2$ waveguide is linearly expanded in a fan-shaped manner with the diagonal of the $a_2 \times b_2$ waveguide as the diameter of the circular arc ϕ_0 (Fig. 4.21a)

$$\phi_0 = \left[a_2^2 + b_2^2\right]^{1/2} \tag{4.104}$$

When the chord length of the circular arc formed by the b_2 transition expands to $2b_2$, it begins to bifurcate from the center at an angle of 90° (Fig. 4.21b). As the distance increases, the bifurcation expands linearly while maintaining the width of each bifurcation arm as b_2 until it becomes a cross-shaped waveguide (Fig. 4.21c). Each arm of the cross-shaped waveguide always maintains the same width b_2 and ϕ_0 remains unchanged, while $a(z)$ decreases to zero and $d(z)$ increases to d_3 (Fig. 4.21c).

The length of the second segment L_2 should be longer than the first segment, and generally 7–15 times that of $\lambda_g/2$ can be taken. The change law of $a(z)$ and $d(z)$ in the second section is:

$$a(z) = a_2 - \frac{a_2}{L_2}z \quad (0 \leq z_2 \leq L_2) \tag{4.105}$$

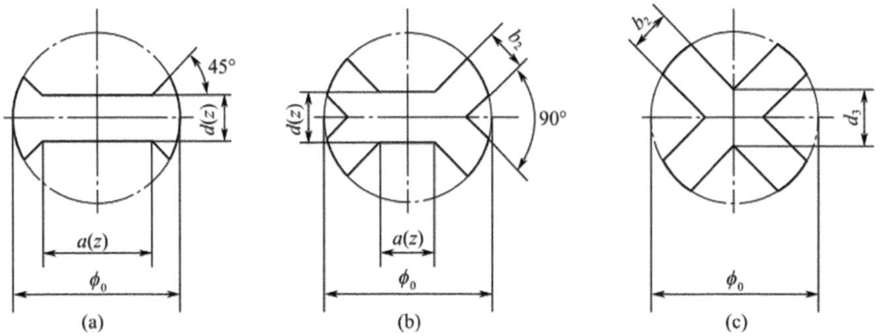

Fig. 4.21 The size change of the second stage of the Marie mode converter, **a** sectoral linear expansion of the rectangular waveguide, **b** the sector starts forks from the center by 90°, **c** Forming a cross-shaped waveguide

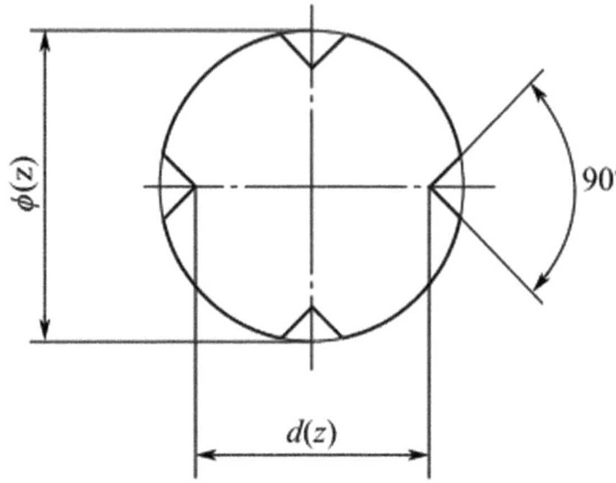

Fig. 4.22 The size change of the third stage of the Marie mode converter

$$d(z) = b_2 + \frac{(\sqrt{2} - 1)b_2}{L_2} z \quad (0 \le z_2 \le L_2) \tag{4.106}$$

The third section starts with a cross-shaped waveguide, linearly increasing the width of the four arms, while the diameter of the circumscribed circle increases from ϕ_0 to ϕ_m (Fig. 4.22).

ϕ_m is determined according to the following principle that the cut-off wavelength of the TE_{01} mode in the circular waveguide with ϕ_m diameter is equal to the cut-off wavelength of the TE_{10} mode in the rectangular waveguide at the input of the converter, that is,

$$\lambda_{c(TE_{10})} = \lambda_{c(TE_{01})} \tag{4.107}$$

From this we can get

$$\begin{aligned} 2a_1 &= 1.64R = 0.82\varphi_m \\ \varphi_m &= \frac{4a_1}{1.64} \end{aligned} \tag{4.108}$$

The diameter ϕ_m may not necessarily meet the requirements of the circular waveguide size for high-power microwave transmission systems. In this case, it should be transitioned to the required size through a transition waveguide. The transition waveguide can have a linear change, or it can change in other functional forms, such as rising cosine function change, Chebyshev function change, etc.

The length L_3 of the third section can be equal to the length L_1 of the first section. In this section, the variations of $d(z)$ and $\phi(z)$ are as follows:

$$d(z) = \sqrt{2}b_2 + \frac{\varphi_m - \sqrt{2}b_2}{L_3} z \quad (0 \le z_3 \le L_3) \tag{4.109}$$

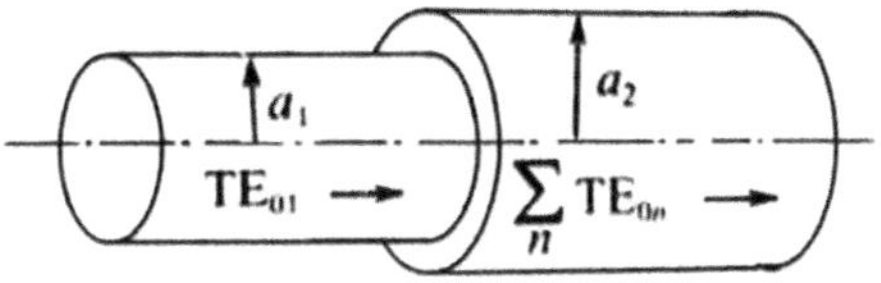

Fig. 4.23 Stepped circular waveguide TE_{01}–TE_{0n} mode converter

$$\phi(z) = \phi_0 + \frac{\phi_m - \phi_0}{L_3} z \quad (0 \le z_3 \le L_3) \tag{4.110}$$

2. **Circular waveguide TE_{0n} mode stepped waveguide converter** [56, 57]

H. Stickel et al. proposed a very simple converter that uses a step change in the overmoded circular waveguide to achieve TE_{01}–TE_{0n} mode conversion, as shown in the structural form in Fig. 4.23. We will introduce the design theory of this converter here.

As shown in Fig. 4.23, a segment of step circular waveguide has radii a_1 and a_2 on either side of the step. Assuming microwaves are injected into the waveguide with radius a_1 from the left end to excite the TE_{01} mode, due to the non-uniformity of the step, all TE_{0n} (n = 1, 2, …) modes with wavelengths not greater than the cutoff wavelength of the TE_{0N} mode will be excited in the waveguide with radius a_2 on the right side of the step, where N is the highest mode that can be transmitted in the waveguide with radius a_2. For the TE_{01} mode, we have

$$E_{\varphi 1} = -\mathrm{j}\frac{\omega\mu}{k_{c1}} H_{01} J_1(k_{c1}r)\mathrm{e}^{-\mathrm{j}(\omega t - \beta_1 z)} \tag{4.111}$$

And for the TE_{0n} mode, we have

$$E_{\varphi n} = -\mathrm{j}\frac{\omega\mu}{k_{cn}} H_{0n} J_1(k_{cn}r)\mathrm{e}^{-\mathrm{j}(\omega t - \beta_n z)} \tag{4.112}$$

where, $k_{c1} = \mu_1/a_1$, $k_{cn} = \mu_n/a_2$, μ_1, μ_n are the first and nth root of the derivative of the Bessel function of zero order. Note that $J_1(kr) = -J_0'(kr)$.

If the non-uniformity introduced by the step is not very large, the amplitude of the reflected wave generated by the step relative to the amplitude of the output TE_{0n} wave can be ignored. For similar reasons, the higher-order TE_{mn} modes with $m \neq 0$ excited in the waveguide on the right side of the step can also be ignored. Therefore, at the boundary of the waveguide step, we should have:

$$E_{\varphi 1} = \sum_{n=1}^{N} E_{\varphi n} \tag{4.113}$$

We take the coordinates of the step boundary as $z = 0$, which can be divided into two parts, namely the part where $r \le a_1$ and the part where $a_1 \le r \le a_2$. Substituting Eqs. (4.111) and (4.112) into Eq. (4.113), obviously

$$\sum_{n=1}^{N} \frac{H_{0n}}{H_{01}} \frac{k_{c1}}{k_{cn}} J_1(k_{cn} r) = \begin{cases} 0, & (z = 0,\ a_1 \le r \le a_2) \\ J_1(k_{c1} r), & (z = 0,\ r \le a_1) \end{cases} \tag{4.114}$$

In the above equation: $H_{0n} k_{c1} / H_{01} k_{cn}$ is the conversion coefficient from TE_{01} mode to TE_{0n} mode.

Multiply both sides of Eq. (4.114) by $J_1(k_{cn} r)$ and integrate both sides from 0 to a_2. Note that the integral on the right side of Eq. (4.114) has the orthogonality of Bessel functions, which means that the integral of all the Bessel function terms with $m \neq n$ is 0. So, the result of the integral becomes [58]:

$$\int_0^{a_2} J_1(k_{c1} r) J_{cn}(k_{cn} r) r \mathrm{d}r = \int_0^{a_2} \frac{H_{0n}}{H_{01}} \frac{k_{c1}}{k_{cn}} J_1^2(k_{cn} r) r \mathrm{d}r \tag{4.115}$$

Using the integral formula, we get:

$$\int_0^{a_2} J_1^2(k_{cn} r) r \mathrm{d}r = \frac{a_2^2}{2} J_0^2(k_{cn} a_2) \tag{4.116}$$

and:

$$\int_0^{a_2} J_1(k_{c1} r) J_1(k_{cn} r) r \mathrm{d}r = \frac{a_2 k_{cn}}{k_{c1}^2 - k_{cn}^2} J_0(k_{cn} a_2) J_1(k_{c1} a_2) \tag{4.117}$$

Substitute Eqs. (4.116) and (4.117) into Eq. (4.115), then

$$\frac{H_{mn}}{H_{m1}} \frac{k_{c1}}{k_{cn}} = \frac{2\mu_n^2}{\alpha\left[\alpha^2 - \mu_n^2\right]} \frac{J_1(\alpha)}{J_0(\mu_n)} \tag{4.118}$$

where,

$$\alpha = \mu_1 \frac{a_2}{a_1} \tag{4.119}$$

When $a = \mu_N$, that is, $a_2/a_1 = \mu_N/\mu_1$, then

$$\frac{H_{0N}}{H_{01}} \frac{k_{c1}}{k_{cN}} = 1 \tag{4.120}$$

This means that the TE_{01} mode is completely converted into the TE_{0n} mode. It can be seen that when $\alpha = \mu_N$, that is,

$$a_2 = a_1 \frac{\mu_N}{\mu_1} \tag{4.121}$$

then 100% conversion efficiency can be achieved. Equation (4.121) is the design basis for the circular waveguide TE_{0n} mode step waveguide converter transformer.

H. Stickel et al. conducted far-field tests on the actually manufactured step over-moded circular waveguide mode converter and obtained the radial distribution of the angular electric field of the TE_{0n} mode. The tests showed that the step waveguide mode converter can achieve the conversion of TE_{01}–TE_{0n} modes, but only relatively low-order modes can achieve ideal results. As the mode order increases, the purity of the obtained modes becomes relatively poor. Specific test results can be found in Sect. 6.3.2 of this book.

The research team led by the authors has also developed step waveguide converters for the TE_{01}–TE_{02} mode, TE_{01}–TE_{03} mode, and TE_{01}–TE_{04} mode of the circular waveguide, and conducted radiation field mode tests. The test results are also presented in Sect. 6.3.2.

4.5.3 Calibration of the Coupler by Network Analyzers

When calibrating a directional coupler used for high-power microwave power measurement with a network analyzer, it should be emphasized that the input to the directional coupler should be the mode for which we need to calibrate the coupling coefficient, and not directly use the coaxial line TEM mode or rectangular waveguide TE_{10} mode output from the network analyzer signal source for calibration. The output signal of the network analyzer signal source must be appropriately transformed into the mode we need before it can be used for calibration.

The technical specifications of directional couplers can generally be calibrated using a network analyzer, and during the calibration process, the various ports of the directional coupler are numbered as shown in Fig. 4.24. Port 1 is the input port, port 2 is the direct pass output port, port 3 is the forward coupled output port, also called the output port, and port 4 is the backward coupled output port, or the isolation port. We use a to represent the normalized voltage of the input wave signal at each port, and b to represent the normalized voltage of the output wave signal at each port.

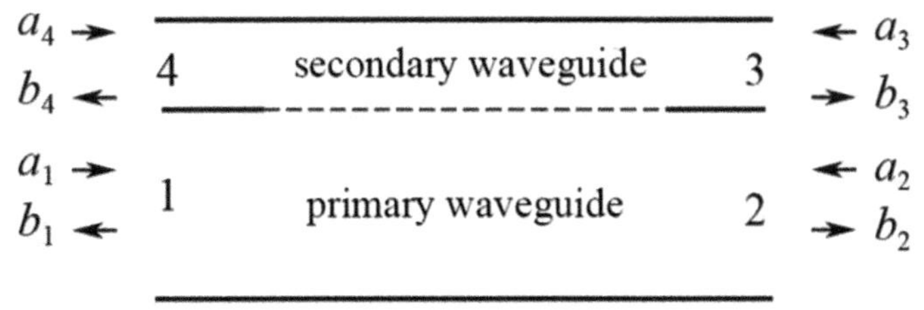

Fig. 4.24 S-parameter network of a directional coupler

(1) **Reflection coefficient and standing wave coefficient**

When measuring the reflection coefficient at port 1, ports 2, 3, and 4 should be connected to matching loads. Similarly, when measuring the reflection coefficient at port 2, all other ports except the one being measured should also be connected to matching loads. By using a network analyzer to measure S_{11} and S_{22} separately, we can obtain the reflection coefficients Γ for ports 1 and 2.

$$\begin{aligned}\Gamma_1 &= S_{11} = \frac{|b_1|}{|a_1|} \\ \Gamma_2 &= S_{22} = \frac{|b_2|}{|a_2|}\end{aligned} \tag{4.122}$$

When using the standing wave coefficient ρ to represent the size of reflection

$$\begin{aligned}\rho_1 &= \frac{1+|S_{11}|}{1-|S_{11}|} \\ \rho_2 &= \frac{1+|S_{22}|}{1-|S_{22}|}\end{aligned} \tag{4.123}$$

For the reciprocal components, we have $\Gamma_1 = \Gamma_2, \rho_1 = \rho_2$.

In network analyzer measurements, the dB value output of S-parameters is often used, and the dB expression of reflection coefficient and standing wave coefficient is

$$\begin{aligned}&S_{11}(\text{Power decibel value}) = 10\lg|\Gamma|^2\ (\text{dB}) \\ &\Gamma = 10^{S_{11}(\text{dB})/20}, \quad \rho = \frac{1+|\Gamma|}{1-|\Gamma|}\end{aligned} \tag{4.124}$$

It can be seen that when using dB values to represent the magnitude of S-parameters, power dB values are generally used. Since it's important to note that the reflection coefficient Γ is always less than 1, so the dB value of S_{11} derived from Γ should be negative. For example, when $S_{11}(\text{dB}) = -20$ dB, it means the reflection coefficient $\Gamma = 0.1$, and the standing wave coefficient $\rho = 1.222$.

(2) **Coupling coefficient**

The coupling coefficient C, expressed as a positive value for directional couplers and any other coupling elements, can be defined as the ratio in dB of the input normalized power at the input port to the output normalized power at the forward coupled output port. Since the coupled output power is always less than the input power at the input port, the obtained coupling coefficient is a positive value. In network analyzer measurements, S_{31} expressed in dB is the negative coupling coefficient, and its negative value represents the positive coupling coefficient. For symmetric networks, since $S_{31} = S_{13}$, the coupling coefficient can also be obtained by measuring S_{13}. When measuring the coupling coefficient, ports 2 and 4 should be connected to matching loads.

$$C = 10\lg\frac{|a_1|^2}{|b_3|^2} = 10\lg\frac{1}{|S_{31}|^2} = -20\lg|S_{31}| = -20\lg|S_{13}| \text{ (dB)} \tag{4.125}$$

(3) **Isolation**

The ratio in dB of the input normalized power at port 1 of directional couplers, power dividers, circulators, microwave bridges, and other components to the output normalized power at port 4 is called isolation. When measured with a network analyzer and with ports 2 and 3 connected to matching loads, the negative values of S_{41} or S_{14} represent isolation. When measuring isolation, ports 2 and 3 should be connected to matching loads.

$$I = 10\lg\frac{|a_1|^2}{|b_4|^2} = 10\lg\frac{1}{|S_{41}|^2} = -20\lg|S_{41}| = -20\lg|S_{14}| \text{ (dB)} \tag{4.126}$$

(4) **Directivity**

The directivity D of a directional coupler is defined as the ratio in dB of the output normalized power at coupled output port 3 to the output normalized power at isolation port 4, or in other words, the ratio in dB of the forward output normalized power to the backward output normalized power in the coupled arm. Directivity can be expressed as the difference in dB between isolation and coupling coefficient, so directivity should be calculated by measuring isolation and coupling coefficient.

$$\begin{aligned} D &= 10\lg\frac{|b_3|^2}{|b_4|^2} = 20\lg|S_{31}| - 20\lg|S_{41}| = 20\lg|S_{13}| - 20\lg|S_{14}| \\ &= -C + I = I - C \text{ (dB)} \end{aligned} \tag{4.127}$$

where, I and C are both positive values.

Chapter 5
Other Electromagnetic-Effect Use in High-Power Microwave Power Measurement

The crystal detector diode power meter, electromagnetic wave pressure effect power meter, Hall effect power meter, Kerr effect power meter, and pyroelectric effect power meter introduced previously in the discussion of conventional repetitive pulse power measurement can also be used to measure high-power single pulse. In addition, scholars from various countries have proposed other methods for measuring high-power single-pulse or low-repetition-rate pulse power, which are also based on various electromagnetic effects of electromagnetic fields. The main ones are:

(a) vacuum detection diode–electron effect power meter.
(b) resistance detector–thermal effect power meter.
(c) capillary energy meter–also a thermal effect power meter, actually a stationary liquid body power meter.
(d) surface acoustic wave power meter–piezoelectric effect power meter.
(e) detectors based on other electromagnetic effects, such as electromagnetic wave pressure effect, Hall effect, magneto-optical effect, pyroelectric effect, etc., can also be used as power detectors for nanosecond single pulses.

In this chapter, we will discuss some special transformers that have been used in high-power microwave measurements, mainly referring to electron effect convertors, thermal effect converters, and electroacoustic converters.

W. Wang et al., *High-Power Microwave Measurement*,
https://doi.org/10.1007/978-981-96-7763-4_5

5.1 Vacuum Diode Power Meter and Capillary Energy Meter

5.1.1 Measurement Method Using Vacuum Detector Diode

The principle of measuring microwave power with a vacuum detection diode is the same as that of a crystal detection diode, but the power capacity of the vacuum detection diode is much larger than that of the crystal detection diode, so it can be used in high-power microwave power measurement. The dynamic range is from a few tenths of a watt to 200 W. If a voltage divider is used, the upper limit can be increased to several hundred kilowatts, but its frequency upper limit is low and can only work in the centimeter wave band, and its volume is large.

The vacuum diode detector is shown in Fig. 5.1 [59]. It uses a former Soviet Union 6D16D high-frequency detection tube for detection, in order to reduce the reflection caused by the insertion of the vacuum diode in the center of the wide side of the WR90 waveguide, the detector uses a three-screw tuner to adjust the match. In order to detect high-power microwave pulses, the output signal of the vacuum diode should be filtered by a specially designed coaxial low-pass filter to strain out the interference signals and display pulse envelope.

By using a high-power microwave source with adjustable output power and output pulse width, researchers can calibrate the vacuum diode detector, that is, obtain the relationship curve between the output voltage of the detector and the input microwave power under different pulse widths. The calibration results show that the microwave pulse power that the detector can withstand can reach about 10 kW, the response

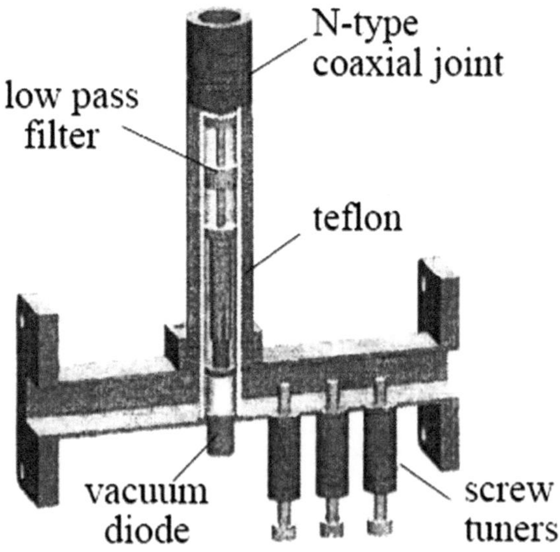

Fig. 5.1 Vacuum diode detector structure

time is less than 2 ns, the detection output voltage is as high as tens of volts, and the dynamic range is large.

Chinese researchers have also conducted experiments on measuring the power density value and its distribution of high-power microwave pulses in the free space radiation field using vacuum diode detectors. The power density measured is up to 100 kW/cm^2. In the X-band, the peak power of the microwave pulse of 10 kW can be directly measured, and the response time is less than 5 ns. In the S and L bands, the peak input power range that can be measured is not less than 20 kW [60].

The vacuum diode high-power microwave pulse measurement system has been verified and used in the measurement of X-band backward wave tubes. In this measurement, the recorded pulse peak voltage is 912 mV, with the pulse width is 22 ns, and the pulse front is 5 ns, and the corresponding microwave pulse peak power input to the detector is 7.2 kW. The peak power density at the horn waveguide antenna feed inlet is 38 kW/cm^2, and the calculated peak output power of the microwave source is 500 MW.

5.1.2 Measurement Method Using Thermionic Diode [61]

Chinese researchers have proposed a new type of high-power pulse microwave measurement detector based on thermal ion diode detection. The diode can withstand microwave pulse peak power up to 100 kW, short response time (less than 2 ns), no need for synchronous signals, and has strong anti-interference ability. Therefore, it is suitable for measuring high-power single-pulse and low-repetition-rate pulse peak power.

1. **Detector working principle and structure**

(1) **Working principle**

The basic structure of the thermal ion diode is shown in Fig. 5.2. It has a simple structure and consists of only two electrodes, cathode and anode. In order to heat the cathode, there should also be a cathode filament. The cathode and anode of the thermal ion diode are placed separately on the central part of the two wide sides of the rectangular waveguide under vacuum condition.

The cathode is heated by the filament and is in a pre-emission state. When a microwave pulse passes through the waveguide, a strong alternating high-frequency electric field $E_W(t)$ is formed in the waveguide, which acts on the cathode and anode of the thermal ion diode. Under the action of the forward electric field, the cathode emits an electron beam, thereby forming a current $I(t)$ in the output circuit. The magnitude of the current is related to the electric field $E_W(t)$. The forward current is further filtered, passed through a sampling resistor, and a voltage is generated on the resistor. This voltage is input to a digital oscilloscope, and the amplitude and waveform of the current $I(t)$ can be measured. Accurate calibration of the detector's sensitivity can achieve accurate measurement of the peak power and power envelope of high-power microwave pulses.

Fig. 5.2 Schematic diagram of the structure of a thermal ion diode

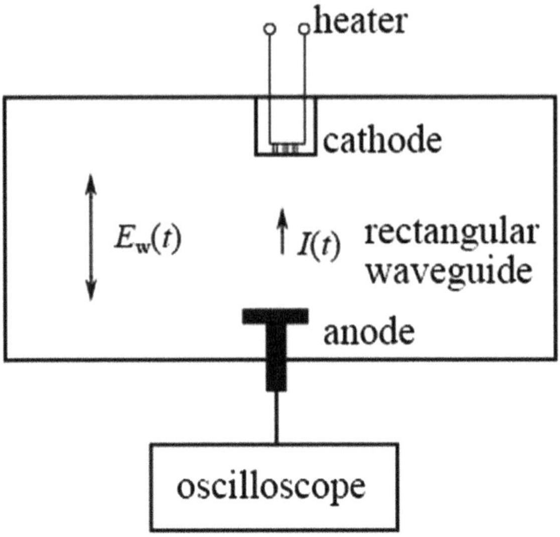

(2) **Structure of the detector**

The structure of the waveguide-type high-power microwave detector based on thermal ion diodes is shown in Fig. 5.3. The mode transmitted in the waveguide is the TE_{10} mode. Since the electric field of the TE_{10} mode is strongest at the center of the wide side of the waveguide, the thermal ion diode is inserted into the hole at the center of the wide side of the waveguide. In order to reduce the influence of reflection or other interference on the thermal ion diode, a waveguide-type matching load is connected to the end of the waveguide to completely absorb the microwave energy passing through the diode and shield other signals from coupling into the transmission waveguide from the end of the waveguide. Since the thermal ion diode is inserted in the waveguide, it will inevitably cause reflection, which increases the standing wave ratio at the input of the waveguide and affects the accuracy of input signal measurement. Therefore, an impedance matching tuner is needed. The simplest and most effective method is to insert an absorbing medium sheet from the narrow side of the waveguide. In addition to eliminating reflection and achieving the matching transmission of microwave signals, the absorbing medium sheet can also attenuate microwave signals. It can absorb forward-transmitted microwave energy to increase the measurement range of the detector, and can also absorb reflected energy from reverse transmission to reduce the possibility of microwave signals oscillating due to back-and-forth reflection. However, when calibrating the thermal ion diode power meter, the attenuation of microwave loss caused by the absorbing medium sheet should be considered. In addition, in order to avoid oscillation interference in the measured microwave detection signal, the output current of the thermal ion diode needs to be filtered. A specially designed low-pass filter should be used to strain out interference signals and to meet the power capacity requirements of the detector.

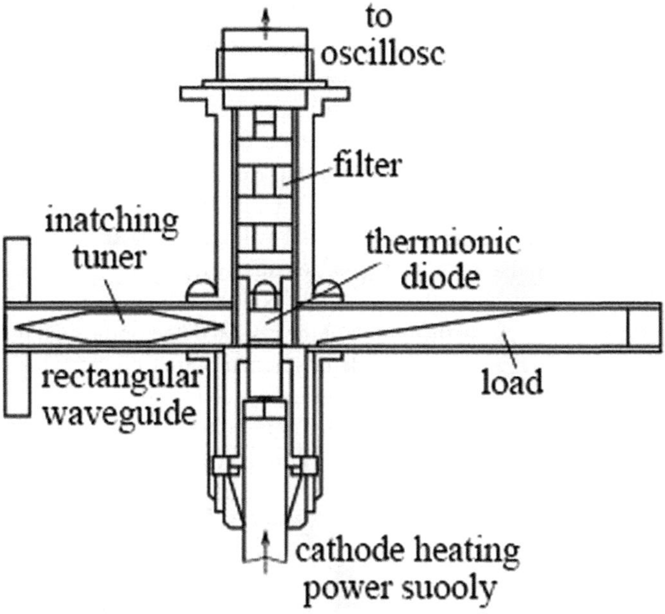

Fig. 5.3 Structure of a waveguide-type thermal ion diode detector

After conducting experiments on graphite, polyester materials, ferromagnetic materials, etc., it was found that ferromagnetic materials perform best as matching tuners for detectors. After multiple adjustments, a matching tuner with a standing wave coefficient of less than 1.1 near the operating frequency point was obtained.

2. **Calibration and experiment of detector sensitivity**

The sensitivity of the detector refers to the corresponding relationship between the microwave input power and the output voltage of the thermionic diode. The calibration system of the detector sensitivity mainly consists of a high-power microwave source, an input power monitoring system, a detector, and an output signal measurement system.

The output pulse width and power of the high-power microwave pulse signal source (such as a klystron) should be adjustable. This microwave pulse is directly input into the thermionic diode detector for detection after passing through an isolator. The detected pulse signal is sent to the oscilloscope to display the voltage size. The microwave power passing through the thermal ion diode is absorbed by the matching load. In order to calibrate the sensitivity of the thermal ion diode, one or two directional couplers can be connected behind the isolator to couple a small part of the microwave signal. This part of the power is detected by the microwave crystal detector and then sent to the oscilloscope for measurement of the output detection voltage. By utilizing the coupling coefficient and attenuation, the output power of the high-power microwave source can be calculated, which corresponds to the input power of the thermal ion diode, and also plays a role in real-time monitoring of microwave power. Adjust the adjustable attenuator to make the signal size sent to the oscilloscope suitable for the oscilloscope measurement range, in order to improve measurement accuracy. The oscilloscope also displays pulse waveforms (pulse width, repetition rate). From this, a relationship can be established between the input microwave power and the detection voltage of the thermal ion diode detector. By changing the microwave power or pulse width, the relationship between the microwave input power and the output signal voltage of the thermal ion diode at different powers and pulse widths, i.e. the sensitivity curve, can be calibrated. The results of this calibration show that the output power of the high-power microwave source obtained by converting the output of the crystal detector through the coupling coefficient and attenuation is smaller than its actual output power. This is because the attenuation caused by the matching tuner introduced by the detector for matching is not included in the calculation. The power input to the thermal ion diode detector actually also undergoes attenuation and loss through the matching tuner, causing the voltage output by the thermionic diode detector to be reduced. Therefore, for more stringent calibration, this loss should be corrected.

The calibration results of the thermal ion detector in the 10 cm wavelength band indicate that the detector has a wide detection power range, with the lower end reaching as low as 1 kW and the upper end exceeding 70 kW. Using this detector for actual measurements of the output power of a high-power microwave radiation source, it was indicated that the output power of the microwave source reached 1.2 GW.

5.1.3 Measurement Method Using Capillary Energy Meter

The capillary energy meter uses the same measurement principle as the glass liquid thermometer to measure microwave energy. This is a high-power microwave power measurement method proposed and applied by research workers in China. The capillary energy meter consists of a microwave transmission waveguide, a temperature sensing bubble, a temperature sensing liquid, a capillary tube connected to the temperature sensing bubble, and a liquid level height measurement device in the capillary tube. Figure 5.4 shows a schematic diagram of the capillary energy meter structure for measuring high-power microwave power in the TM_{01} mode of a circular waveguide using alcohol as the temperature sensing liquid.

1. **Sensitivity of capillary energy meter**

Assuming that the microwave energy injected into the waveguide is fully absorbed by the energy meter, we can calculate the relationship between the microwave energy absorbed by the liquid and the rise height of the liquid surface in the capillary tube based on the physical characteristics of the temperature sensing liquid.

The relationship between the energy absorbed by the liquid and the temperature rise is

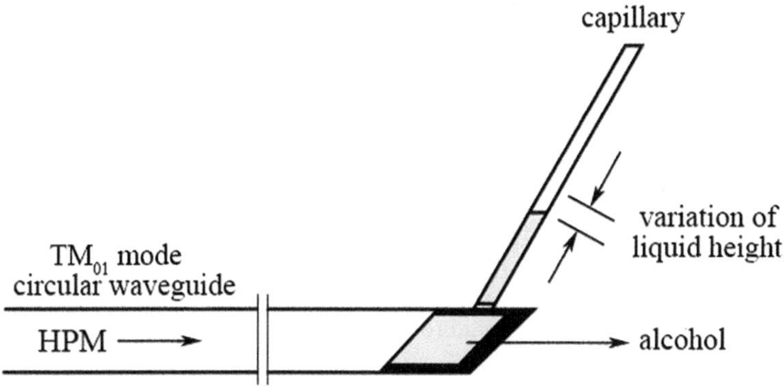

Fig. 5.4 Schematic diagram of the X-band capillary energy meter structure

$$Q = c_p \rho V_1 \Delta T \text{ (Joule)} \quad (5.1)$$

where,

Q is the absorbed microwave energy (J),

c_p is the specific heat capacity of the temperature sensing liquid at constant pressure (J/g K);

ρ is the density of the temperature sensing liquid (g/cm^3);

V_1 is the volume of the temperature sensing liquid before absorbing microwaves, that is, the volume of the temperature sensing bubble (cm^3) (for more accurate calculation, it should also include the volume corresponding to the original liquid level height in the capillary tube. If the volume of the capillary tube is much smaller than the volume of the temperature sensing bubble, the original liquid volume in the capillary tube can also be ignored);

ΔT is the temperature rise of the liquid after absorbing microwaves (K).

When a temperature sensing liquid absorbs microwave energy, its temperature rises and its volume expands. Since the volume of the temperature sensing bubble is fixed, the expansion of the liquid volume can only be reflected by the rise of the liquid level height in the capillary tube. We introduce the concept of volume expansion coefficient to describe the relationship between the temperature rise of the liquid and the volume expansion. When the temperature of the liquid changes by 1 °C, the ratio of its change in volume to its volume at 0 °C is called the volume expansion coefficient of the liquid, expressed as β (1/° K). When the temperature is not very high, β can be directly calculated according to the following equation:

$$\beta = \frac{V_T - V_0}{V_0 T}; \quad \mathrm{V_T} = V_0(1 + \beta T) \quad (5.2)$$

where, V_0 is the volume of the temperature sensing liquid at 0 °C, V_T is its volume at temperature T.

Since β is generally very small, when the temperature is not very high, β can be directly calculated based on the volume before and after the object expands, without the need to calculate the volume at 0 °C, that is,

$$\beta = \frac{V_2 - V_1}{V_1 \Delta T} = \frac{\Delta V}{V_1 \Delta T} \quad (5.3)$$
$$\Delta V = V_2 - V_1 = \pi r_m^2 \Delta l$$

where,

V_2 is the volume of the temperature sensing liquid after absorbing microwaves,
V_1 is the volume before absorbing microwaves,
ΔT is the temperature rise of the liquid after absorbing microwaves,
r_m is the inner radius of the capillary tube,
Δl is the height of the liquid level rise in the capillary tube.

From the above three equations, it is not difficult to obtain

$$D = \frac{\Delta l}{Q} = \frac{\beta}{C_p \rho \, \pi r_m^2} \quad (5.4)$$

This equation reflects the relationship between the microwave energy absorbed by the liquid and the height of the liquid level rise in the capillary tube, that is, the sensitivity D of the capillary energy meter.

It should be noted that the rise height Δl of the capillary liquid level refers to the rise height measured when the capillary tube is placed vertically. In addition, the top of the capillary tube should not be sealed, in order to avoid air compression when the liquid rises, which will cause resistance to the free rise of the liquid level. If sealing is required, the necessary measures should be taken to exhaust the air before sealing.

The sensitivity of the capillary energy meter can be calculated based on the characteristic parameters and geometric dimensions of the liquid used by the capillary energy meter. Then, based on the rise height of the liquid level in the capillary tube, the microwave energy Q absorbed by the liquid can be calculated, and then the microwave power can be calculated from Q based on the pulse width and repetition frequency of the microwave.

2. **Application of capillary energy meter**

(1) **Selection of liquid**

Table 5.1 gives the relevant characteristics of three common liquid materials.

$\frac{\Delta l}{Q}$ (mm/J)	r_m		
	0.1 mm	0.2 mm	0.5 mm
Water	1.6	0.4	0.06
Alcohol	19.0	4.7	0.76
Mercury	2.7	0.7	0.11

Table 5.1 The physical properties of water, alcohol, and mercury and their sensitivity at different r_m

Liquid	Density ρ/(g/cm^3)	Specific heat capacity c_p/(J/g K)	Volume expansion β/K^{-1} coefficient
Water	1.0	4.18	0.207×10^{-3}
Alcohol	0.79	2.38	1.12×10^{-3}
Mercury	13.5951	0.1394	0.16×10^{-3}

According to Table 5.1, when alcohol is used as the temperature sensing liquid, the sensitivity of the capillary energy meter is the highest. Therefore, in practical applications of capillary energy meters, alcohol is the best choice as the temperature sensing liquid. For example, when the inner radius of the capillary tube is $r_m =$ 0.48 mm, according to the data in Table 5.1, the sensitivity of the energy meter when using alcohol can be calculated to be 0.823 mm/J.

(2) **Matching between the temperature sensing bubble and the capillary energy meter**

The temperature sensing bubble of the capillary energy meter is generally made of quartz glass. In order to ensure that microwave energy can be absorbed as much as possible by the temperature sensing liquid in the temperature sensing bubble, the connection between the temperature sensing bulb (including the liquid inside) and the microwave transmission waveguide should be matched as far as possible. This problem often depends on the shape of the temperature sensing bubble and the placement method of the temperature sensing bubble in the waveguide. In the capillary energy meter, the function of the temperature sensing bubble is actually completely similar to that of the water load in the liquid calorimeter; the difference is that the water load is used in the flow-type liquid power meter, while the temperature sensing bubble is used in the static-type liquid power meter. Therefore, the two inlet and outlet pipes of the water load should be replaced with a capillary tube of the temperature sensing bubble. In order to accurately read the change of the liquid level in the capillary tube, the capillary tube must be placed vertically to facilitate the measurement of the rise height of the liquid level after being heated. If the capillary tube is inclined relative to the vertical direction, the inclination angle should be corrected when calculating the rise height of the liquid level due to thermal expansion, that is, converted into the change of height and volume in the vertical direction.

In order to check the matching degree between the temperature sensing bubble and the waveguide connection, it is necessary to measure the standing wave ratio of the energy meter. Since the capillary energy meter relies on the thermal effect of the temperature sensing liquid to realize power measurement, it depends only on the size of the microwave energy absorbed by the liquid, which is not directly related to the microwave mode. Therefore, as long as the temperature sensing bubble is connected to the waveguide to achieve matching, the ideal absorption of microwave energy can be achieved. Of course, the quality of the matching is also related to the mode to a certain extent, but for most cases, the shape and placement method of a temperature sensing bubble that matches well can be considered to have little difference in standing wave ratio for different modes. In the case of relatively simple mode excitation, direct measurement of the standing wave ratio of the capillary energy meter using the high-power microwave mode to be measured is of course a more ideal choice.

If the temperature sensing bubble cannot absorb all the microwave energy transmitted by the waveguide, it will likely cause some microwave reflection, resulting in an increase in standing wave ratio. Therefore, the direct measurement of the standing wave ratio of the capillary energy meter can reflect of the efficiency η of the temperature sensing bubble in absorbing microwave energy, that is,

$$\eta = 1 - \left(\frac{\rho - 1}{\rho + 1}\right)^2 \tag{5.5}$$

where, ρ is the standing wave ratio of the measured capillary energy meter.

The schematic diagram of the system used by Chinese research workers for measuring the standing wave coefficient of the capillary energy meter used to measure the TM_{01} mode power of the circular waveguide is shown in Fig. 5.5. The system uses the TM_{01} mode exciter to generate the mode of the capillary energy meter when actually measuring the high-power microwave power, which makes the measurement of the standing wave coefficient more accurate. After measuring ρ and calculating η, Eq. (5.1) should be corrected.

(3) **Calibration of the capillary energy meter**

Before using the capillary energy meter for power measurement, its sensitivity needs to be calibrated. The calibration system is shown in Fig. 5.6.

(a) In order to calibrate the sensitivity under conditions as close to actual measurement as possible, the signal source cannot use the output signal of the network analyzer. Its power is too small and the sensitivity calibration error is too large, which is not suitable for actual testing under high-power conditions. In addition, continuous wave signals are not conducive to the conversion of energy and

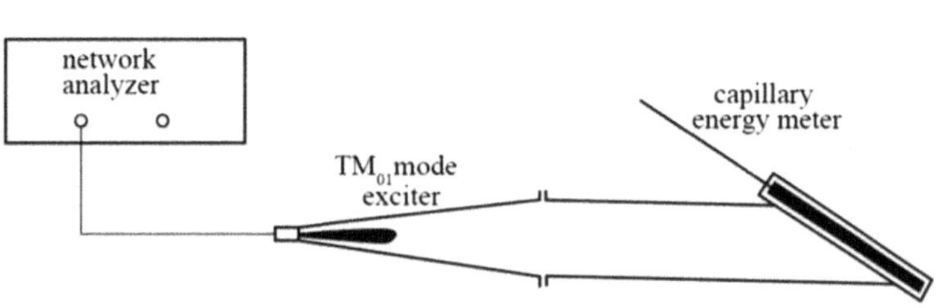

Fig. 5.5 Schematic diagram of the standing wave coefficient measurement system for capillary energy meter

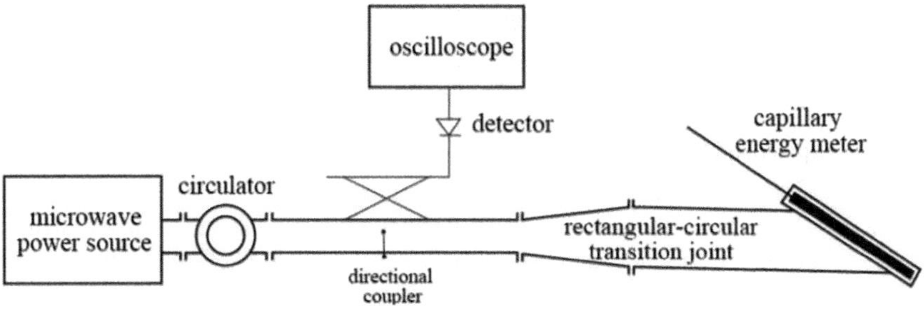

Fig. 5.6 Schematic diagram of the capillary energy meter calibration system

power. Therefore, a pulsed power source with a relatively high power level (such as a hundred kilowatts or even megawatts) should be used as the measurement signal.

(b) Since the capillary energy meter responds to the size of microwave energy, not the microwave mode, as long as the temperature sensing bubble of the capillary energy meter can absorb microwave energy as much as possible without reflection, its sensitivity will mainly depend on the size of the microwave energy, while the impact of the microwave mode is relatively small. In this way, we can avoid converting the output mode of the signal source used for calibration into the high-power microwave mode to be measured, and only need to make the energy meter match the signal source as much as possible to ensure that the energy meter can absorb as much microwave energy as possible from the output of the signal source. Therefore, only the rectangular-circular convertor is used in the calibration system to transform the rectangular waveguide output by the microwave source into the circular waveguide of the capillary energy meter, and no mode conversion element is connected again.

In order to improve the matching performance of the calibration system, the standing wave of the rectangular-circular transition waveguide can be measured and the size can be adjusted to improve the matching performance. The most direct way to measure the standing wave of the rectangular-circular transition waveguide is to connect a matching load for circular waveguide to the circular waveguide end, the rectangular waveguide end or it is already a standard waveguide, or it can be transformed into a standard waveguide through a transition waveguide, so that it can be connected to the network analyzer for measurement. However, the manufacture and performance identification of the circular waveguide matching load will encounter certain difficulties. Another method for measuring the standing wave of the rectangular-circular transition waveguide is to use two completely identical rectangular-circular transition waveguides for back-to-back measurement. At this time, the circular waveguide ends of the two rectangular-circular transition waveguides are directly connected, and a small section of the same size of straight circular waveguide needs to be added during the measurement process. The two rectangular waveguide ends can be respectively connected to the network analyzer and the matching load for measurement. The specific measurement and calculation methods can be referred to the relevant literature.

When measuring back-to-back, S_{21} can also be measured at the same time to check the transmission efficiency of the rectangular-circular transition waveguide. S_{21} includes both the reflection loss of the rectangular-circular transition waveguide and its own high-frequency loss, so it more accurately reflects the efficiency of transmitting microwave energy to the energy meter.

(c) The capillary energy meter connected to the measurement system and the capillary energy meter connected to the calibration system should be the same energy meter, so that their sensitivity is the same and no measurement error is generated.

(4) **Power measurement of high-power microwaves**

If the pulse width of the high-power microwave is τ(s), the repetition frequency is F(Hz), the sensitivity of the energy meter is $D = \Delta l/Q$(mm/J), where the rise height of the capillary tube liquid level during measurement is Δl, then the microwave power is

$$P = \frac{\Delta l/D}{\tau F} \tag{5.6}$$

But it should be noted that Eq. (5.6) has a limitation, that is, the high-power microwave source only works for 1 s. Because power is obtained by the ratio of microwave energy to the time it acts on the energy meter. In formula (5.6), F represents the number of microwave pulses generated in 1 s, τ represents the working time of each pulse. Thus, τF represents the total working time of the microwave in 1 s (unit is s). Also, $\Delta l/D$ is the energy generated when the microwave acts on the energy meter (joule value), so the ratio of the two is power.

In practical applications, the working time of microwaves is generally not easy to control or accurately determine, which brings uncertainty to power measurement, because as the working time of microwaves increases, the heat absorbed by the energy meter increases correspondingly, and the liquid level of the capillary tube rises continuously, but this change is not linear. Energy meters lose heat through channels such as conduction, radiation, and air convection, and the higher the temperature, the faster the heat loss. This causes the rise of the capillary liquid level to exhibit nonlinear characteristics and reach equilibrium after a certain period of time, and the liquid level no longer rises. In Sect. 2.2.1, we analyzed the heat balance process of a dry calorimeter and pointed out that after the calorimeter reaches thermal equilibrium, and after being calibrated, the steady-state temperature rise can be used as a measure of the measured power. Similar analysis can also be applied here to analyze the capillary energy meter.

In general, it is best to use the capillary energy meter for power measurement of single-pulse microwaves. At this time, $F = 1$ in the denominator of Eq. (5.6), and since the pulse width is known, the pulse width represents the total working time of the microwave pulse, so the power can be calculated more accurately based on the measured energy.

However, due to the low response of alcohol to heat and the relatively long thermal equilibrium process, the sensitivity of the capillary energy meter is relatively low, which makes the measurement accuracy of the capillary energy meter relatively low, thereby limiting its widespread use.

5.2 Power Measurement Using Resistive Sensor

Resistive sensors can actually be classified as a type of semiconductor thermistor that we introduced in Sect. 5.2.3. They are all semiconductor materials and undergo changes in resistance value when heated. However, the semiconductor thermistor we introduced in Sect. 5.2.3 emphasizes its thermal response sensitivity, so it is called a thermistor; while the resistive sensor to be discussed here emphasizes its response speed and power capacity for heat, to suit the measurement of high-power ultra-short pulse microwave.

The use of resistive sensors for high-power microwave measurement was first proposed by M. Dagys, Z. Kancleris and others from the Microwave Laboratory of the Semiconductor Physics Research Institute in Vilnius, Lithuania, and the main research work was also completed by the laboratory.

5.2.1 Resistive Sensor

1. **Introduction to resistive sensor** [62]

The resistivity of semiconductor materials is very sensitive to temperature. The well-known thermistor is a semiconductor element made using this characteristic, but its power capacity is very low and can only be used for small-power microwave measurement. The semiconductor resistance material that can withstand high-power microwaves is the resistive sensor. Although the power meter made using them is also a type of thermal effect power meter, the two are different in working principle. The principle of the thermistor is the overall temperature rise of the semiconductor material after absorbing microwave energy, which causes the material resistance to increase. In contrast, for the resistive sensor, the carriers in the material, namely the electrons, are heated by the microwave electric field, which causes the electrons to scatter irregularly during the migration process, so that the mobility decreases rapidly with the increase of temperature, that is, the resistance increases. Since the resistance of the entire sensor is large enough and the duration of the microwave pulse is very short, the effect of the microwave electric field can only cause the thermal effect of the electrons, leading to an increase in resistance. Moreover, it cannot cause the temperature of the entire sensor to change promptly, which in turn would change the resistance. This phenomenon was discovered more than 50 years ago. When there is a strong electric field, the current through the semiconductor material will no longer obey Ohm´s law, indicating that its resistance has changed. This is because in a strong electric field, electrons gain extra energy, and these heated electrons will be scattered more by the lattice, resulting in an increase in resistance.

In terms of microwave pulse measurement, the resistive sensor is also different from the diode detector. Point contact or Schottky diodes are commonly used to measure small power (not exceeding several hundred milliwatts) microwave pulses. When the peak power of the pulse is relatively high, it is necessary to couple out small power through a probe or small hole (directional coupler), or reduce the microwave pulse power through an attenuator, so that the measured power entering the detector is below the maximum power capacity allowed by the detector. The application of coupler or attenuator with a large attenuation amount will inevitably reduce the measurement accuracy, increase the volume and weight of the measurement system, and make the control of the measurement system more complicated; while the resistive sensor has a very large output signal range (up to tens of volts), so it can directly measure high-power pulses without the need for directional couplers or attenuators.

The thermal effect power meter we introduced in Chap. 2, including the thermistor power meter, can only measure the continuous wave microwave power or the average power of continuous pulse microwaves. Since we cannot ignore the thermal inertia, the thermal response time is far beyond the duration of the microwave pulse, so they are not suitable for measuring the pulse power of single pulses or low-repetition-frequency pulses. In the resistive sensor composed of n-type semiconductor Si, the heating of electrons is a very fast process, with a characteristic time on the order of a few picoseconds. Therefore, the actual response time of the resistive sensor depends only on the rise time of the bias voltage applied to the resistive sensor, and this mainly depends on the resistance of the resistive sensor.

The phenomenon that the resistance of the resistive sensor changes under the action of a microwave electric field was first discovered on n-type germanium semiconductor materials. Under the strong electric field of several hundred volts per centimeter, the volt-ampere characteristics of the semiconductor do not obey Ohms law, indicating that its resistance has changed. n-type silicon semiconductor materials are now more commonly used because they can withstand higher temperatures (200 °C), that is, stronger microwave electric fields. A microwave electric field of about 1 kV/cm can cause a 1% temperature change. In the waveguide of the X-band, this is equivalent to about 1 kW of microwave power.

Currently, the operating frequency range of the resistive sensor can reach 1–40 GHz. The pulse power range of the measurement is: in the waveguide, 100 kW in the S-band, 70 kW in the X-band, and 3 kW in the Ka-band. The response time is: 2.5 ns in the S-band, 0.5 ns in the X-band, and 0.2 ns in the Ka-band. It can be seen that the resistive detector is very suitable for measuring the pulse power of high-power microwaves with extremely short pulses.

2. **Sensitivity of the resistive sensor** [63]

The change in resistance of the resistive sensor quantitatively reflects the microwave pulse power transmitted in the waveguide, so the sensitivity ζ of the resistive sensor can be defined as

$$\zeta = \frac{\Delta R/R}{P} \tag{5.7}$$

where, *ΔR/R* is the relative change in resistance value of the semiconductor element in the resistive sensor, namely the semiconductor element, under the action of the microwave electric field, and *P* is the microwave pulse power transmitted in the waveguide. In order to measure the change in resistance, the resistance detector must be connected to a DC power supply or a pulse power supply with a pulse width much wider than the microwave pulse width, in order to measure the resistance size and its change of the resistance detector according to $R = U_0/I$. This external power supply can also be referred to as DC bias, denoted as U0. When using a pulse power supply, the magnitude of Δ*R/R* should be calculated by taking the average of current values of the microwave electric field passing through the resistance sensor within one pulse width.

Under low-power conditions, the average electric field strength in the resistive sensor is quite small. Therefore, when expanding the change in the resistance of the resistance sensor with respect to the average electric field in the sensitive element of the resistance sensor (referring to the semiconductor material in the power probe of the resistance sensor) into a power series, only the first term can be considered. Thus, in the linear region, *ΔR/R* in Eq. (5.7) can be expressed as

$$\frac{\Delta R}{R} = \beta^*(f)\langle E_m^2 \rangle \tag{5.8}$$

where, $\langle E_{\mathrm{m}}^2 \rangle$ is the average value of the square of the microwave electric field amplitude in the sensitive element of the resistive sensor, which is proportional to the power of the average electric field; $\beta^*(f)$ is the effective warm-electron coefficient, which is defined as the deviation of the voltage-current characteristic from Ohm's law and is a function of frequency. It can be well expressed by Eq. (5.9), which considers the electron thermal inertia when the resistance changes in the microwave electric field, that is,

$$\beta^*(f) = \beta\left[\frac{1}{2} + \frac{1}{1 + (2\pi f \tau_\varepsilon)^2}\right] \tag{5.9}$$

where, f is the frequency of the microwave field, τ_ε is a phenomenological energy relaxation time, β is the warm-electron coefficient in the DC bias electric field. By substituting Eqs. (5.8) and (5.9) into Eq. (5.7), we can get

$$\zeta = \frac{\beta}{2}\left[1 + \frac{2}{1 + (2\pi f \tau_\varepsilon)^2}\right]\frac{\langle E_m^2 \rangle}{P} \tag{5.10}$$

The power P in the microwave transmission line is proportional to $E_0{}^2$, where E_0 is the electric field amplitude at the center of the empty waveguide, so ζ should be proportional to $\langle E_{\mathrm{m}}^2 \rangle / E_0^2$. By substituting the power expression of the transmission mode in the waveguide into formula (5.10), we can obtain the relationship between ζ and $\langle E_{\mathrm{m}}^2 \rangle / E_0^2$. For n-Si semiconductors, Z. Kancleris et al. by measuring the appropriate $\beta^*(f)$ over a fairly wide frequency range obtained $\tau_\varepsilon = 2.9 \times 10^{-12}$ s at room temperature.

M. Dagys and Z. Kanc1eris reported that the warm-electron coefficient β of n-Si semiconductors is as follows at 300 K temperature when the electric field is along the <111> direction of the crystal: when the resistivity is 5 Ω cm, $\beta = 6.0 \times 10^{-8}$ cm²/V²; when the resistivity is 20 Ω cm, $\beta = 6.2 \times 10^{-8}$ cm²/V²; when the resistivity is 200 cm, $\beta = 6.7 \times 10^{-8}$ cm²/V².

In the calculation formula of ζ, only $\langle E_{\mathrm{m}}^2 \rangle$ is an unknown term, which can be used to calculate the field component of the transmission mode in the waveguide by the finite difference time domain method (FDTD, and then the average value of the electric field amplitude in the sensitive element of the resistive sensor can be obtained.

5.2.2 *Power Probe of Resistive Sensor*

1. **Rectangular waveguide**

(1) **TE$_{10}$ mode probe [64]**

The thermal effect of electrons is the only reason for the change in resistance of the resistive sensor. In order to record the change in resistance of the resistive sensor, we must apply a certain DC bias voltage U_0 to the resistive sensor and output the pulse signal generated by the resistive sensor due to the microwave pulse through capacitive coupling. After applying a high-power microwave pulse signal to the resistive sensor, the pulse amplitude U_s output by the power probe can be represented as

$$U_s = U_0 \frac{\Delta R}{R} \tag{5.11}$$

The actual structure of the probe is shown in Fig. 5.7a. The resistive sensor is made into a rectangular rod shape, with ohmic contacts formed at both ends, and its resistivity is about 20–200 Ω cm. ohmic contact refers to the contact between metal and semiconductor, and the resistance value of the contact surface is much smaller than the resistance of the semiconductor itself, so that when a voltage is applied to the component, most of the voltage drop is in the active region of the semiconductor rather than on the contact surface. The main measures to achieve ohmic contact are to perform high doping or introduce a large number of recombination centers in the end

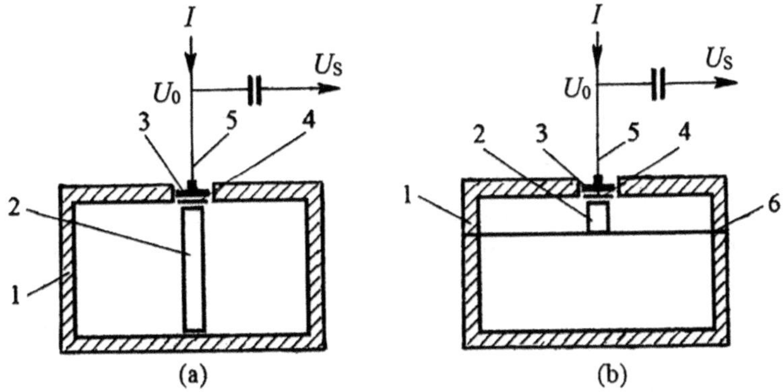

Fig. 5.7 Rectangular waveguide TE_{10} mode resistive sensor power probe **a** resistive sensor is placed throughout the waveguide as the power probe, **b** Thin-film type power probe. 1—waveguide, 2—resistive sensor semiconductor device, 3—insulation gasket, 4—metal gasket, 5—output lead, 6—metal film

surface layer of the semiconductor. Then place the semiconductor element between the two wide sides of the rectangular waveguide (parallel to the narrow side), with one end grounded and directly in contact with the waveguide wide side, and the other end insulated from the waveguide and connected to the measurement circuit.

To improve the thermal characteristics of the resistive sensor and extend its range of measuring high-power pulse power, a thin-film type TE_{10} mode power probe (Fig. 5.7b) has been developed. The resistive sensor is placed between a metal film and the waveguide wide side, and the length of the semiconductor element in the resistive sensor is about 1/10 of the height of the waveguide narrow side. The advantages of the thin-film type resistive sensor probe are: first, it extends the power measurement level, because the resistive sensor no longer bears the entire high-frequency voltage between the waveguide wide sides, but only bears the microwave power of about 1/10 passing through the waveguide, thereby enabling the power passing through the entire waveguide to be increased. Second, because the height of the semiconductor element is reduced, which improves the thermal characteristics of the resistive sensor. Third, the resistive sensor probe only occupies a small part of the waveguide cross-section, so even if the characteristic resistance of the resistive sensor is relatively small, the reflection of microwave power transmission will be reduced. This sensor has been used in laboratories in Russia, Sweden, the United States, etc. to measure nanosecond-level high-power microwave pulses in the S-band to Ka-band. The power capacity of this power probe has been measured to be 3 MW/m^2 using the far-field radiation measurement method.

From Eq. (5.11), it can be seen that the signal output from the resistive sensor will increase linearly with the DC voltage applied to the resistive sensor. However, the DC voltage cannot be increased too much. Otherwise it will cause the resistive sensor to overheat. To overcome this difficulty and obtain a larger output signal, a synchronous pulse source can be used instead of a DC bias source. This pulse source generates a current pulse with a width that is approximately 300 microseconds wider than the measured signal pulse. Its amplitude can be adjusted to the required size. At the time of 280 microseconds, it triggers a high-power microwave pulse, and the output signal is indicated by an oscilloscope. However, this type of resistive sensor requires a synchronous trigger signal to function properly, making real-time online measurements, especially long-distance measurements, relatively difficult. Additionally, due to the problem of triggering switch dithering in most pulse power sources, this sensor cannot guarantee capturing every microwave pulse signal, especially the pulse signals under repetitive frequency operation.

(2) **Sensitivity of the TE_{10} mode probe** [63]

Using the finite difference time domain method, the average microwave electric field amplitude $\langle E_m \rangle$ in the semiconductor sensitive element of the resistive sensor can be obtained. For ease of calculation, we take its dimensionless value $\langle E_m / E_0 \rangle$ normalized to the electric field amplitude E_0 in the empty waveguide. Since E_0 can be expressed as

$$E_0^2 = \frac{4P\sqrt{\mu_0/\varepsilon_0}}{ab\sqrt{1-(f_c/f)^2}} \tag{5.12}$$

where, a and b are the wide and narrow dimensions of the rectangular waveguide, f_c is the cutoff frequency of the TE_{10} mode. Using Eq. (5.7) to (5.12), we obtain the sensitivity of the linear region of the resistive sensor's output signal to changes in input power as

$$\zeta(f) = \frac{4\beta\sqrt{\mu_0/\varepsilon_0}}{ab\sqrt{1-(f_c/f)^2}}\left[\frac{1}{2}+\frac{1}{1+(2\pi f\tau_\varepsilon)^2}\right]\frac{\langle E_m^2 \rangle}{E_0^2} \tag{5.13}$$

$\langle E_m \rangle$ is calculated through the FDTD method.

2. **Circular waveguide**

(1) **Power probe [65]**

The power probe of the resistive sensor in the circular waveguide is shown in Fig. 5.8.

The n-Si semiconductor sensitive element is closely attached to the inner wall of the circular waveguide, with a thin metallic film coated on the top and an insulating film coated on the bottom. It is connected to the external circuit (bias input and signal output) through holes on the insulating film, forming the power probe of the circular waveguide resistive sensor. Its ideal form should be as shown in Fig. 5.8a. However, in reality, this ideal solution is difficult to achieve because it is challenging to fabricate a small section of n-Si sensitive element in a circular shape, with its arc fitting the wall of the circular waveguide. The most likely approach, as shown in Fig. 5.8b, it is formed by two pieces of semiconductor slices of rectangular resistive

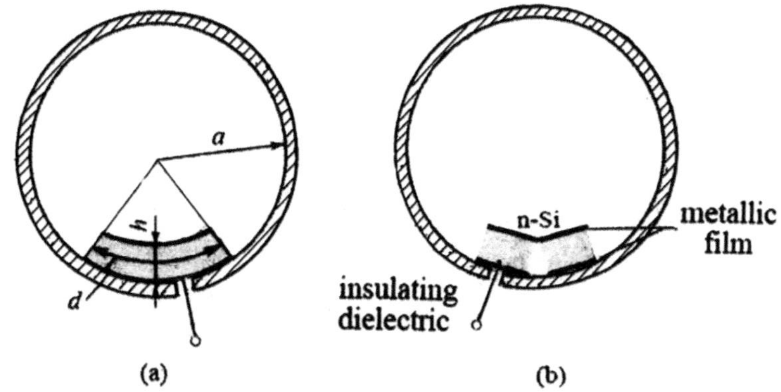

Fig. 5.8 Cross-sectional view of resistive sensor power probe in circular waveguide **a** ideal form, **b** practical application form

sensors placed closely together. Their tops are connected together by a thin metal film (electrically shorted), while the bottom of one semiconductor slice is connected to the inner wall of the circular waveguide, and the bottom of the other semiconductor slice is isolated insulated from the waveguide. These two pieces form a parallel circuit and are connected to the conductor through holes in the insulating film, allowing the resistive sensor to be biased with DC voltage and output pulse signals. In this way, the two resistive sensors are in series for DC current, but in parallel for microwave electric fields, a design that can be fully realized using semiconductor technology.

(2) **Sensitivity**
(a) TE_{11} mode [65]

The sensitivity of the resistive sensor is defined in Eqs. (5.7) and (5.8). By using the expression of the field components of the TE_{11} mode in the circular waveguide. Then, by integrating the Poynting vector of the TE11 mode over the entire waveguide cross-section, we can calculate its power. Finally, this power can be substituted into Eq. (5.7) to obtain the sensitivity of the TE_{11} mode.

$$\zeta = \frac{\beta^*}{\pi a^2}\left[J_0^2(\mu_{11}) + \left(1 - \frac{2}{\mu_{11}^2}\right)J_1^2(\mu_{11})\right]^{-1} \frac{\sqrt{\mu_0/\varepsilon_0}}{\sqrt{1-(\lambda/\lambda_c)^2}}\left(\frac{\langle E_m\rangle}{E_0}\right)^2 \quad (5.14)$$

where,

a is the radius of the circular waveguide,

E_0 is the maximum amplitude of the TE11 mode field in the circular waveguide,

λ and λ_c are the wavelength and cutoff wavelength of the TE_{11} mode microwave signal in free space and in the circular waveguide, respectively,

J_0 and J_1 are the zero-order and first-order Bessel functions of the first kind,

μ_{11} is the first root of the derivative of J_1, $\mu_{11} = 1.841$.

(b) TE_{01} mode [66]

By substituting the power expression of the TE_{01} mode in the circular waveguide into the sensitivity definition Eq. (5.7) and using Eq. (5.8), we can obtain the sensitivity of the resistive sensor TE_{01} mode in the circular waveguide.

$$\zeta = \frac{2\beta^*}{\pi a^2}\left(\frac{J_1(\mu_{11})}{J_0(\mu_{01})}\right)^2 \frac{\sqrt{\mu_0/\varepsilon_0}}{\sqrt{1-(\lambda/\lambda_c)^2}}\left(\frac{\langle E_m\rangle}{E_0}\right)^2 \tag{5.15}$$

where, λ and λ_c are the wavelength and cutoff wavelength of the TE_{01} mode in free space and in the circular waveguide, respectively, $\mu_{01} = 3.832$ is the first root of the derivative of J_0, the remaining symbols are defined the same as those in Eq. (5.14).

3. **Response time of the resistive sensor**

The response time of the resistive sensor to microwave pulses can be measured using time domain reflectometry (TDR). TDR utilizes the transmission of a signal in a certain transmission line. When there is an impedance change (non-uniformity) in the transmission path, a portion of the signal is reflected, while the rest continues to transmit along the line. By measuring the time from the reflection point to the signal output point in the time domain, the position of the impedance change point in the transmission path can be calculated. The impedance of the resistive sensor is different before and after receiving microwave pulses, causing different reflections. By measuring the time difference between the two different reflections, the response time to microwave pulses can be determined.

When measuring, the bias voltage of the resistive sensor is provided by a pulse with a rise time in the sub-nanosecond range. By measuring the time interval between the initial pulse and the reflected pulse, the response time of the resistive sensor can be calculated. The measurement results for different bands are given in Table 5.2 [64].

From the data in Table 5.2, it can be seen that the response time of the resistive sensor to microwaves generally does not exceed 10 microwave oscillation period.

Table 5.2 Response time of the resistive sensor

Frequency band	Response time τ/ns	Ratio of response time τ to microwave oscillation period T (τ/T)
S	2.5	8
C	2	10
X	0.5	5
Ka	0.4	7
Ku	0.2	7

5.2.3 Power Measurement Method Using Resistive Sensor

1. **Measurement of the relationship between sensitivity and signal frequency** [67]

The measurement of the sensitivity of the resistive sensor can be carried out using the measurement system shown in Fig. 5.9. The frequency-adjustable microwave signal source outputs an average power of about 200mW. Using a small-power microwave source can effectively improve the frequency response measurement of the resistive sensor. The microwave signal is modulated using a modulator, and a modulation frequency of 10 kHz is selected to avoid additional changes in the resistance due to lattice heating. In general, the signal applied to the resistive sensor is the sum of the microwave signal and the modulation signal, so the change in its resistance is jointly caused by the thermal effects of the electrons and the Joule heating of the crystal lattice. Since lattice heating depends only on the characteristic resistance of the n-Si semiconductor used to make the resistive sensor and decreases with increasing modulation frequency, in order to avoid the heating of the crystal lattice affecting the output signal, the modulation frequency must be chosen high enough. At this point, the influence of Joule heating on the signal (or resistance change) can be ignored. The power of the microwave source applied to the resistive sensor is measured using an average power meter, and the signal output by the resistive sensor is amplified using a phase-locked amplifier to improve measurement accuracy.

By measuring the signal output from the lock-in amplifier and the power transmitted to the resistive sensor, the sensitivity of the resistive sensor can be represented as,

$$\zeta = \frac{\pi U_m}{\sqrt{2} U_0 P} \tag{5.16}$$

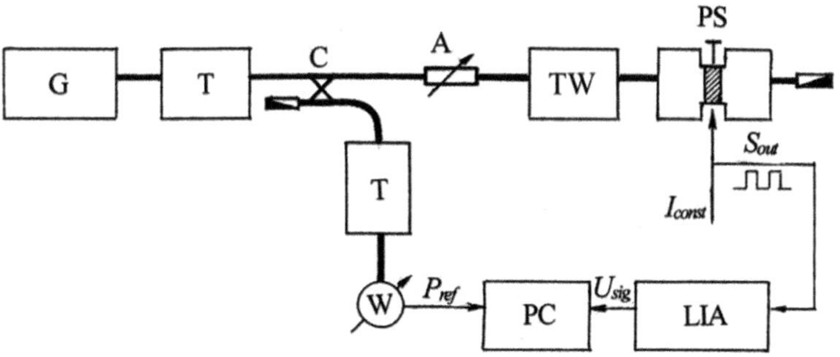

Fig. 5.9 Resistive sensor sensitivity measurement system G—microwave signal generator, T—coaxial–waveguide convertor, TW—rectangular waveguide–double-ridged waveguide convertor, C—directional coupler (DC-10-20 dB), A—variable precise attenuator, RS—resistive sensor, W—reference power RS output, PC—computer, LIA—lock-in amplifier

where, U_m is the output signal amplified by the lock-in amplifier, U_0 is the DC bias voltage applied to the sensitive element.

By changing the operating frequency of the microwave signal source and then measuring the sensitivity, we can finally obtain the relationship curve between sensitivity and signal frequency. If the output power of the microwave signal source is relatively small and the power coupled out by the directional coupler is too small, we can also use a power divider to directly split out a portion of the power from the signal source for measurement.

2. **Measurement of microwave power transmitted in waveguide** [63,67]

When measuring the relationship between power and the output signal of the resistive sensor, we should use a high-power microwave signal source, typically using a magnetron, as shown in the measurement system in Fig. 5.10. The resistive sensor is directly connected to the output of the magnetron, with a pulse width of 4 μs and a repetition frequency of 25 Hz. The reference power probe is coupled to the output waveguide of the directional coupler, and the output power of the magnetron is adjusted by changing the attenuation amount L of the precision attenuator. By measuring the output signal of the resistive sensor at different output powers, we can obtain the relationship between the output signal and the microwave pulse power. The DC bias voltage is set to 10 V and directly applied to the resistive sensor. The output signal can be measured with or without an amplifier, and the amplification factor of the amplifier is adjusted as needed.

By measuring the output signal U_s of the resistive sensor, the change in resistance $\Delta R/R$ can be calculated according to the following formula,

$$\frac{\Delta R}{R} = \frac{U_s}{U_0} \frac{1 + R/R_a}{1 - U_s R/U_0 R_a} \tag{5.17}$$

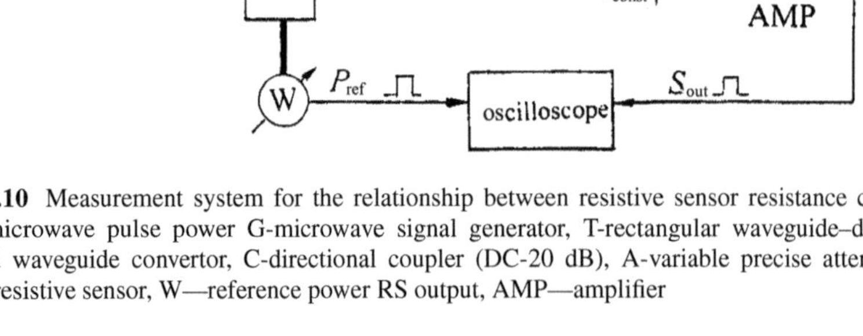

Fig. 5.10 Measurement system for the relationship between resistive sensor resistance change and microwave pulse power G-microwave signal generator, T-rectangular waveguide–double-ridged waveguide convertor, C-directional coupler (DC-20 dB), A-variable precise attenuator, RS—resistive sensor, W—reference power RS output, AMP—amplifier

where, R_a is the input impedance of the amplifier or oscilloscope. Since R_a is much larger than the resistance R of the resistive sensor, the relative change in resistance of the resistive sensor is relatively easier to determine. If we ignore the second factor on the right side of Eq. (5.17), we get Eq. (5.11).

The relationship between $\Delta R/R$ and the microwave pulse power P can be approximated as a quadratic polynomial of $\Delta R/R$ in a fairly wide range of power, that is

$$A\frac{\Delta R}{R} + B\left(\frac{\Delta R}{R}\right)^2 = P \tag{5.18}$$

where, the coefficients A and B are determined based on experimental measurement data of the relationship between $\Delta R/R$ and P. If we only consider the linear region of Eq. (5.18), that is, ignore the quadratic term, and then compare it with Eq. (5.7), we can obtain

$$\zeta = \frac{1}{A} \tag{5.19}$$

3. **Measurement of microwave power in the radiation space**

The measurement of high-power microwave pulse power using a resistive sensor can be performed using a radiation far-field measurement system, as shown in Fig. 5.11. The measurement is carried out in a microwave darkroom, and the resistive sensor is directly connected to the receiving antenna, eliminating the need for a directional coupler or attenuator, thus avoiding additional measurement errors. This system can measure pulse power density up to several megawatts per square meter.

The sensitivity of the radiation far-field measurement system can be expressed as:

$$\chi = \frac{4\pi l^2 U_s}{G_s P U_0} \tag{5.20}$$

where,

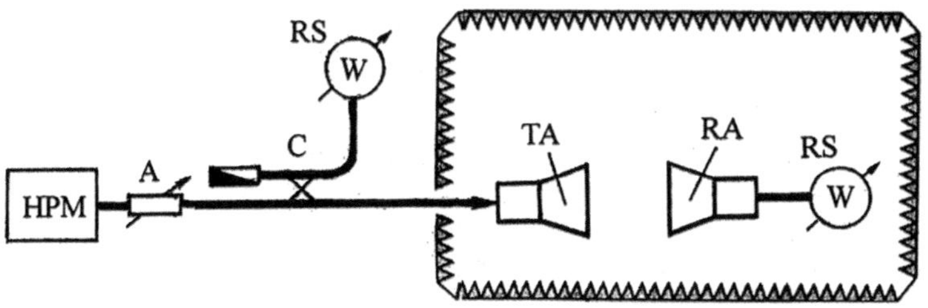

Fig. 5.11 Schematic diagram of the system for high-power microwave pulse power measurement using resistive sensor. HPM—high-power microwave source, A—variable precise attenuator, RS—resistive sensor, W—reference power RS output, TA—radiating antenna, RA—receiving antenna

l is the distance from the radiating antenna to the measuring resistive sensor,
U_s is the signal output from the resistive sensor,
G_s is the gain of the radiating antenna,
P is the microwave power transmitted to the radiating antenna,
U_0 is the bias voltage of the resistive sensor.

5.2.4 Application of Resistive Sensor in China

1. **P-type semiconductor resistive sensor** [68]

In China, in-depth research has also been conducted on high-power microwave pulse power measurement using resistive sensors. In the 3 cm band, Chinese researchers use a resistive sensor made of p-type semiconductor Si, which is different from the n-Si type proposed by M. Dagys and Z. Kancleris. Under the action of a strong microwave electric field, the energy of its charge carriers increases transiently and transitions back to the top of the valence band, resulting in a decrease in the momentum diffusion time of the charge carriers and an increase in the bulk resistance of the sensitive element of the resistive sensor. The peak microwave power is then characterized by the change in bulk resistance.

P-type Si semiconductors can be used to make resistive sensors of various types within the frequency range of 3–30 GHz, with a response time of less than 2 ns, capable of withstanding power nearly 6 orders of magnitude higher than ordinary microwave detectors, mainly limited by waveguide breakdown. The response time of their resistivity to microwave electric fields is on the order of picoseconds, capable of withstanding electric field intensities greater than 50 kV/cm. The pulsed bias voltage applied to the resistive sensor is 18 V with a pulse width of 500μs.

The power probe structure of the BJ-100 standard waveguide resistive sensor developed by Chinese researchers is shown in Fig. 5.12. The waveguide consists of two parts, connected as a whole using positioning pins and fixing screws. In the waveguide, two copper foils are fixed parallel to the wide side and symmetrically to the center of the narrow side, with a thickness of 0.1 mm and a spacing of 2 mm between them. Two copper foils should be kept parallel, such a structure ensures that the smallest disturbance to the distribution of the polarized electric field of the TE_{10} modes within the waveguide. The sensitive element of the detector is fixed at the center of the two copper foils, with dimensions of 4.0 mm × 2.0 mm × 1.0 mm, and its two end faces should have good ohmic contact with the copper foils. One copper foil has good electrical contact with the waveguide side wall (narrow side) as the ground electrode, i.e., it is at the same potential as the outer conductor of the coaxial cable fixed on the top of the waveguide. The other copper foil is insulated from the waveguide and is connected to the inner conductor of the coaxial cable as the output for the resistive sensor's detection signal and the input for the bias voltage. The connection points of the detector probe's electrodes should produce as small parasitic capacitance and inductance as possible.

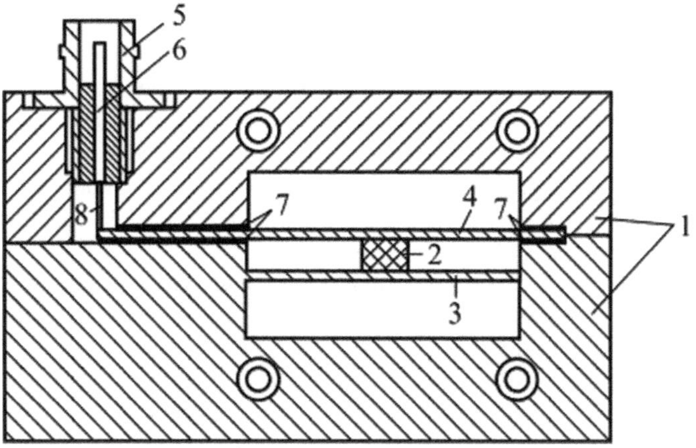

Fig. 5.12 Power probe structure of p-type Si semiconductor detector 1—waveguides; 2—p-type Si semiconductor sensitive components; 3, 4—copper foil; 5—coaxial cable joint outer conductor; 6—coaxial inner conductor; 7—insulation medium; 8—coaxial transmission line

This sensor operates in the range of 8.2–12.5 GHz, and the test results show that the response time of the sensor is at least less than 1.7 ns. It can reliably operate under a driving power supply of 100 V and 80us, and can withstand microwave power of over 200 kW.

2. **N-type semiconductor resistive sensor**
(1) **Terahertz band resistive sensor [69]**

The more successful work of Chinese researchers is the development of an overmoded waveguide 0.3–0.4 THz microwave pulse sensor using n-Si semiconductor. In the terahertz band, the standard waveguide is too small, and the sensitive element of the resistive sensor is even smaller, making it difficult to manufacture and even more challenging to place and fix it in the waveguide. Therefore, they proposed a solution to transition the standard waveguide to an overmoded waveguide, thereby increasing the waveguide size (see Fig. 5.13).

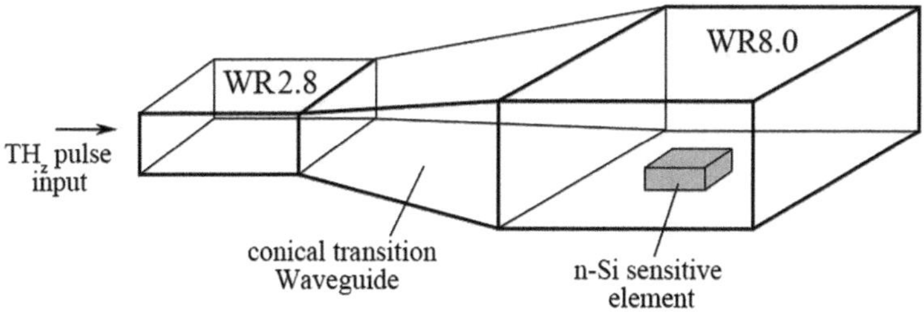

Fig. 5.13 Power probe of the terahertz pulse overmoded waveguide resistive sensor

The power probe of this resistive sensor is similar to the power probe of the resistive sensor in a circular waveguide (see Fig. 5.8b). The two pieces of rectangular sensitive elements of the resistive sensor are directly attached to the bottom center of the overmoded waveguide, with their tops connected by a metal film to form a short circuit. At the bottom, one sensitive element is coated with a metallic film and is connected to the grounding of the wide side of the waveguide, while another sensitive element is insulated from the wide side of the waveguide but connected to a bias source through a cable and outputs a signal.

The distribution of electromagnetic field components, voltage standing wave ratio, and average electric field in the power probe is obtained using the three-dimensional finite difference time domain method. By adjusting the length, width, height, and characteristic resistance of the sensitive element of the resistive sensor, the performance of the resistive sensor is optimized. Its relative sensitivity is about 0.24 kW^{-1}, with fluctuations not exceeding $\pm 22\%$, and the maximum voltage standing wave ratio in the 0.3–0.4 THz frequency range is 2.74. With the use of overmoded waveguide, such a resistive sensor can be manufactured and assembled smoothly.

(2) **X, S-band resistive sensor** [70]

After improving the manufacturing process of the n-type semiconductor resistive sensor, Chinese researchers designed a new bias power supply and sensor structure, increasing the detection sensitivity of the sensor by an order of magnitude. The X-band resistive sensor detects an output pulse signal amplitude of 9 V at a microwave power of 60 kW, while the S-band resistive sensor reaches an output pulse signal amplitude of 10 V at 600 kW, suitable for high-power microwave power measurements with a pulse width of 10 to 500 ns.

The structure of the sensor (power probe) is shown in Fig. 5.14, and the entire probe is directly installed inside the waveguide flange, forming an integrated structure, making the processing more convenient and the structure more compact, with no weld seams inside the waveguide and lower microwave transmission loss. The electrodes are made of thin copper sheets, and the plane of the copper sheet inside the waveguide remains perpendicular to the microwave electric field to minimize its impact on the microwave field. The electrode is connected to the inner conductor of the BNC connector, with the BNC connector serving as the input for bias and the output interface of the sensor.

The results of the S-band relativistic backward wave tube measurement using the power probe are shown in Fig. 5.15, and the measurement is carried out in the case of a short pulse in the order of nanoseconds and a wide pulse of 200 ns, and a good pulse waveform is obtained, and after calibration the power of the backward wave tube can be obtained by conversion.

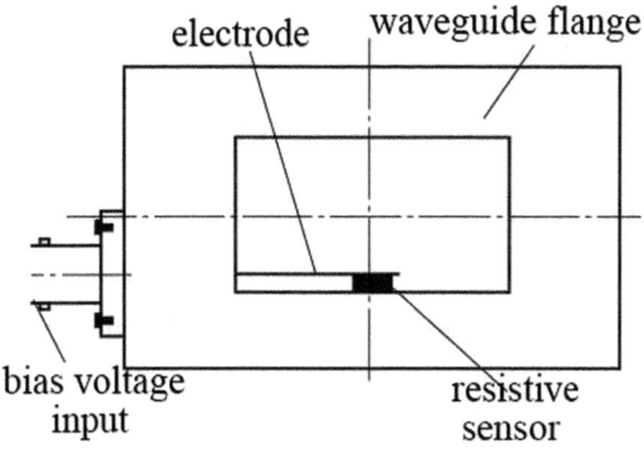

Fig. 5.14 The power probe structure of an S-band n-type semiconductor resistance detector

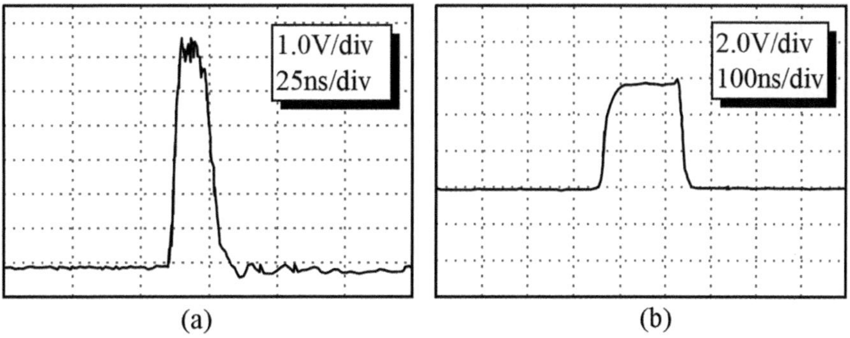

Fig. 5.15 Pulse waveforms of relativistic backwave tube detected by the resistive sensor **a** short pulse waveform, **b** wide pulse waveform

5.3 Electroacoustic Method for the Power Measurement

5.3.1 Fundamental Principles of Electroacoustic Method

In the methods for high-power microwave power measurement introduced in this chapter, except for the resistor detection method, the rest can only output weak detection signals, resulting in low signal-to-noise ratio, and poor anti-noise ability, affecting measurement accuracy and resolution. In addition, the detectors are relatively large in size and when they are directly placed in the waveguide, they unavoidably cause interference to the waveguide field. In contrast, electroacoustic method measurement can largely overcome these shortcomings.

1. **Surface acoustic wave**

Surface acoustic wave is a new technology that emerged in the late 1960s, which is based on an interdisciplinary combination of acoustics and electronics. Currently, surface acoustic wave technology has been widely used in various disciplines and technical fields such as seismology, astronomy, signal processing, aerospace, petroleum exploration, non-destructive testing, electronic tags, etc.

When an alternating electric field is applied through a transducer to a piezoelectric material, it will excite elastic waves caused by material deformation in the piezoelectric body. This function of the piezoelectric body is called the inverse piezoelectric effect. Conversely, elastic waves in the piezoelectric body can also excite electrical signals, known as the positive piezoelectric effect. Elastic waves will generate surface acoustic waves (SAW) propagating along the surface of the piezoelectric surface. The amplitude of the surface acoustic wave rapidly decreases with increasing depth into the substrate. The vibration frequency of the surface acoustic wave is the same as the frequency of the applied alternating electrical signal. Compared to the bulk acoustic waves propagating within a solid medium, SAW has two significant characteristics: high energy density and slow propagation speed.

The main piezoelectric materials include single crystal quartz, $LiNbO_3$, $LiTaO_3$, $Bi_{12}GeO_{20}$, piezoelectric ceramics.

5. **2.Surface acoustic wave devices**

The working principle of surface acoustic wave devices is as follows: two acoustic-electric transducers are made on the piezoelectric substrate. The transducer at one end of the substrate (input transducer) converts the input electrical signal into an acoustic signal through the inverse piezoelectric effect. This acoustic signal propagates along the surface of the substrate and is ultimately converted back into an output electrical signal by the transducer (output transducer) at the other end of the substrate. The overall function of the surface acoustic wave device is achieved through various processing of the acoustic signals that propagate on the piezoelectric substrate and utilizing the characteristics of the acousto-electric transducer.

It can be seen that the key to the production and application of surface acoustic wave devices is the transducer. Currently, the most effective transducer is the interdigital one.

The basic structure of the surface acoustic wave device is to use semiconductor integrated circuit planar technology. A 0.1um thick aluminum film is deposited on the polished surface of the piezoelectric substrate, and then a mask pattern is used for lithography to make the input transducer and output transducer.

The superiority of the surface acoustic wave device is as follows:

(a) Surface acoustic waves have extremely slow propagation speed and short wavelengths. Therefore, in the same frequency range, the size of the surface acoustic wave device is much smaller than that of the corresponding electromagnetic wave device, and the weight is also greatly reduced. For example, the delay obtained with a 1 km long microwave transmission line can be achieved with a surface acoustic wave delay line with a transmission path of only 1 m. It can be seen that the use of surface acoustic wave technology can achieve ultra-miniaturization of electronic devices. A surface acoustic wave device is often only a few millimeters in size, and the line width of the transducer is only 0.25–0.2 μm.
(b) Since surface acoustic waves propagate along the solid surface at an extremely slow speed (generally 3000–4000 m/s), which is 70,000 to 100,000 times slower than the speed of electromagnetic waves, time—varying signals can be completely presented on the surface of the crystal substrate at a given instant. Therefore, when the signal travels between the input and output ends of the device, it is easy to sample and transform the signal, such as compressing and widening pulse signals, encoding and decoding, and convolving two arbitrary analog signals, etc.

Due to the relatively long delay between the signal output and input in the surface acoustic wave device, the noise signal has been attenuated during this period, making the device highly noise-resistant.

In addition, in many cases, the performance of surface acoustic wave devices far exceeds the level that even the best electromagnetic wave devices can achieve. For example, surface acoustic waves can be used to make pulse compression filters with a time-bandwidth product greater than 5000. In the UHF (ultra-high frequency, frequency range of 300 to 3000 MHz) band, resonant cavities with a Q value exceeding 50,000 can be created, and filters with a frequency response non-flatness of only ± 0.3 to ± 0.5 dB and an out-of-band suppression of over 40 dB can be fabricated.

(c) Since the surface acoustic wave device is made using semiconductor planar processing on single crystal materials, it has good consistency and repeatability, is easy to mass-produce, and when using certain single crystal materials or composite materials, the surface acoustic wave device has extremely high temperature stability.
(d) The surface acoustic wave device has strong radiation resistance and a large dynamic range, up to 100 dB. This is because it utilizes elastic waves on the crystal surface without involving the process of electron migration.

The main characteristics of the surface acoustic wave device are excellent frequency selectivity (selectable frequency range from 10 MHz to 3 GHz), small input–output impedance error, low transmission loss, good electromagnetic interference resistance, high reliability, small size, light weight, and the ability to achieve multiple complex functions. The shortcomings of SAW lie in the fact that the required substrate is made of single crystal material, making its price relatively high, and the directional

Fig. 5.16 Structure of an interdigital transducer

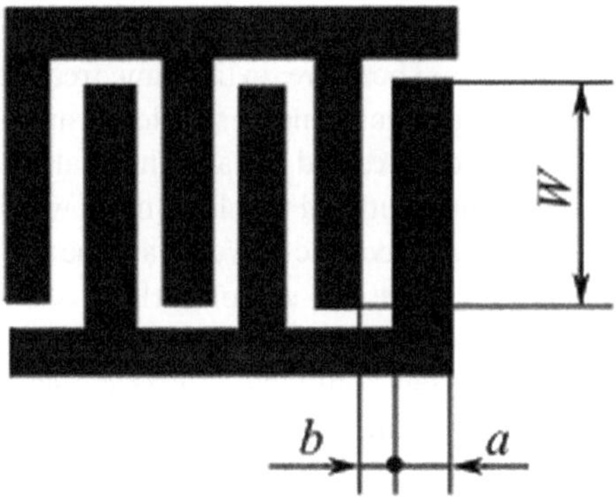

cutting, polishing, and manufacturing processes for the substrate are demanding. Due to the strict requirements of substrate crystallization and manufacturing accuracy, its cost is high. Due to process limitations, the operating frequency upper limit of the surface acoustic wave device is limited to 2–3 GHz at present.

3. **Transducer** [71]
(1) **Linear interdigital transducer**

A device that can convert acoustic–electrical energy is called a transducer. Various schemes for exciting surface acoustic waves have been proposed, such as wedge-shaped transducers, comb transducers, etc., the conversion efficiency of these methods is low and it's difficult to obtain high-frequency surface acoustic waves, so they are impractical. So far, the most widely used and effective transducer is the interdigital transducer (IDT), and its basic structure is shown in Fig. 5.16.

The IDT is composed of several metal thin film electrode strips deposited on a piezoelectric substrate material, and these electrode strips are arranged in a cross configuration, and connected at both ends by a bus-bar wire. Its shape is like two rows of intersecting fingers laid flat, so it's called an interdigital electrode. The interdigital electrode is used as both a transmitting transducer to excite surface acoustic waves and as a receiving transducer to receive surface acoustic waves and excite electrical signals; thus, it is reversible. When an AC signal of the appropriate frequency is applied to the transmitting IDT, an electric field distribution appears inside the piezoelectric substrate, as shown in Fig. 5.17. This electric field can be decomposed into two components, E_v and E_u, perpendicular and parallel to the surface, respectively. Due to the inverse piezoelectric effect of the substrate, this electric field causes a deformation of the material between the finger electrodes, resulting in mass point displacement. E_u causes the mass point to produce compression and expansion displacement parallel to the surface, while E_v causes vertical shear displacement up and down perpendicular to the surface. As the applied signal changes periodically, this

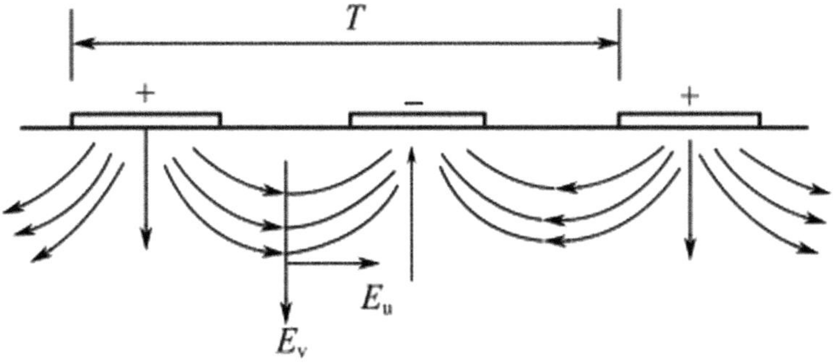

Fig. 5.17 The electric field distribution when a microwave signal is applied to the IDT

displacement will also change periodically, thus generating surface acoustic waves propagating along both sides of the IDT, with a frequency equal to the frequency of the applied electrical signal. The useless wave of one side can be absorbed by a high-frequency lossy medium, while the SAW on the other side propagates to the receiving IDT, and with the aid of the direct piezoelectric effect, the SAW is converted into an electrical signal output again.

(2) **Nonlinear grid transducer**

The grid consists of a series of parallel and equidistant electrode strips deposited on the substrate material. It can be obtained by two methods: the first method is to use semiconductor integrated circuit technology to deposit aluminum (or gold) on electrostrictive material or nonlinear piezoelectric material substrates, and then use photolithography to shape the metal film into a grid shape. In this case, the area of the substrate material not covered by the metal film becomes the area for exciting surface acoustic waves. The second method is to deposit electrostrictive material or nonlinear piezoelectric material strips on a dielectric substrate as the grid. This grating structure can serve as both a transducer for sound-to-electric energy conversion and as a reflector for reflecting surface acoustic waves.

When the grid is used as a reflector, the most typical application is to form a surface acoustic wave resonator, which can be made into a high-Q resonator in the microwave band with a quality factor of tens of thousands. By placing a grid reflector at each end of two interdigital transducers, with one interdigital transducer serving as the input port and the other as the output port, the output signal is amplified by an external amplifier and fed back to its input. As long as the gain of the amplifier can compensate for the losses of the resonator and its connecting wires, and satisfy certain phase conditions, the resonator can oscillate. The oscillation frequency of the surface acoustic wave resonator will change with factors such as temperature and deformation of the piezoelectric substrate material, so it can be used as a sensor to measure various physical quantities.

5.3.2 Electroacoustic Method for Pulse Power Measurement of High-Power Microwave [72, 73]

1. **Power probe**

At the beginning of this section, we have pointed out the shortcomings of the various methods for measuring the power of high-power microwave pulses that have been described previously, and these shortcomings can be overcome to some extent by measuring the power of high-power microwave pulses by electroacoustic method.

The probe for measuring the electric field intensity of MW level high-power microwave pulses using electroacoustic method is shown in Fig. 5.18. The basic component of the probe is a Y-shaped lithium niobate piezoelectric material plate, on which two electroacoustic transducers are deposited. Transducer 1 of Fig. 5.18 is a nonlinear grid transducer as the input, which is excited to generate a pulsed sound wave under the action of a microwave electric field, and the pulse width of the sound wave depends on the length of the transducer and the propagation speed of the sound wave, i.e., L_1/v, and its carrier frequency is determined by the spatial period h_1 of the grid. The signal of the surface acoustic wave propagating along the dielectric plate is received by a linear interdigital transducer 4, re-exciting the pulsed electrical signal, and is displayed by a coaxial line input storage oscilloscope. The amplitude of the output signal is related to the strength of the microwave electric field, so the input transducer should be placed at the strongest place of the electric field. Due to the low propagation velocity of the surface acoustic wave, the surface acoustic wave generated by the input transducer can only reach the output transducer after a certain delay time, during which the surrounding radio interference has enough time to be attenuated and will not interfere with the effective output signal of the probe, that is, this power probe has a high noise rejection ability.

The connection method between the electroacoustic power probe and the waveguide is shown in Fig. 5.18. The input transducer of the probe extends into the waveguide. Due to its size is extremely small, which is only a few tenths of a millimeter, so it minimizes the probe's disturbance to the microwave field to the greatest extent possible.

2. **Experimental characteristics**

The connection between the electroacoustic power probe and the waveguide is shown in Fig. 5.19, where, 2, 3, 4 constitute the power probe shown in Fig. 5.18. The experiment was carried out with a relativistic device with an output power of 4–5 MW, a pulse width of 500 ns and a wavelength of 3.9 cm. The amplitude of the acoustic signal excited by the input transducer made of electrostrictive material film is proportional to the square of the electric field strength E^2, and this pulse signal may be delayed by 100 μs and converted into an electrical signal output by the output transducer.

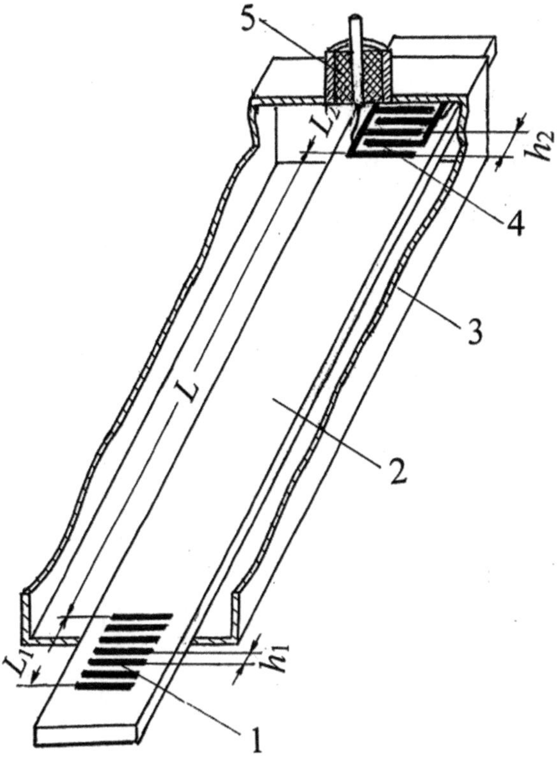

Fig. 5.18 A probe for measuring the electric field strength of MW-scale high-power microwave pulses by electroacoustic method 1—nonlinear grid transducer, 2—dielectric plate, 3–acoustic waveguide (shielding shell), 4–linear interdigital transducer (IDT), 5–coaxial wire

The calibration curve obtained from the experiment, that is, the relationship curve between the amplitude of the output electrical signal and the electric field strength, is shown in Fig. 5.20. After calibrating the oscilloscope, the relationship between the output power of the relativistic device and the output signal of the power probe can be obtained, thus, the power can be determined from the output signal amplitude.

The characteristics of the acoustoelectric measurement method are as follows.

(a) High power capacity. The power capacity is mainly limited by the waveguide itself, and the experimental system has conducted power measurements of 5 MW in the 3.9 cm band.
(b) High sensitivity and it causes very little interference with the microwave field.
(c) It has strong anti-interference ability. Due to the time delay between the input transducer and the output transducer, during this delay process, various interference signals have already been attenuated.
(d) The manufacturing process is conducive to integration and standardization, with high repeatability and reliability.
(e) It has a small volume and light weight.
(f) Currently, the working frequency of the surface acoustic wave device is mainly limited to the centimeter wave band.

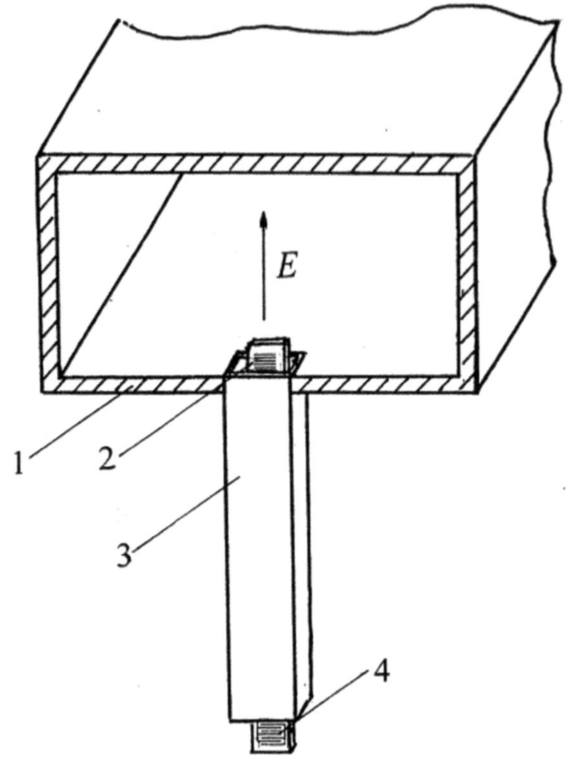

Fig. 5.19 Connection between the electroacoustic power probe and the waveguide. 1—microwave waveguide, 2—nonlinear input transducer, 3—acoustic waveguide, 4—linear output transducer

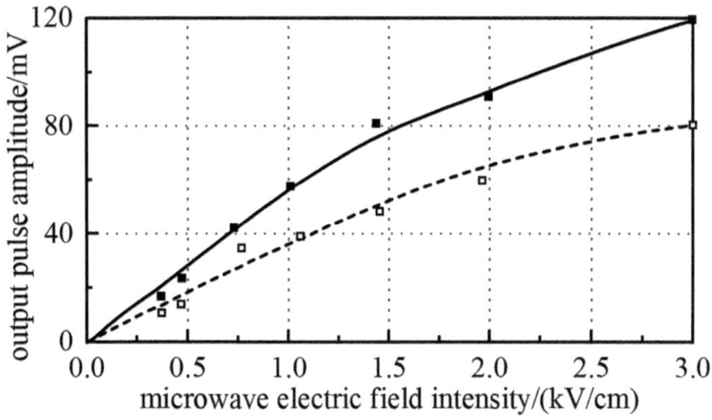

Fig. 5.20 The relationship between the output electrical signal and the electric field strength measured obtained from the experiment 1—bulk acoustic wave, delay 21 μs, 2—surface acoustic wave, delay 13 μs

Chapter 6
Mode Field Measurement of High-Power Microwave Mode

6.1 Introduction

In Chap. 1, we introduced the necessity of mode recognition and analysis (mode measurement) in high-power microwave measurement, and briefly introduced the main methods of mode measurement. From this chapter on, we will discuss various methods of mode measurement in detail.

In high-power microwave transmission systems, overmoded waveguides are usually used to transmit microwave power. Overmoded waveguide refers to a waveguide whose size is larger than required size of the fundamental mode that can be transmitted at a certain frequency, so that it can transmit the high-order modes output by the high-power microwave source. Or simply put, at a certain frequency, the waveguide that can transmit higher-order modes than the fundamental mode is called an overmoded waveguide, because in such a waveguide, the size of the waveguide required to transmit the high-order mode is larger than that of the fundamental mode waveguide. Therefore, all modes (including the fundamental mode) that the order is lower than that of the high-order mode can be transmitted in such a waveguide, so it is called an overmoded waveguide. Especially in high-power microwave systems, the higher the order of mode output by the high-power microwave source, the more severe the overmoding required for the waveguide that can transmit the mode is. Sometimes, in order to transmit higher power, the waveguide size may be even larger than that required by the high-order working mode output by the high-power microwave source, so that parasitic modes higher than the high-order working mode can also propagate.

The main reason why HPM adopts overmoded waveguide transmission is that:

(1) **Most of the microwaves output by HPM sources are often high-order modes**

For example, the gyrotrons often work on TE_{0n}, TE_{12}, TE_{13}, TE_{21} and other modes, while the millimeter-wave gyrotrons work more on high-order whispering gallery lateral modes TE_{mn} (usually $m \geq 10, n \geq 2, m \gg n$), and the vircator generally outputs

W. Wang et al., *High-Power Microwave Measurement*,
https://doi.org/10.1007/978-981-96-7763-4_6

TM_{0n} multimode, etc. Therefore, only overmoded waveguides that can transmit these high-order modes can be used as output systems for these devices.

(2) **To meet the needs of increasing system power capacity**

Since the output power of HPM sources is all above the level of hundreds of megawatts, or even tens of gigawatts, such power levels will produce extremely high high-frequency voltages in ordinary standard single-mode waveguides, causing waveguide breakdown and interrupting the transmission of microwave energy. In order to avoid breakdown and increase the power capacity of the waveguide system, it is necessary to enlarge the waveguide size, that is, to use overmoded waveguides.

The application of overmoded waveguides in HPM systems will inevitably bring about the problem of multimode transmission, that is, in the waveguide system, in addition to the main mode, there will be multiple other modes, that is, parasitic modes or miscellaneous modes. On the one hand, because some HPM devices themselves do not output pure single-mode output, but often contain some miscellaneous modes, such as gyrotrons, vircator, etc., on the other hand, even if the HPM source outputs a single mode, any non-uniformity in the transmission of overmoded waveguide systems, such as the small variation of waveguide size, shape and axis, the connection of waveguides, the existence of coupling probes or small holes, etc., will excite other modes whose order is lower than the HPM source output mode, and even higher order parasitic modes may be excited when the waveguide size allows. Therefore, in overmoded waveguide systems, multimode transmission is almost inevitable, that is, parasitic modes (miscellaneous modes) will more or less exist.

Since there are multiple modes coexisting in HPM transmission systems, according to the principle of orthogonalization of regular waveguide modes, except for some higher-order modes that have already been cut off by the waveguide size and will be attenuated and unable to continue transmission during the process, the remaining modes can propagate independently in the overmoded waveguide with their own power. Therefore, in order to understand the mode composition in high-power systems and the relative power content of each mode, it becomes necessary to identify, analyze, and measure these transmission modes.

In Chap. 1, we divided mode recognition methods into two categories: mode field measurement and mode spectrum measurement. Each method includes multiple specific measurement methods [74, 75]. In this chapter, we will discuss mode field measurement methods in detail, and introduce mode spectrum measurement methods in Chap. 7. The so-called mode field measurement refers to a method of directly measuring the radiation field or waveguide field of a high-power microwave system, and identifying and analyzing mode composition according to the field distribution obtained by the measurement. Its most characteristic is to measure directly the mode field structure. This method mainly includes image display method, radiation field measurement method, thermal imaging analysis method, and waveguide field measurement method.

6.2 Image Display Method

The method of displaying the electric field structure of the microwave mode transmitted by the microwave system directly in a planar image to determine the mode is called the image display method. The image display method is a method of displaying the mode field structure by using microwave energy to change the color or light emission of certain materials or components. Therefore, according to the different materials or components used, the image display method has several different specific methods.

The image display method is intuitive, simple, convenient, and is not affected by frequency, so it is suitable for any band. Its disadvantage is that it generally only applies to the identification of single-mode or a small number of modes coexisting. When there are more coexisting modes, the image overlap is often blurred, making it difficult to distinguish each mode. In addition, this method cannot be used for online measurement, that is, it cannot be used to measure under the normal connection state between the microwave system and the working load. That is to say, in order to obtain the mode image, the microwave source must be disconnected from the working load (a matching load can be connected to the terminal when needed) so that the microwave energy can be used for imaging. Of course, the image display method cannot give the relative power of the mode, and can only roughly judge the relative electric field strength based on the degree of color change or light emission brightness in different areas of the image.

6.2.1 Scorched Spot Method

A piece of white paper with a relatively good texture is directly sandwiched between the output waveguide connection flanges of the high-power microwave source, and the microwave energy is used to burn and form scorched spots on the paper. The position and degree of the scorched spot correspond to the distribution and strength of the microwave electric field, that is, the field structure. Therefore, according to the shape and degree of the scorched spot, the transmission mode of the microwave system can be identified. The terminal of the output waveguide should be connected to a matching load to prevent the reflection wave from affecting the accurate formation of the image. This is the simplest and most feasible method that gyrotron developers often used in the early days. This method requires that the microwave source output has sufficient average power and a certain ablation time: if the microwave power is too small, no matter how long the ablation time is, the total energy is insufficient, and it is impossible to leave a scorched spot on the paper; after reaching a certain microwave average power, the required ablation time is inversely proportional to the power. Generally speaking, it is appropriate to use a power density of several watts to tens of watts per square centimeter of average power density, and the time is tens of seconds to several minutes. If the power is too large or the time is too long, the paper

Fig. 6.1 Scorched spot of the output modes TE_{01}, TE_{13}, and TE_{53} of gyrotrons

will be burned through, and the scorched spot will spread, which cannot correctly reflect the distribution of the field, and vice versa, it is impossible to form a clear and complete field structure image on the paper.

The scorched spot method cannot obtain a very clear mode image, and is basically only suitable for the identification of single-mode and relatively low-order modes. When there are multiple modes or higher-order modes, the overlap of the mode scorched spots and the spread of the scorched position will make the mode image blurred, making it difficult to identify.

Figure 6.1 shows the scorched spot images of the working modes of the gyrotrons developed by the University of Electronic Science and Technology of China. They work on 15 GHz (TE_{01} mode), 37.5 GHz (TE_{13} mode), and 65 GHz (TE_{53} mode), respectively. It can be seen that the identification of the TE_{53} mode is already quite difficult [74, 75].

6.2.2 Thermosensitive Paper (Film) Method

By using the characteristic of specially made thermosensitive paper or liquid crystal film that changes color when heated, people place them at the output port of the microwave system or at a certain distance from the port. Under the action of microwave energy, they will change color and form a thermosensitive image. The microwave mode can be determined based on the degree of color change and the distribution of the color change area. Thermosensitive paper and liquid crystal film have high sensitivity, so they can quickly identify the transmission mode of small and medium power microwave systems. However, their overload capacity is poor. They are easily burned out when the power is slightly larger or the microwave action time is slightly longer. Compared with the scorched spot method, the spatial resolution of the mode image obtained by thermosensitive paper and liquid crystal film is slightly higher, and the image is clearer. Another advantage of thermosensitive paper and liquid crystal film is that their color change can disappear after a period of time after the microwave power irradiation stops, so they can be reused. Obviously, this is also a simple, low-cost pattern recognition method.

1. **Reconstruction of the mode components** [76]

Generally speaking, when the image formed by the thermosensitive film is formed by the combined action of the microwave electric fields of multiple modes, it will be more difficult to identify these mode types, especially their composition components. To solve this problem, it is necessary to rely on the theoretical calculation of the mode field and the comparison of the thermosensitive image, or use the artificially set mode composition to reconstruct the electric field image through theoretical calculation, and compare it with the actual measured image. When the two are basically the same, it can be considered that the set mode composition is the actual mode composition.

After we have obtained the thermosensitive image of a certain microwave mode, in order to determine the mode composition of the image and the relative components of each mode, we can assume that the image contains N determined modes, and its composite electric field can be represented in polar coordinates.

$$E(r,\phi) = \left|\sum_{m=1}^{N} A_m \exp(\mathrm{j}\varphi_m)\boldsymbol{e}_m(r,\phi)\right| \tag{6.1}$$

where, $\boldsymbol{e}_m$ is the normalized electric field of the m-th mode; A_m and φ_m are the amplitude and phase of the m-th mode, respectively. Thus, determining a mode requires two parameters, amplitude and phase. Obviously, the temperature rise at each point on the thermosensitive image is proportional to the microwave power at that point in space, that is, the square of the electric field strength $|E|$ at that point. Therefore, it can be considered that

$$P(r,\varphi) \propto |E(r,\varphi)|^2 = \left|\sum_{m=1}^{N} A_m \mathrm{e}^{\mathrm{j}\phi_m}\boldsymbol{e}_m(r,\varphi)\right|^2 \tag{6.2}$$

If we assume that a certain thermosensitive image is formed by N modes set in advance, we read $2N$ temperature rise values on the image (use infrared camera to connect to computer for reading), which correspond to the microwave power or correspond to $|E|^2$. Therefore, when calculated as relative quantities, these $2N$ data can be directly substituted into Eq. (6.2) as $P(r, \phi)$ to obtain $2N$ equations. Since we only need to consider the relative size of the various mode fields, we can generally set the amplitude of the main mode to 1 and the phase to 0°, so that only $(N-1)$ modes need to be determined. Therefore, we only need to establish $2(N-1)$ equations.

Solve the system of equations consisting of these $2(N-1)$ equations, and obtain the amplitude and phase of N modes in Eq. (6.1). Then, using these N modes (including the main mode) and Eq. (6.2), the electric fields of these N modes are superimposed and reconstructed into a modes image. If the obtained image is very close to the measured thermosensitive image in an acceptable degree, it means that the N modes we assumed at the beginning are correct. The amplitude of each mode obtained is already a normalized value relative to the amplitude of the main mode (because we have set the amplitude of the main mode to 1), so the A_m^2 value of each other mode is the power of the mode relative to the main mode. Of course, the A_m^2 values of all modes can be added together as the relative power value of the total power, and then the relative amount of each mode, including the main mode, in the total power is calculated separately.

Due to the low spatial resolution of the image display method, the temperature we read from the thermosensitive image is often the average value in a certain small area, which will bring some errors to the above mode reconstruction process. For this purpose, we can directly replace the temperature measurement of a certain point with the average temperature of a certain area, that is, with the average value of the square of the relative electric field strength or the average value of the relative power. At the same time, we can directly take the ratio of the two measurement values to obtain the equation. According to the specific structure of the image, we can use the following two methods to read the average value of the relative power. Method 1: On the thermosensitive image, for example, we can measure the ratio of the square of the relative electric field strength from r_1 to r_2 and from r_3 to r_4 on two different circles, that is,

$$\frac{\int_{r_1}^{\mathrm{r}_2}\int_0^{2\pi} P(r,\phi)\mathrm{d}r\mathrm{d}\phi}{\int_{r_3}^{\mathrm{r}_4}\int_0^{2\pi} P(r,\phi)\mathrm{d}r\mathrm{d}\phi} \propto \frac{\int_{r_1}^{\mathrm{r}_2}\int_0^{2\pi} |E(r,\phi)|^2\mathrm{d}r\mathrm{d}\phi}{\int_{r_3}^{\mathrm{r}_4}\int_0^{2\pi} |E(r,\phi)|^2\mathrm{d}r\mathrm{d}\phi} \tag{6.3}$$

Method 2: We can also take the ratio of the square of the relative electric field strength of two different sectors with the same radius a, such as between φ_1 and φ_2 and between φ_3 and φ_4,

$$\frac{\int_{\phi_1}^{\phi_2}\int_0^{a} P(r,\phi)\mathrm{d}r\mathrm{d}\phi}{\int_{\phi_3}^{\phi_4}\int_0^{a} P(r,\phi)\mathrm{d}r\mathrm{d}\phi} \propto \frac{\int_{\phi_1}^{\phi_2}\int_0^{a} |E(r,\phi)|^2\mathrm{d}r\mathrm{d}\phi}{\int_{\phi_3}^{\phi_4}\int_0^{a} |E(r,\phi)|^2\mathrm{d}r\mathrm{d}\phi} \tag{6.4}$$

Substituting Eqs. (6.1) or (6.2) into Eqs. (6.3) or (6.4), using the result obtained from Eqs. (6.3) or (6.4) as the theoretical value, and read the measured value on the thermosensitive image in the corresponding area of Eqs. (6.3) or (6.4). i.e. after setting N mode combinations, selecting $2N$ areas to read $2N$ measurement values. Since Eqs. (6.3) or (6.4) can only establish an equation by the ratio of two measurement values, of course, we can compare them with the same measurement value, so $2N$ measurement values can only be established $2(N - 1)$ equations. By solving this system of equations, the relative amplitude and phase of $(N - 1)$ modes are obtained, and then the mode image of $(N - 1)$ modes is reconstructed. The correctness of the theoretical image is judged by comparing it with the actual measured image.

The temperature rise value mentioned above, that is, the square of the relative electric field strength value or the relative power value, can be obtained by combining infrared camera and computer technology, which can significantly improve the image resolution.

2. **Thermosensitive paper mode image** [77]

Figure 6.2 shows the mode change image of each stage of a three-stage mode converter obtained by German scholars using thermosensitive paper. This mode converter is designed to obtain a microwave wave beam with near-linear polarization when using the output power of the gyrotron for plasma heating in controlled nuclear fusion research. It works at 28 GHz, and the output mode of the gyrotron is TE_{02} mode, and subsequently it is converted into the required output mode through $TE_{02} \rightarrow TE_{01} \rightarrow TE_{11} \rightarrow HE_{11}$. From the figure, it can be seen that the axial symmetry of the TE_{02} mode is deviates slightly from the theoretical situation, which indicates that there is a small part of asymmetric parasitic mode in the output mode. This is caused by the impure output of the gyrotron and the inhomogeneity of the 90-degree bent waveguide in the output system, but its content is only about 1%.

3. **Liquid crystal film mode image**

(1) **Output mode of gyrotron FUIVA** [76]

Japanese scholars at Fukui University used a liquid crystal film to display the output mode image of the gyrotron FUIVA. This is a TE_{23} high-order mode gyrotron, which works at a frequency of 294.7 GHz and an output power of 11W. The mode images experimentally measured under microwave action times of 5, 6, and 7 s are shown in Fig. 6.3. In order to estimate the content of the main parasitic mode TE_{22} mode output by the gyrotron, they assumed that the image is formed by TE_{23} and TE_{22} modes, and then the measurement values of three circular rings are taken on the image. Two equations are established using a method similar to Eq. (6.3). By solving this system of equations, the relative power ratio of TE_{23} and TE_{22} modes is 91%–94% and 6%–9%, respectively, and their phase difference is π. The TE_{23} mode with 93% relative power and TE_{22} mode with 7% relative power and a phase difference of π are used to reconstruct the mode image, which is shown in Fig. 6.4. Compared with the measured image, the two are quite consistent.

(2) **The experiment of TE02 mode gyrotron** [76]

Japanese scholars at Fukui University placed a liquid crystal film at a certain distance from the front end of another gyrotron output waveguide with the main mode of TE_{02}. According to the different microwave irradiation times, the image of the TE_{02} mode

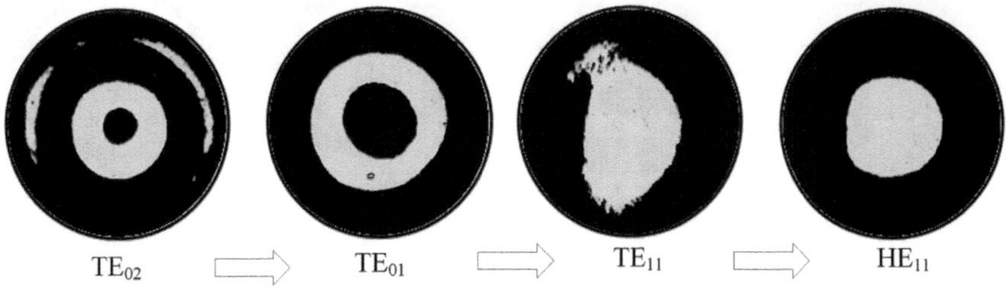

Fig. 6.2 Thermosensitive paper display of mode change in a mode converter

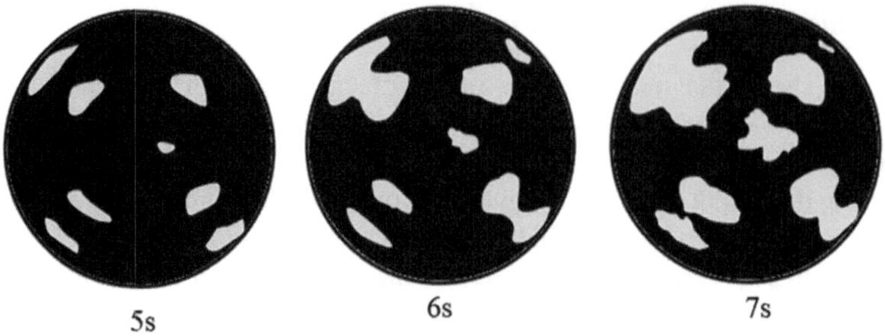

Fig. 6.3 The liquid crystal plane film image of the output mode of the gyrotron FUIVA

Fig. 6.4 The theoretically reconstructed image

was obtained when it worked at 207.5 GHz and an output power of 2.2W (Fig. 6.5). It is not difficult to see from the figure that the microwave irradiation time of 5 s is more appropriate, 6 s is too long, and the image resolution is reduced, while 3 s is obviously insufficient and cannot form a complete image. Even with the microwave irradiation time of 4–5 s, the obtained image does not completely conform to the field structure of the TE_{02} mode.

Japanese scholars assume that the difference between the image and the field structure of the TE_{02} mode is caused by the parasitic TE_{22} mode. They took two circular rings on the image and established a relative intensity equation using Eq. (6.3). Then, another relative intensity equation was established based on two sectors of the same

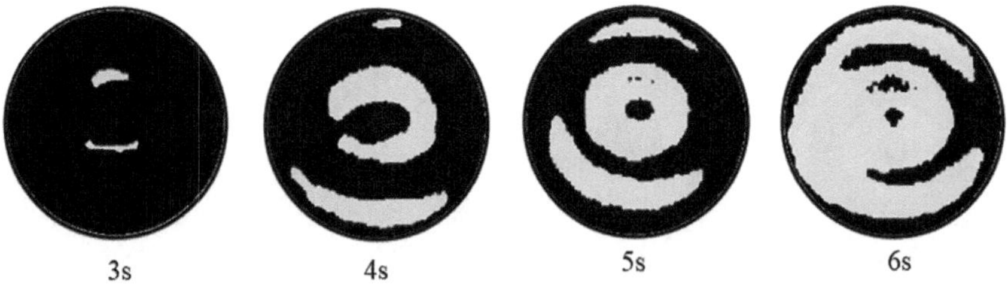

Fig. 6.5 The liquid crystal display image of the output mode of another gyrotron under different microwave irradiation time

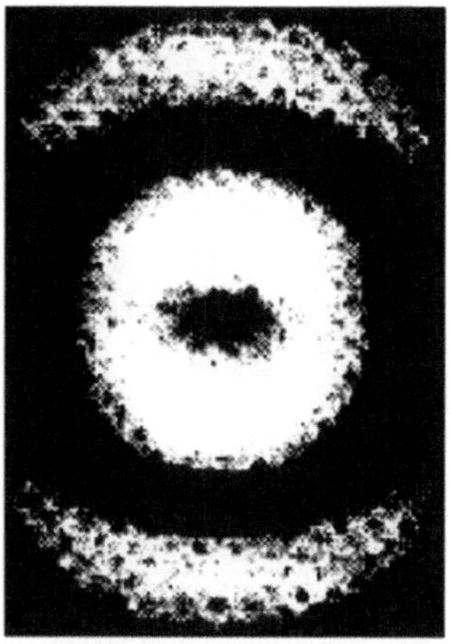

Fig. 6.6 The theoretically reconstructed image

radius on the image in Fig. 6.5, according to Eq. (6.4). By solving the two equations, the relative power of TE_{02} and TE_{22} modes is 85–87% and 13–15%, respectively, and their phase difference is 0°. They used the TE_{02} mode with 86% relative power and the TE_{22} mode with 14% relative power and a phase difference of 0° to reconstruct the comprehensive image of the two modes, as shown in Fig. 6.6. It can be considered that the reconstructed comprehensive image is in good agreement with the measured image.

4. **Cobalt chloride impregnated paper mode image** [54]

Japanese scholar Kenichi Noda proposed various rectangular-circular mode exciters (see Sect. 4.5.1) and, using cobalt chloride impregnated paper, obtained images of the output modes of these exciters. The cobalt chloride impregnated paper is placed at the port of the circular waveguide. The microwave electric field acts on the paper, where the paper is heated and dried, and the cobalt chloride turns blue where the electric field is strong, and the part that is not heated by the microwave electric field becomes light pink, thereby displaying the mode image, which can be photographed and recorded. Figure 6.7 shows the mode images of TE_{11}, TE_{01}, TE_{41}, TE_{31}, TM_{11}, and TE_{12} modes generated by these exciters.

6.2.3 Neon Tube Array Method

Microwaves can excite gas discharge tubes to emit light. By using this phenomenon, we can realize the optical display of the mode diagram. The specific method is to radiate microwave energy into space through a waveguide port or horn, and to place a flat display device composed of a large number of regularly arranged neon tubes in front of it. When the radiated microwave power acts on the neon tube array, the neon tube will emit light. The distribution of neon tube light and the brightness

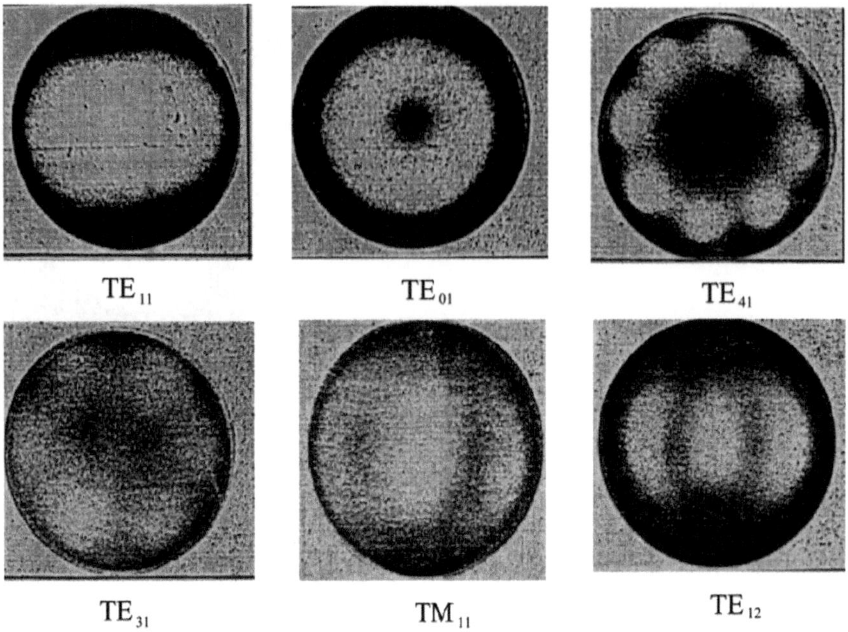

Fig. 6.7 The output mode field distribution diagram of various mode exciters proposed by Kenichi Noda

reflect the spatial structure distribution of the electric field, thereby determining the radiation field mode. Since the neon tube array can be made large enough, it can be placed at a distance from the microwave radiation port, so this method is suitable for mode identification of relativistic devices operating in single-pulse mode and having considerable power. However, in the far-field region, the distribution of the radiation field and the mode field in the system will be different, but in the case of a single mode, it is completely feasible for mode recognition.

Scholars from the Institute of Applied Physics of the former Soviet Academy of Sciences and the Institute of High Current Electronics used this method to successfully display the output modes TE13 and TE01 of relativistic orotron and relativistic backward wave tube (BWT) (Fig. 6.8). Among them, the relativistic orotron works in the 8 and 5 mm bands, and the output power is 120 $\pm$ 20 MW and 50 $\pm$ 20 MW, respectively, with a pulse width of 6–8 ns [78].

Fluorescent lamps arranged closely together can also be used to display the mode diagram because fluorescent lamps will also emit light under microwave electric field excitation, and the brightness of the emitted light is related to the intensity of the microwave irradiation, thereby displaying the structure of the mode field. Figure 6.9 shows the fluorescent lamp display image of the TE_{11} mode [79]. Since the volume of the fluorescent lamp is much larger than that of the neon tube, the mode diagram

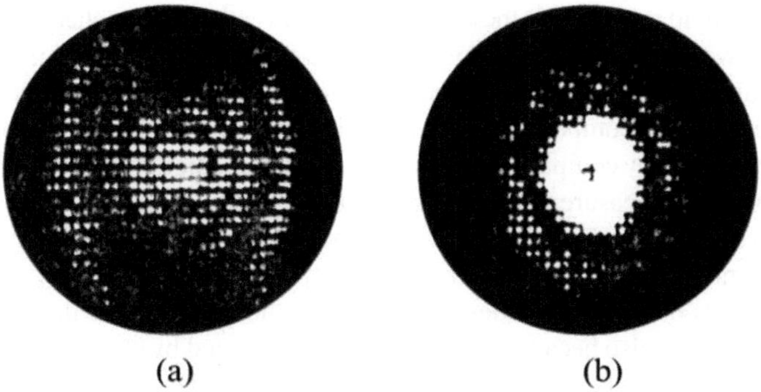

Fig. 6.8 The neon tube array display image of the output mode of relativistic devices, **a** TE13 mode image output by relativistic orotron, **b** TE_{01} mode image output by relativistic backward wave tube

Fig. 6.9 The mode image of the TE_{11} mode displayed using a fluorescent lamp

displayed by it will obviously be much rougher than that displayed by the neon tube array, and it is not suitable for displaying more complex mode structure diagrams, so it is not practical.

6.3 Radiation Field Measurement Method

6.3.1 Overview

1. **Radiation field mode measurement**

Open the waveguide terminal of the microwave system or connect it to a radiation horn to radiate the high-power microwave mode field into space, and then measure the spatial radiation field. By analyzing the measured radiation field distribution diagram (radiation pattern), the mode composition in the system can be obtained, which is the basic principle of radiation field mode recognition. In the case of a single mode or only two modes, the mode can be directly determined according to the field distribution of the radiation field measured in the φ direction or θ direction in the spherical coordinate system. However, when there is multi-mode radiation, the radiation pattern of the radiation field obtained will become more complicated, and it is difficult to directly determine the mode composition from the radiation diagram. The mode composition can be determined by comparing it with the theoretical radiation field or by numerical calculation. If the measurement of the radiation field is accurate enough, we can also quantitatively analyze the relative power of each mode. However, in reality, accurate measurement of the radiation field is very difficult. This is not only because the radiation fields of multiple modes overlap each other, making the field distribution very complex, but also because this complexity is affected by changes in the relative power ratio between the modes and the phase difference between them. In addition, the receiving horn used for measurement always has a certain aperture size, and we can actually only measure the average value of the radiation field on the effective cross section of the receiving horn, so it is impossible to truly measure the field value at each point in space.

The radiation field measurement method can theoretically perform multi-mode recognition. Since the radiation field of the microwave mode is known mathematically, as long as the measurement of the microwave radiation field distribution is accurate enough, no matter how many modes are simultaneously radiated, it should be possible to analyze the mode composition by comparing the theoretical radiation field distribution with the measured distribution diagram. However, in reality, due to the influence of some actual factors, the radiation field itself cannot strictly conform to the theoretical distribution, and there will be considerable errors in the measurement of the radiation field. We cannot truly measure the field value point by point, thereby obscuring a large number of details in the radiation field distribution. In addition, when there are many modes, it is necessary to repeatedly set the mode, establish multiple equations based on the field distribution diagram obtained by measurement, and then solve these equations simultaneously to obtain the theoretically reconstructed mode field distribution diagram, and compare it with the measured image. This process must be repeated until the two are consistent, and the entire process is very complex and time-consuming. These limit the number of modes that the radiation field measurement method can recognize and the accuracy of recognition. Therefore, this method is very inconvenient for mode recognition when there are many modes at the same time.

For cases where there are fewer modes (1–2) and the mode field structure is relatively simple (low-order symmetric mode), the radiation field measurement method is a simple and effective mode discrimination method that is widely used.

2. **Coordinate system**

The spherical coordinate system (R, θ, φ) used in the radiation field measurement method is shown in Fig. 6.10. In actual measurement, an open rectangular waveguide or rectangular horn is placed at a distance of R from the radiation port as a receiving antenna, and the back end is connected to the basic mode rectangular waveguide and crystal detector.

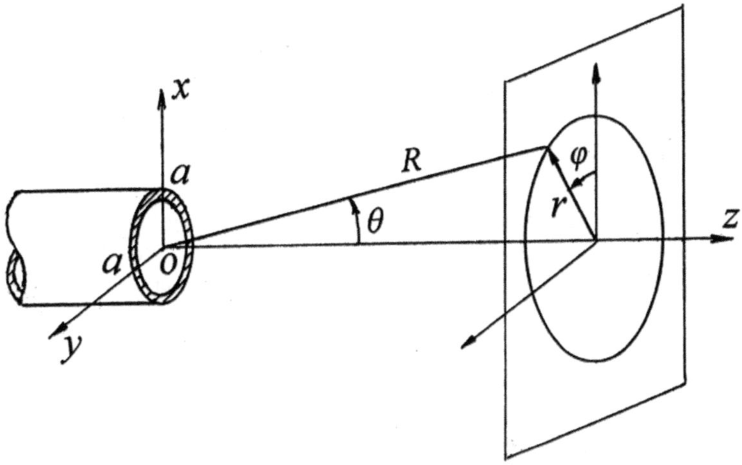

Fig. 6.10 The coordinate system of the radiation field measurement

It is important to note that regardless of the position of the receiving antenna in the radiation field, what is measured is the electric field component that is perpendicular to the wide side of the open waveguide or receiving horn. The result obtained will be proportional to the square of the measured electric field component within the square law range of the crystal detector. In other words, the wide side of the open waveguide or receiving horn should be oriented perpendicular to the electric field component we wish to measure.

Square values of the resulting electric field component are proportional to the microwave power, so the relative power of the radiation field at each point can be plotted as a curve to obtain the radiation pattern, which is the field distribution diagram.

6.3.2 Direct Discrimination Method

The method of directly judging the mode in the transmission system based on the radiation pattern obtained by radiation field measurement is called the direct discrimination method. This method is generally only applicable to mode recognition of high-power microwave systems with a single mode or only one parasitic mode, such as gyrotron and mode converters. In particular, this method is more common for circular electrically symmetric modes. Internationally, this method has also been widely used in units engaged in high-power microwave research.

1. **Mode converter of stepped circular waveguide**

The stepped circular waveguide mode converter is one of the simplest mode converters. It is composed of two circular waveguides with different diameters directly connected, forming a step at the connection. By correctly designing the diameters of the two circular waveguides, the transformation from TE_{01} mode to TE_{0n} mode can be achieved. We have already given the design theory of this mode converter in Sect. 4.5.2 [56, 57].

(1) **Mode image of the converter developed by H. Stickel, etc.**

Figure 6.11 shows the far-field radiation measurement diagram of the input and output modes of a set of stepped circular waveguide mode converters. This is the measured result of the TE_{01}–TE_{0n} circular waveguide mode converter designed by German scholars. The ordinate in the figure represents the relative power size, which is proportional to $|E_{\varphi}|^2$, and the abscissa is the angle θ deviated from the axis (z-axis in Fig. 6.10). This indicates that the receiving horn is moving in the xoz plane shown in Fig. 6.10, and the wide side of the horn is parallel to the x-axis. We can directly determine the type of the main mode from the figure, and further observation can also determine the main parasitic mode. For example, in the output distribution diagram of electric field of TE_{02} mode and TE_{03} mode, there is a relatively obvious TE_{01} input mode component. In the TE_{04} output mode, there is obviously not an inconsiderable TE_{03} mode component. Similarly, in the output modes of TE_{05} and TE_{06} both indicate the presence of TE_{04} mode.

(2) **Mode image of the converter developed by the University of Electronic Science and Technology of China**

According to the theory introduced in Sect. 4.5.2, the author's research group also developed a stepped circular waveguide mode converter from TE_{01} mode to TE_{02},

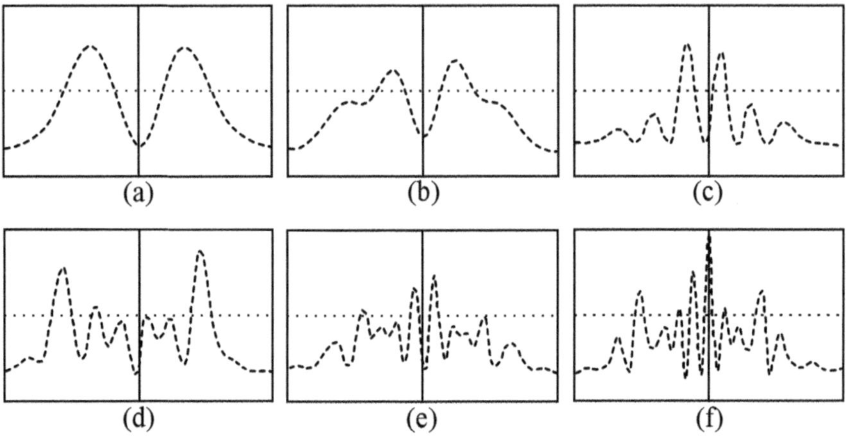

Fig. 6.11 The far-field radiation pattern of the input and output modes of the stepped circular waveguide mode converter measured at 132 GHz, **a** input TE_{01} mode, **b** output mode of TE_{01}–TE_{02} converter, **c** output mode of TE_{01}–TE_{03} converter, **d** output mode of TE_{01}–TE_{04} converter, **e** output mode of TE_{01}–TE_{05} converter, **f** output mode of TE_{01}–TE_{06} converter

TE_{03}, and TE_{04} modes. In order to obtain the TE_{01} mode in the circular waveguide, we also made a Marie-type cross-shaped waveguide converter (Sect. 4.5.2), which converts the TE_{10} mode in the standard rectangular waveguide into the circular waveguide TE_{01} mode. The radiation field measurement results of these converters are shown in Fig. 6.12.

From the figure, it can be seen that the symmetry of the TE_{01} mode field distribution output by the Marie-type cross-shaped waveguide mode converter is not ideal, which leads to the lack of symmetry of the obtained TE_{02}, TE_{03}, and TE_{04} mode field distributions. The asymmetry of the TE_{01} mode field distribution is caused by the fact that the symmetry of the four arms of the electroforming core mold is not satisfactory. This occurs when machining is carried out on the cross waveguide part of the cross waveguide mode converter, resulting in the imperfect symmetry of the TE_{01} mode field distribution.

2. **Mode image of other circular waveguide mode converter** [77]

A series of circular waveguide mode converters were proposed by M. Thumm et al. at the University of Stuttgart in Germany, and these converters were measured and

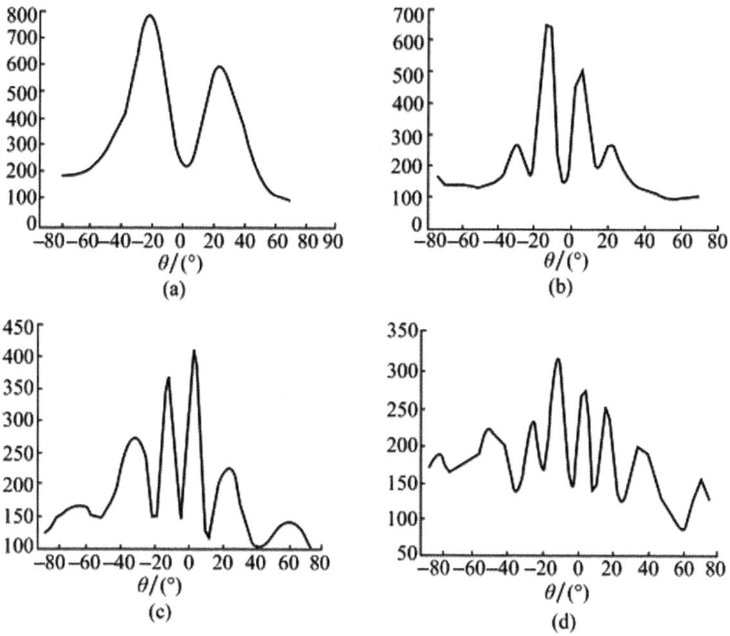

Fig. 6.12 The radiation field measurement results of the Marie-type cross-shaped waveguide mode converter and the stepped circular waveguide mode converter produced by the University of Electronic Science and Technology of China, **a** output TE_{01} mode of Marie-type cross-shaped waveguide converter, **b** output TE_{02} mode of converter, **c** output TE_{03} mode of converter, **d** output TE_{04} mode of converter

analyzed using the far-field measurement method to obtain many important parameters such as the output mode composition, the relative content of each mode, and the conversion efficiency of the converter. We only take the corrugated waveguide mode converter as an example, and some measurement results are used to illustrate the application of the far-field measurement method in the mode determination and analysis of mode converters.

When the inner radius a of the circular waveguide changes according to the following rule:

$$a(z) = a_0\left[1 + \varepsilon_0 \sin(2\pi z / \lambda_B)\right] \tag{6.5}$$

the transformation from the circular waveguide TE_{0n} mode to the $TE_{0n'}$ mode can be realized, which is called the corrugated waveguide mode converter (Fig. 6.13). In general, only $n' = n\text{-}1$ can be achieved. In the formula, a_0 is the average radius of the waveguide; $a_0\varepsilon_0$ is the amplitude of the waveguide radius change; λ_B is the beat wavelength of the TE_{0n} mode and the $TE_{0n'}$ mode, that is,

$$\lambda_B = l\frac{\lambda_{0n}\lambda_{0n'}}{|\lambda_{0n} - \lambda_{0n'}|} \quad (1 = \pm 1, \pm 2 \cdots) \tag{6.6}$$

where, λ_{0n} and $\lambda_{0n'}$ are the waveguide wavelengths of the TE_{0n} mode and the $TE_{0n'}$ mode in a uniform circular waveguide with a_0 as the radius, respectively.

At $f = 28$ GHz and $a_0 = 31.75$ mm, a series of far-field patterns obtained by changing the length of a TE_{01}–TE_{02}–TE_{01} cyclic converter (an integer multiple of λ_B) are shown in Fig. 6.14. The figure is the result of measuring under small power. The length of the converter in the figure increases from $0\lambda_B$ to $13\lambda_B$, and the far-field distribution is measured once for each additional length of one λ_B. It can be seen from the figure that the converter has a cycle period of $6\lambda_B$, $0\lambda_B$ is the input end, inputting TE_{01} mode, after $6\lambda_B$ length, it is converted to TE_{02} mode, and after another $6\lambda_B$ length, it is converted back to TE_{01} mode at $12\lambda_B$. In fact, the entire mode converter can be regarded as the connection of two mode converters: from $0\lambda_B$ to $6\lambda_B$ is the first segment of the converter, which takes the TE_{01} mode as the input mode and the TE_{02} mode as the output mode; from $6\lambda_B$ to $12\lambda_B$ is the second segment of the converter, TE_{02} mode becomes the input mode, and TE_{01} mode becomes the output mode.

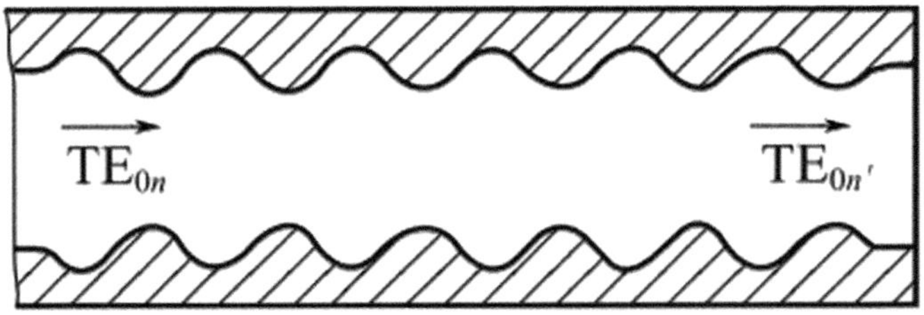

Fig. 6.13 Corrugated waveguide mode converter

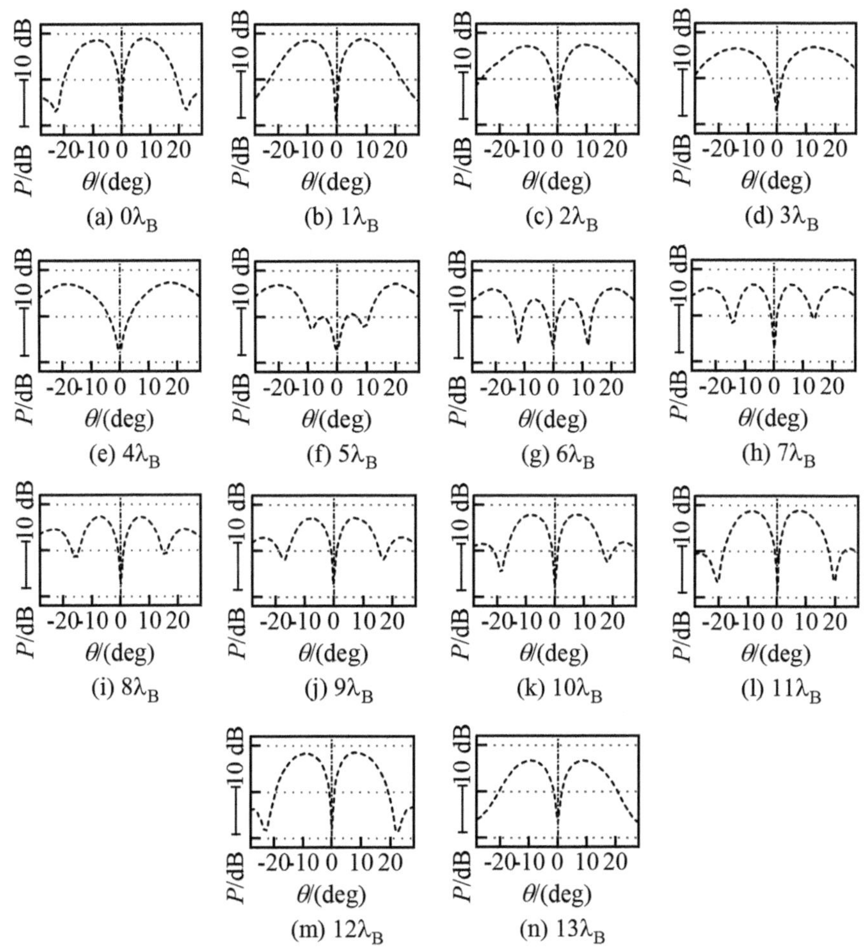

Fig. 6.14 Series of measured far-field radiation pattern obtained by changing the length of a TE01–TE02–TE01 periodic cyclic converter under small power measurement at 28 GHz

The measured conversion efficiency of the TE_{02}–TE_{01} conversion is 98.5%. The numerical solution indicates that the conversion coefficient from the TE_{02} mode to the TE_{01} mode reaches 99.2%. In the output microwave, the TE_{02} mode accounts for only 0.3%, and the TE_{03} mode accounts for 0.4%. The corrugated waveguide mode converter can also realize the mode conversion of TM_{0n}–$TM_{0n'}$, which is similar to the TE_{0n}–$TE_{0n'}$ conversion. Generally, it can only be converted in the way of $n' = n - 1$, and it can be converted into TM_{01} mode after $(n - 1)$ conversions.

M. Thumm et al. also proposed a series of other circular waveguide mode converters, such as the tortuous waveguide mode converter (TE_{01}–TE_{11}, TM_{01}–TM_{11}), the corrugated waveguide mode converter (TE_{11}–HE_{11}, TM_{11}–HE_{11}), the curved waveguide mode converter (TE_{01}–TM_{11}), and the S-shaped curved waveguide converter (TM_{01}–TE_{11}). Their measurement can be entirely carried out using the same method. For example, Fig. 6.15 is a far-field pattern of the TE_{01}–TE_{02}–TE_{03}–TE_{04} corrugated waveguide mode converter under small power measurement at 28 GHz; Fig. 6.16 is a far-field H-plane and E-plane measurement diagram of the folded waveguide TE_{01}–TE_{11} mode converter working at 70 GHz. Since we are mainly discussing mode recognition here, mode conversion is not the focus of this book, so the specific shapes and design methods of these converters will not be discussed in detail. Readers can read Refs. [5, 55] for more information.

It can be seen that the far-field measurement method can be used to conveniently judge and analyze the modes and conversion results before and after the mode converter conversion.

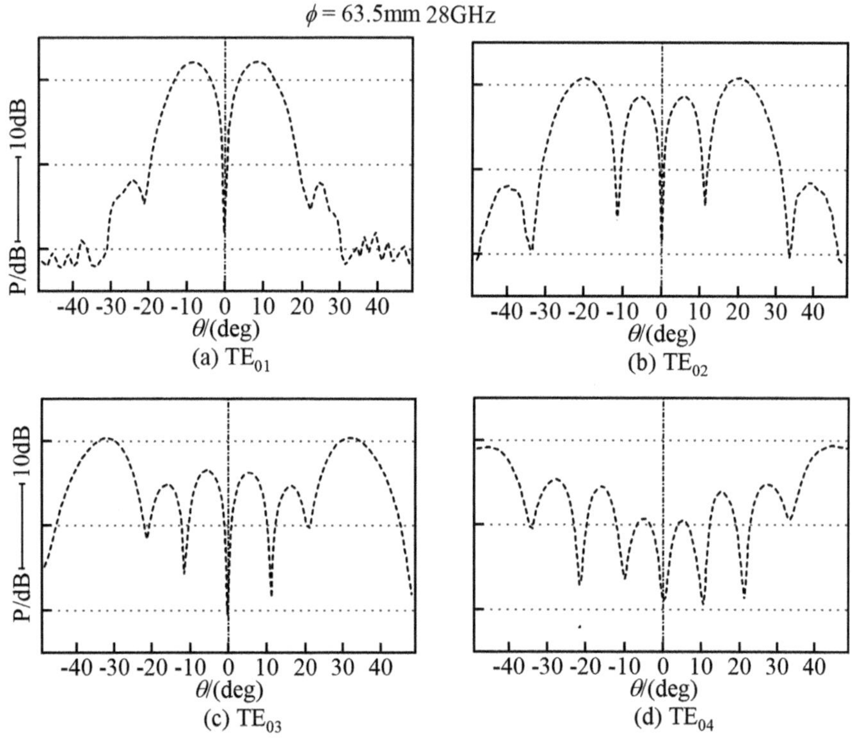

Fig. 6.15 Far-field pattern directional diagram of TE_{01}–TE_{02}–TE_{03}–TE_{04} corrugated waveguide mode converter under small power measurement

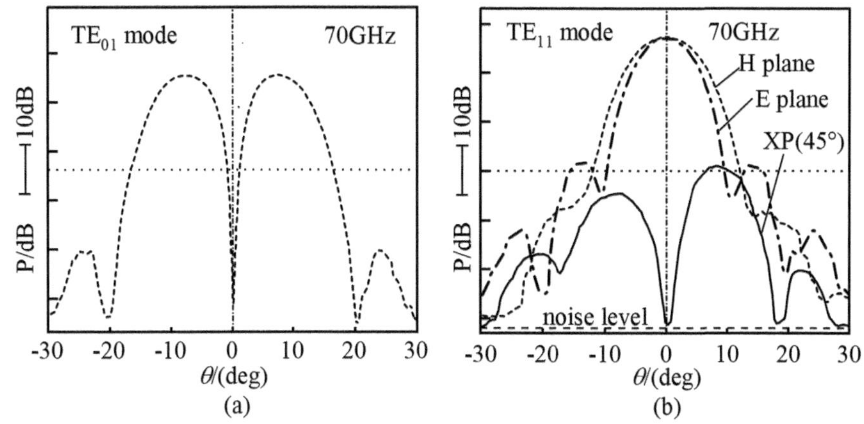

Fig. 6.16 Far-field H-plane and E-plane mode pattern diagram of TE_{01}–TE_{11} tortuous waveguide mode converter under small power measurement

6.3.3 *Numerical Calculation Method [80, 81]*

1. **Theoretical basis**

The far-field radiation patterns of the TE_{mn} and TM_{mn} modes in circular waveguides at a distance R from the radiation port ($R > 2D^2/\lambda$, where $D = 2a$, a is the radius of the circular waveguide, and λ is the free space wavelength) have been given in Eqs. (3.2)–(3.5). For convenience of analysis, we rewrite them in the following form.

TE_{mn} mode:

$$
\begin{aligned}
E_r &= 0 \\
E_\theta &= \mathrm{j}^{m+1}\left(\frac{m\omega\mu}{2R}\right) U_{mn}(\theta)\cos\left[m\phi + (m-1)\frac{\pi}{2}\right]\mathrm{e}^{-\mathrm{j}kR} \\
E_\phi &= -\mathrm{j}^{m+1}\left(\frac{ka\omega\mu}{2R}\right) V_{mn}(\theta)\sin\left[m\phi + (m-1)\frac{\pi}{2}\right]\mathrm{e}^{-\mathrm{j}kR}
\end{aligned}
\tag{6.7}
$$

where,

$$
\begin{aligned}
U_{mn}(\theta) &= \left[1 + \frac{\beta_{mn}}{k}\cos\theta + \Gamma\left(1 - \frac{\beta_{mn}}{k}\cos\theta\right)\right]\frac{J_m(k_c a)J_m(ka\sin\theta)}{\sin\theta} \\
V_{mn}(\theta) &= \left[\frac{\beta_{mn}}{k} + \cos\theta - \Gamma\left(\frac{\beta_{mn}}{k} - \cos\theta\right)\right]\frac{J_m(k_c a)J'_m(ka\sin\theta)}{[1 - (k\sin\theta/k_c)^2]}
\end{aligned}
\tag{6.8}
$$

where,

k is the free space wave number,

J_m is the m-th order Bessel function of the first kind, and J'_m is its derivative,

β_{mn} is the longitudinal propagation constant (phase constant), $k_c^2 = k^2 - \beta_{mn}^2$,

Γ is the reflection coefficient of the wave at the radiation port, for the transmission modes of $\beta_{mn}\, a > 1$, Γ is generally small and negligible.

$$k_c = \frac{\mu'_{mn}}{a} \tag{6.9}$$

where, μ'_{mn} is the n-th zero point of J'_m, i.e. $J'_m(\mu'_{mn}) = 0$, When $m = 0$, $E_\theta = 0$, that is, the radiation field of the TE_{0n} mode only has the E_φ component.

TM_{mn} mode:

$$\begin{aligned} E_r &= 0 \\ E_\phi &= 0 \\ E_\theta &= -\mathrm{j}^{m+1}\frac{ka}{2R}Q_{mn}(\theta)\cos m\phi\ \mathrm{e}^{-\mathrm{j}kR} \end{aligned} \tag{6.10}$$

where

$$Q_{mn}(\theta) = k_c\left[\frac{\beta_{mn}}{k} + \cos\theta + \Gamma\left(\frac{\beta_{mn}}{k} - \cos\theta\right)\right]\frac{J'_m(k_c a)J_m(ka\sin\theta)}{\sin\theta\left[1 - (k_c/k\sin\theta)^2\right]} \tag{6.11}$$

$$k_c = \frac{\mu_{mn}}{a} \tag{6.12}$$

where μ_{mn} is the n-th zero point of J_m, that is, $J_m(\mu_{mn}) = 0$, and other symbols have the same meaning as TEmn mode.

The method of numerical calculation for mode recognition is as follows: each possible mode can be determined by its amplitude A and phase α on the waveguide end face. That is, in the far-field radiation, in the case of multi-mode coexistence, if considering the phase difference ϕ_{mn} between them, the radiation field of the combination of several TE_{mn} modes with power P_{mn} can be written in the following form at point P:

$$\begin{aligned} E_r &= 0 \\ E_\theta &= \sum_{mn}\mathrm{j}^{m+1}A_{mn}\frac{m}{2R}U_{mn}(\theta)\sin\left[m\left(\phi + \frac{\pi}{2}\right)\right]\mathrm{e}^{-\mathrm{j}(kR+\varphi_{mn})} \\ E_\phi &= \sum_{mn}\mathrm{j}^{m+1}A_{mn}\frac{ka}{2R}V_{mn}(\theta)\cos\left[m\left(\phi + \frac{\pi}{2}\right)\right]\mathrm{e}^{-\mathrm{j}(kR+\varphi_{mn})} \end{aligned} \tag{6.13}$$

where

$$A_{mn} = \frac{1}{J_m(\mu'_{mn})}\left[\frac{\varepsilon_m\eta}{\pi}P_{mn}\frac{k}{(\mu'^2_{mn})\beta_{mn}}\right]^{1/2} \tag{6.14}$$

where, $\eta = 377\ \Omega$ is the free space wave impedance, P_{mn} is the transmission power of the TE_{mn} mode, when $m = 0$, $\varepsilon_m = 1$; when $m \neq 0$, $\varepsilon_m = 2$.

It's clear that given a far-field point P (R, φ, θ), for the case of multimode coexistence, as long as the amplitude A_{mn} and phase ϕ_{mn} of each mode can be obtained, the entire radiation field can be determined. In order to solve the values of A_{mn} and ϕ_{mn}, a set of mode combinations is first set, such as some N different TE_{mn} modes, resulting in a total of $2N$ undetermined parameters. Then, on the measured radiation pattern (e.g., $|E_\varphi|^2$ distribution), the $|E_\varphi|^2$ values at $2N$ different θ angles are read, and using the expression for the far-field radiation E_φ (6.13), a system of $2N$ algebraic equations is established, which can be solved using numerical computation to obtain the $2N$ undetermined parameters. Then, based on this set of obtained parameters, N different modes are formed, and their combination of far-field radiation patterns is theoretically calculated. If the calculated pattern matches the measured pattern within an acceptable error range, it can be considered that the assumed N modes are reasonable. If they do not match, a new set of mode combinations should be established, and the above calculation process should be repeated.

When actually reading relative power data from the pattern, specific angles should be chosen, such as θ angles where $ka\sin\theta = \mu'_{mn}$, in order to greatly simplify the computation. This is because: when $ka\sin\theta = \mu'_{mn}$, the numerator term $J'_m(ka\sin\theta)$ and the denominator term 1-($k\sin\theta/k_c$)2 in $V_{mn}(\theta)$ are both 0 for the TE_{mn} mode, and using L'Hôpital's rule, its value can be determined; for all other modes with the same m but different n, only the numerator term $J'_m(ka\sin\theta)$ continues to be 0, while the denominator term, due to the different k_c, is no longer 0, therefore, $V_{mn}(\theta)$ is 0, i.e., E_φ is 0. This means that for each TE_{mn} mode, there will be a specific θ_{mn} angle where only that TE_{mn} mode exists in the radiation field, and all other TE modes with the same m but different n do not exist in the radiation field. In the single-mode judgment method described below, we will further discuss these special angles.

From the above introduction, it is not difficult to see that although the numerical computation method for far-field radiation can theoretically identify the simultaneous existence of multiple modes and obtain the relative power ratios between the modes, in practice, this is a very complex and time-consuming process, and it is difficult to achieve complete consistency between the theoretically calculated radiation pattern and the measured pattern. This inconsistency becomes more pronounced, especially when there are many modes. Additionally, the inability to perform strict point-by-point measurements on the radiation pattern itself makes this method even more difficult to be widely applied in practice.

2. **Calculation example**

Reference [81] provides an example of using numerical computation to determine the mode composition. Suppose the far-field radiation pattern of a high-power microwave source has been measured (Fig. 6.15a). We assume that this pattern is composed of the radiation fields of 5 modes: TE_{01}, TE_{02}, TE_{03}, TE_{11}, and TE_{12}. In this case, we should establish 10 equations to solve for these 5 modes. To simplify the calculation, we should choose specific angles θ_i ($i = 1, 2, 3, 4, 5$) when reading power data. These angles are $ka\sin\theta_1$, $ka\sin\theta_2$, $ka\sin\theta_3$, $ka\sin\theta_4$, $ka\sin\theta_5$, where $ka\sin\theta_1$, $ka\sin\theta_2$, and

$ka\sin\theta_3$ are the first, second, and third zero points of the derivative of the zero-order Bessel function; $ka\sin\theta_4$ and $ka\sin\theta_5$ are the first and second zero points of the derivative of the first-order Bessel function. As for the angle φ, we symmetrically sample the points at $\varphi = \pm 90°$. This way, we will actually read power values $P(R, \theta, \varphi)$ at the following ten points: P_1 $(R, \theta_1, 90°)$, P_2 $(R, \theta_1, -90°)$, P_3 $(R, \theta_2, 90°)$, P_4 $(R, \theta_2, -90°)$, P_5 $(R, \theta_3, 90°)$, P_6 $(R, \theta_3, -90°)$, P_7 $(R, \theta_4, 90°)$, P_8 $(R, \theta_4, -90°)$, P_9 $(R, \theta_5, 90°)$, P_{10} $(R, \theta_5, -90°)$.

Furthermore, we assume that the amplitudes of TE_{01}, TE_{02}, TE_{03}, TE_{11}, and TE_{12} at the waveguide radiation port are A_1, A_2, A_3, A_4, A_5, and their phases are $\phi_1, \phi_2, \phi_3, \phi_4, \phi_5$. This allows us to write the following 10 algebraic equations:

$$
\begin{aligned}
P_1 &= 0.5(A_1^2g_{11}^2 + A_4^2g_{41}^2 + A_5^2g_{51}^2) - A_1A_4g_{11}g_{41}\sin(\varphi_1 - \varphi_4) \\
&\quad - A_1A_5g_{11}g_{51}\sin(\varphi_1 - \varphi_5) + A_4A_5g_{41}g_{51}\cos(\varphi_4 - \varphi_5) \\
P_2 &= 0.5(A_1^2g_{11}^2 + A_4^2g_{41}^2 + A_5^2g_{51}^2) + A_1A_4g_{11}g_{41}\sin(\varphi_1 - \varphi_4) \\
&\quad + A_1A_5g_{11}g_{51}\sin(\varphi_1 - \varphi_5) + A_4A_5g_{41}g_{51}\cos(\varphi_4 - \varphi_5) \\
P_3 &= 0.5(A_2^2g_{22}^2 + A_4^2g_{42}^2 + A_5^2g_{52}^2) - A_2A_4g_{22}g_{42}\sin(\varphi_2 - \varphi_4) \\
&\quad - A_2A_5g_{22}g_{52}\sin(\varphi_2 - \varphi_5) + A_4A_5g_{42}g_{52}\cos(\varphi_4 - \varphi_5) \\
P_4 &= 0.5(A_2^2g_{22}^2 + A_4^2g_{42}^2 + A_5^2g_{52}^2) + A_2A_4g_{22}g_{42}\sin(\varphi_2 - \varphi_4) \\
&\quad + A_2A_5g_{22}g_{52}\sin(\varphi_2 - \varphi_5) + A_4A_5g_{42}g_{52}\cos(\varphi_4 - \varphi_5) \\
P_5 &= 0.5(A_3^2g_{33}^2 + A_4^2g_{43}^2 + A_5^2g_{53}^2) - A_3A_4g_{33}g_{43}\sin(\varphi_3 - \varphi_4) \\
&\quad - A_3A_5g_{33}g_{53}\sin(\varphi_3 - \varphi_5) + A_4A_5g_{43}g_{53}\cos(\varphi_4 - \varphi_5) \\
P_6 &= 0.5(A_4^2g_{40}^2 + A_5^2g_{50}^2) + A_4A_5g_{40}g_{50}\cos(\varphi_4 - \varphi_5) \\
P_7 &= 0.5(A_1^2g_{14}^2 + A_2^2g_{24}^2 + A_3^2g_{34}^2 + A_4^2g_{44}^2) \\
&\quad + A_1A_2g_{14}g_{24}\cos(\varphi_1 - \varphi_2) + A_1A_3g_{14}g_{34}\cos(\varphi_1 - \varphi_3) \\
&\quad - A_1A_4g_{14}g_{44}\sin(\varphi_1 - \varphi_4) + A_2A_3g_{24}g_{34}\cos(\varphi_2 - \varphi_3) \\
&\quad - A_2A_4g_{24}g_{44}\sin(\varphi_2 - \varphi_4) - A_3A_4g_{34}g_{44}\sin(\varphi_3 - \varphi_4) \\
P_8 &= 0.5(A_1^2g_{14}^2 + A_2^2g_{24}^2 + A_3^2g_{34}^2 + A_4^2g_{44}^2) \\
&\quad + A_1A_2g_{14}g_{24}\cos(\varphi_1 - \varphi_2) + A_1A_3g_{14}g_{34}\cos(\varphi_1 - \varphi_3) \\
&\quad + A_1A_4g_{14}g_{44}\sin(\varphi_1 - \varphi_4) + A_2A_3g_{24}g_{34}\cos(\varphi_2 - \varphi_3) \\
&\quad + A_2A_4g_{24}g_{44}\sin(\varphi_2 - \varphi_4) + A_3A_4g_{34}g_{44}\sin(\varphi_3 - \varphi_4) \\
P_9 &= 0.5(A_1^2g_{15}^2 + A_2^2g_{25}^2 + A_3^2g_{35}^2 + A_5^2g_{55}^2) \\
&\quad + A_1A_2g_{15}g_{25}\cos(\varphi_1 - \varphi_2) + A_1A_3g_{15}g_{35}\cos(\varphi_1 - \varphi_3) \\
&\quad - A_1A_5g_{15}g_{55}\sin(\varphi_1 - \varphi_5) + A_2A_3g_{25}g_{35}\cos(\varphi_2 - \varphi_3) \\
&\quad - A_2A_5g_{25}g_{55}\sin(\varphi_2 - \varphi_5) - A_3A_5g_{35}g_{55}\sin(\varphi_3 - \varphi_5) \\
P_{10} &= 0.5(A_1^2g_{15}^2 + A_2^2g_{25}^2 + A_3^2g_{35}^2 + A_5^2g_{55}^2) \\
&\quad + A_1A_2g_{15}g_{25}\cos(\varphi_1 - \varphi_2) + A_1A_3g_{15}g_{35}\cos(\varphi_1 - \varphi_3) \\
&\quad + A_1A_5g_{15}g_{55}\sin(\varphi_1 - \varphi_5) + A_2A_3g_{25}g_{35}\cos(\varphi_2 - \varphi_3)
\end{aligned}
$$

$$+ A_2A_5g_{25}g_{55}\sin(\varphi_2 - \varphi_5) + A_3A_5g_{35}g_{55}\sin(\varphi_3 - \varphi_5) \tag{6.15}$$

where:

$$g_{ji} = \begin{cases} G_{01}(\theta_i,\phi = 90^\circ)_{j=1} \\ G_{02}(\theta_i,\phi = 90^\circ)_{j=2} \\ G_{03}(\theta_i,\phi = 90^\circ)_{j=3} \\ G_{11}(\theta_i,\phi = 90^\circ)_{j=4} \\ G_{12}(\theta_i,\phi = 90^\circ)_{j=5} \end{cases} \tag{6.16}$$

$$G_{mn}(\theta, \phi) = V_{mn}(\theta)\sin\left[m\phi + (m-1)\pi/2\right]$$

The far-field radiation pattern measured in the experiment, as shown in Fig. 6.17a, is obtained by a pure TE_{01} mode that radiates to the far field after passing through a 90° bent waveguide. Due to the short tapered waveguide at the input end of the bent waveguide and the inherent elliptical distortion of the bent waveguide itself, the pattern exhibits a strong asymmetry with respect to the centerline of the radiating waveguide. During the measurement, the axis of the bent waveguide lies in the *yoz* plane (Fig. 6.10), and the receiving antenna measures the E_x component point-by-point in a straight line parallel to the axis. For the ideal TE_{01} mode, the measured E_x component should result in a symmetric radiation pattern. The asymmetrical measured pattern can only be fitted using the numerical calculation method described above. We assume that the pattern is formed by the radiation fields of the aforementioned TE_{01}, TE_{02}, TE_{03}, TE_{11}, and TE_{12} modes. Data is read from the experimental pattern curve, and by solving equation set (6.15), the relative amplitudes and phases of these 5 modes are determined. The results show the relative strengths between the modes as $TE_{01}/TE_{02}/TE_{03}/TE_{11}/TE_{12} = 0.08/0.23/0.35/0.19/0.10$. Using the data obtained for the 5 modes, the theoretically calculated far-field pattern is shown in Fig. 6.17b, which closely approximates the measured pattern but still requires adjustment of the mode composition to achieve a more satisfactory fit. The errors in this method mainly stem from the reasonableness of the initial assumption about the modes and the measurement errors of the experimental pattern, as well as the random errors in reading data from the pattern.

6.3.4 Single-Mode Discrimination Method [81, 82]

1. **Theoretical basis of single-mode discrimination**

Clearly, using the numerical calculation method for mode recognition is a very complex and cumbersome process. Reference [81] proposes another mode recognition method: the single-mode discrimination method, which is relatively more convenient than the numerical calculation method. The application of this method is based on the following characteristics of the radiation pattern:

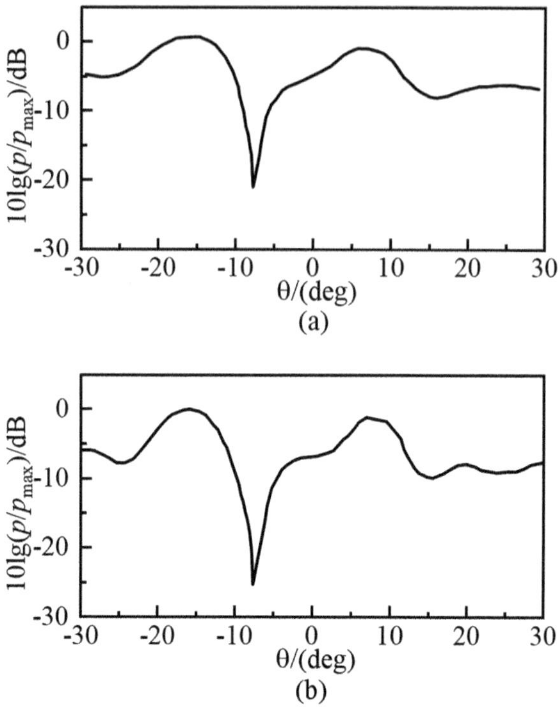

Fig. 6.17 Far-field mode pattern of multi-mode combination, **a** experimentally measured mode pattern: **b** theoretically calculated mode pattern

(1) **In a series of modes of TE_{mn} or TM_{mn} with the same m value but different n values, there exists a specific θ angle for each n value, at which only the radiation of that mode is non-zero, while the radiation of all other modes with the same m value but different n values is zero.**

In the E_φ expression in Eq. (6.7), for the TE_{01} mode, if we take $ka\sin\theta_1 = \mu'_{01} = 3.8317$, then in the $V_{01}(\theta_1)$ expression (6.8), the numerator $J'_0(3.8317) = 0$, and at this point due to $k_c = \mu'_{mn}/a$, $[1-(k\sin\theta_1/k_c)2] = 0$, the denominator also becomes 0, which becomes an indeterminate form where both the numerator and denominator are 0, and we can use L'Hôpital's rule to find the definite value of this ratio. At this point, for all TE_{0n} modes other than TE_{01}, the numerator in the $V_{0n}(\theta_1)$ expression is still 0, but the denominator is not 0, so $V_{0n}(\theta_1) = 0$ ($n \neq 0$), meaning they have no radiation field E_φ at θ_1. In a more general case, we can conclude that for each TE_{mn} mode, there will be a special angle θ_{mn}, at which only that TE_{mn} mode has a radiation field E_φ, while all other TE_{mn} modes with the same m but different n have no E_φ radiation field.

For any TM_{mn} mode, there are similar characteristics. In the $Q_{mn}(\theta)$ expression (6.11) for the E_θ component in Eq. (6.10), when $ka\sin\theta = \mu_{mn}$, both the numerator and denominator are 0, and we can similarly use L'Hôpital's rule to find the definite value. But at this point, for all TM_{mn} modes with the same m but different n, the numerator is 0 and the denominator is not 0, so $Q_{mn}(\theta) = 0$, meaning they do not have an E_θ radiation at that θ_{mn} angle.

So, for both TE and TM modes, in a series of modes with the same *m* but different *n*, there is a specific θ angle for each *n*, at which only that mode has a non-zero radiation field, while all other modes with the same *m* but different *n* have a radiation field of 0. This is a characteristic of the radiation field.

(2) Far-field radiation patterns of TE_{mn} or TM_{mn} modes with different *m* values exhibit symmetry in certain special directions in space

In the E_φ expression in Eq. (6.7), for the sum of radiation field at angles φ, $\varphi + 90°$, $\varphi + 180°$, and $\varphi + 270°$, the radiation fields for TE_{1n} and TE_{2n} modes will be zero. This means that the $\sin[m\varphi + (m\text{-}1)\pi/2]$ term will be zero when $m = 1$ or $m = 2$, and when φ takes the four values mentioned. Substituting these values into the E_φ expression and adding the results together gives the following:

$$E_\phi(\phi)(+E_\phi(\phi + 90°) + E_\phi(\phi + 180°) + E_\phi(\phi + 270°) \tag{6.17}$$

for TE_{1n} and TE_{2n} modes, the sum is 0. At the same time, for all TM_{mn} modes, according to Eq. (6.10), E_φ is already equal to 0. This indicates that the sum of the radiation fields at these four angles will not include contributions from TE_{1n}, TE_{2n}, and all TM_{mn} modes.

This conclusion holds for any value of φ, meaning that even if we don't know the specific direction of the waveguide system, i.e., the orientation of the electric vector of the TE_{1n} mode at the radiation port center, we can still use the above method to exclude contributions from the TE_{1n}, TE_{2n}, and TM_{mn} modes. If the output of a high-power microwave source only contains TE_{0n}, TE_{1n}, TE_{2n}, and TM_{mn} modes, then the combined signal represented by Eq. (6.13) will only contain the TE_{0n} mode.

Similar characteristics can be found for other TE or TM modes. In short, far-field radiation patterns of TE_{mn} or TM_{mn} modes with different m values exhibit different symmetries relative to certain directions in space. In other words, we can always find some angles where the superposition of the radiation fields E_φ or E_θ can exclude radiation from a certain type of mode with different m values. This is another characteristic of the radiation field.

(3) Distinguishing modes with different *n* values

In the above, we have distinguished the TE_{0n} mode separately. If further distinction is needed for the different modes within the TE_{0n} mode with different *n* values, it can be achieved by placing the receiving antenna at specific θ angles for measurement. For example, when the signal combination is:

$$E_\phi(\phi, \theta_1) + E_\phi(\phi + 90°, \theta_1) + E_\phi(\phi + 180°, \theta_1) + E_\phi(\phi + 270°, \theta_1) \tag{6.18}$$

And when $ka\sin\theta_1 = 3.8317$, then in the combined signal above, the E_φ component of the TE_{02}, TE_{03} modes is all zero, containing only the contribution of the TE_{01} mode.

2. **Single-mode judgment method for various modes**

The single-mode judgment method for the far-field radiation is based on the three basic characteristics mentioned above. We take the modes where $m = 0, 1, 2$ and n is any positive integer as an example to explain the application of this judgment method. In the measurement system, a fundamental mode rectangular waveguide is connected behind the receiving antenna, and the signals are combined using a magic T, and the combined signal is input into a spectrum analyzer.

(1) **Judgment of TE$_{0n}$ mode**

The radiation field E_φ is measured at four angles φ, φ+90°, φ+180°, φ+270°, where φ is an arbitrary value, and then the outputs are detected after being superimposed, that is

$$p(\theta) = \left\langle \left[E_\varphi(\theta, \varphi) + E_\varphi(\theta, \varphi + 90^\circ) + E_\varphi(\theta, \varphi + 180^\circ) + E_\varphi(\theta, \varphi + 270^\circ) \right]^2 \right\rangle \tag{6.19}$$

where, $p(\theta)$ represents relative power. According to the earlier analysis, $p(\theta)$ will only have contributions from the TE$_{0n}$ mode (with $m = 0, 1, 2$), and all other TE$_{mn}$ modes with $m \neq 0$ and all TM$_{mn}$ modes have been excluded. In Eq. (6.19), the symbol⟨⟩represents averaging over time, and the square and time average operations actually represent the detection process.

To further determine the value of n from the TE$_{0n}$ mode, we just need to select specific angles for θ. As mentioned earlier, when the chosen θ satisfies $ka\sin\theta_1 = \mu'_{01} = 3.8317$, the value obtained from Eq. (6.19) is the relative power $p(\theta_1)$ of the TE$_{01}$ mode, and when θ satisfies $ka\sin\theta_2 = \mu'_{02} = 7.0156$, $ka\sin\theta_3 = \mu'_{03} = 10.1735$, ..., the obtained $p(\theta_2)$, $p(\theta_3)$ are the contributions of the TE$_{02}$, TE$_{03}$, etc. modes, respectively.

(2) **Judgment of TE$_{1n}$ mode**

If E_φ is measured only at φ=φ and φ+180°, then the combined result

$$p(\theta) = \left\langle \left[E_\varphi(\theta, \varphi) - E_\varphi(\theta, \varphi + 180^\circ) \right]^2 \right\rangle \tag{6.20}$$

will only contain the contribution of the TE1n mode and not the power of the TE$_{0n}$, TE$_{2n}$, and TM$_{mn}$ modes. Here, the value of φ can be any value. Furthermore, taking θ values that satisfy $ka\sin\theta_1 = \mu'_{11} = 1.8412$, $ka\sin\theta_2 = \mu'_{12} = 5.3314$, $ka\sin\theta_3 = \mu'_{13} = 8.5363$, ... will respectively represent the contributions of the TE$_{11}$, TE$_{12}$, TE$_{13}$, ... modes.

(3) **Judgment of TE$_{2n}$ mode**

If the measured values of E_φ at four angles φ=φ, φ+90°, φ+180°, φ+270° are recombined as:

$$p(\theta) = \left\langle \left[E_\varphi(\theta, \varphi) - E_\varphi(\theta, \varphi + 90^\circ) + E_\varphi(\theta, \varphi + 180^\circ) - E_\varphi(\theta, \varphi + 270^\circ) \right]^2 \right\rangle \tag{6.21}$$

Then in $p(\theta)$ there will only be the contribution of the TE_{2n} mode, excluding those of the TE_{0n}, TE_{1n}, and TM_{mn} modes. Now taking θ as $ka\sin\theta_1 = \mu'_{21} = 3.0542$, $ka\sin\theta_2 = \mu'_{22} = 6.7061$, $ka\sin\theta_3 = \mu'_{23} = 9.9695$, ... will respectively represent the contributions of the TE_{21}, TE_{22}, TE_{23}, ... modes.

(4) **Judgment of TM_{0n} mode**

By measuring the E_θ component at angles $\varphi = 45^\circ$ and $\varphi = -135^\circ$, the combined signal

$$p(\theta) = \left\langle \left[E_\theta(\theta, \varphi = 45^\circ) + E_\theta(\theta, \varphi = -135^\circ) \right]^2 \right\rangle \tag{6.22}$$

will only contain the relative power of the TM_{0n} and TE_{2n} modes, and not the contribution of the TM_{1n}, TM_{2n}, TE_{0n}, and TE_{1n} modes. We can use the judgment Eq. (6.21) from (3) above to determine the contribution of the TE_{2n} mode. After we determine $p(\theta)$ for the TE_{2n} mode, by subtracting this contribution from Eq. (6.22), we can get the contribution of the TM_{0n} mode. Taking $ka\sin\theta_1 = \mu_{01} = 2.4048$, $ka\sin\theta_2 = \mu_{02} = 5.5201$, $ka\sin\theta_3 = \mu_{03} = 8.6537$, ... will respectively help in determining the existence of the TM_{01}, TM_{02}, TM_{03}, ... modes.

(5) **Judgment of TM_{1n} mode**

The above measured values of E_θ are recombined as

$$p(\theta) = \left\langle \left[E_\theta(\theta, \varphi = 45^\circ) - E_\theta(\theta, \varphi = -135^\circ) \right]^2 \right\rangle \tag{6.23}$$

Then $p(\theta)$ will only contain the contributions of the TM_{1n} and TE_{1n} modes, and not the contributions of the TM_{0n}, TM_{2n}, TE_{0n}, and TE_{2n} modes. From the previous judgment Eq. (6.20) for the TE_{1n} mode, we can already determine the contribution of the TE_{1n} mode. By subtracting the contribution of the TE_{1n} mode from Eq. (6.23), the contribution of the TM_{1n} mode can also be determined. Then taking $ka\sin\theta_1 = \mu_{11} = 3.8317$, $ka\sin\theta_2 = \mu_{12} = 7.0156$, $ka\sin\theta_3 = \mu_{13} = 10.1735$, ... will respectively decide the existence of the TM_{11}, TM_{12}, TM_{13}, ... modes.

(6) **Judgment of TM_{2n} mode**

Taking the signal combination as

$$p(\theta) = \left\langle \left[E_\theta(\theta, \varphi = 0^\circ) + E_\theta(\theta, \varphi = 180^\circ) \right]^2 \right\rangle \tag{6.24}$$

then it will only contain the contributions of the TM_{2n} and TM_{0n} modes. We have already determined the contribution of the TM_{0n} mode from the previous judgment

Eq. (6.22), so only the relative power of the TM_{2n} mode remains. Then, when taking values of θ such that $ka\sin\theta_1 = \mu_{21} = 5.1356$, $ka\sin\theta_2 = \mu_{22} = 8.4172$, $ka\sin\theta_3 = \mu_{23} = 11.6198...$, we can respectively determine the existence of the TM_{21}, TM_{22}, TM_{23}, ….

From the above introduction, it is not difficult to see that in the radiation field measurement method, only the direct discrimination method is widely used because it is relatively simple and practical. The numerical calculation method and single mode judgment method are quite complex, and in actual radiation field measurements, due to the limitation of the receiving antenna size, it is not possible to measure accurately point by point, resulting in low spatial resolution. Additionally, the receiving antenna itself affects the radiation field, further reducing the measurement accuracy. Therefore, the measurements at specific angles may not strictly conform to the theoretical mode combinations, making mode identification difficult to achieve accurate results or even completely ineffective. Due to these reasons, the practicality of these two methods is limited, with only a certain application in the University of Stuttgart, Germany, and very few other institutions have adopted them.

6.4 Analysis Methods of Thermal Image

The thermal imaging analysis method also relies on the obtained microwave mode radiation image for mode recognition. Unlike image display methods and radiation field measurement methods, it uses microwave thermal effect materials with high sensitivity, microwave thermoelectric effect components or materials to make target planes. When the microwave radiation reaches the target plane, it does not directly produce a visible image, but instead forms a thermal image composed of a temperature distribution or directly converts the thermal image into current distribution or voltage distribution. Then, an infrared camera and computer are used to process and obtain a visual image of the radiation field structure.

By changing the point-by-point measurement of the microwave system's radiation field to a one-time imaging, it overcomes the limitations of the receiving antenna's aperture size on spatial resolution of measurement and the measurement errors caused by the radiation field's changes (instability of the radiation source) during the measurement process. The characteristics of thermal imaging analysis are:

(a) It creates a one-time plane image using microwave thermal effect and electromagnetic effect, which can be completed in a short time or even instantaneously.
(b) It is an indirect imaging method, the microwave effect produces a thermal distribution image, or the voltage distribution or current distribution image obtained by the thermal image again. These are invisible images and require secondary processing to form visible images. This is the main difference between thermal imaging and image display methods. The latter also uses

microwave electrothermal effects to produce images, but they are directly visible images.

(c) Thermal imaging is easily processed directly through infrared cameras and computers, allowing for better indication of the fine structure of the radiation field and facilitating multi-mode recognition and analysis of the relative proportion of power between mode components.

When using the radiation field measurement method, obtaining a complete radiation field distribution requires point-by-point measurement. The instability of the radiation source and a certain aperture size of the receiving antenna during the measurement process seriously affect the accuracy and precision of the radiation image measurement, especially for high-power microwave sources. Due to the poor repeatability between microwave pulses, relying on many times single pulse emissions to complete the radiation field measurement is obviously more difficult and unreliable.

Obtaining radiation images through thermal imaging analysis can overcome the drawbacks of the radiation field measurement method, due to the highly sensitive nature of the electrothermal components and materials to microwaves. The energy of just a single pulse can form a thermal image. This approach can obtain a more detailed radiation field structure without the need for lengthy point-by-point measurements. Furthermore, computer processing can directly yield richer or even three-dimensional image information.

6.4.1 Analytical Method of Infrared Image

1. **The experiment at Tsukuba University in Japan** [80]

Using a microwave absorber plate (ECCOSORB AN72 produced by Emerson & Cuming Company or other microwave absorbing materials) as the microwave radiation target, placed it at a certain distance from the radiation source (far-field or near-field position). When the microwaves radiated from the radiation source are absorbed by the absorber plate, this causes the temperature of the absorber plate to rise, thus forming a two-dimensional temperature distribution, i.e., a thermal image. Using an infrared camera, this thermal image can be recorded, turning the invisible thermal image into a visible image. Alternatively, the images recorded by the camera can be directly input into a computer for analysis, allowing for the drawing of two-dimensional or even three-dimensional amplitude distribution maps, isothermal lines (isovalue) distribution maps, and so on, as needed.

In many high-power microwave experiments, infrared cameras are often used to obtain images of the radiation pattern. It has a very wide temperature dynamic range, which can not only record the temperature changes of 0.02–4 °C caused by continuous wave radiation power below 1W, but also capture the distribution images of temperatures in the highest tens of degrees Celsius temperature generated by pulse powers ranging from several tens of kilowatts to 200 kW. ECCOSORB AN72 absorber plate is a multi-layered impregnated carbon polyurethane plastic plate, which is lightweight and flexible, capable of responding to a microwave with a wide frequency range. The relationship between the absorber plate temperature increase and the time-averaged value of power is almost linear within the conventional power bandwidth range. The time constant for the temperature relaxation of the absorption plate is approximately a few seconds, and the main heat dissipation is believed to be accomplished through convective cooling by air. The thermal diffusion along the surface of the absorber plate is much smaller than the thermal diffusion caused by air convection, which is advantageous for capturing the thermal distribution structure in the thermal image. However, thermal diffusion will limit the spatial resolution in the central region of the radiation pattern, where the power density will change rapidly within a small range.

In Fig. 6.18a, Japanese scholars used thermal imaging to obtain the near-field radiation distribution of the input mode TE_{01} mode when developing a five-stage stepped circular waveguide TE_{01}–TE_{02} mode converter. The mode is obtained by converting the rectangular waveguide TE_{10} mode into the circular waveguide TE_{01} mode, and the TE_{10} mode is generated by the output of a klystron operating at a frequency of 41 GHz and outputting power less than 1W. The radiation distribution was recorded at a distance of 20 cm from the waveguide radiation port using an absorber plate (ECCOSORB AN72, Emerson & Cuming Company) and an infrared camera (Japan Avionics TVS-2200), and then processed by a computer. From the image, it is evident that there are small protrusions on either side of the two main peaks, indicating that there is a small parasitic TE_{02} mode component in this TE_{01} mode. The obtained TE_{01} mode is input into the TE_{01}–TE_{02} mode converter to generate the TE_{02} mode. The power content of the TE_{01} mode can be adjusted by changing the number of steps in the stepped circular mode waveguide converter. The output port diameter of the mode converter is taperedly converted to a circular waveguide of 37 mm, then connected to a uniform circular waveguide with a diameter of 37 mm, and the flat incision at the terminal of the uniform circular waveguide serves as the radiating port. The output radiation power distribution of the TE_{02} mode is shown in Fig. 6.18b–d. The position of the peak on the image indicates that the main mode is the TE_{02} mode. However, when the length of the uniform circular waveguide of diameter 37 mm at the output end of the mode converter is changed, the power density ratio between the peaks on the image changes. Figure 6.18b–d represent the radiation situation at different lengths of the uniform circular waveguide.

2. **The experiment at Osaka University in Japan** [83]

The experiment conducted by universities including Osaka University in Japan involved using quasi-optical techniques to focus Gaussian beams in the research of high-power millimeter-wave gyrotron output beams. They captured images of the beams at different positions using ECCOSORB AN absorption plates and an infrared camera. The experiment was conducted using a 60 GHz gyrotron. After passing through the output window of the gyrotron, they connected a tapered waveguide, reducing the diameter of the circular waveguide from 63.5 mm to 25.6 mm.

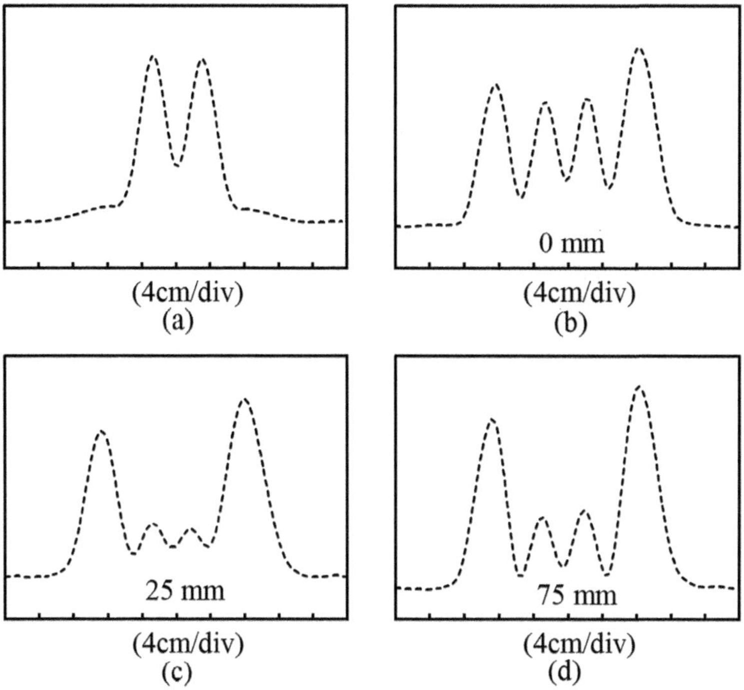

Fig. 6.18 Near-field thermal images of multi-step TE_{01}–TE_{02} mode converter input TE_{01} mode and output TE_{02} mode, **a** input mode TE_{01} of mode converter, **b** TE_{02} output mode when the length of the uniform circular waveguide at the output end is 0 mm, **c** TE_{02} output mode when the length of the uniform circular waveguide is 25 mm, **d** TE_{02} output mode when the length of the uniform circular waveguide is 75 mm

Then, they directly connected a circular waveguide with a semi-incision end as the radiation antenna. The radiated microwave beam is output after passing through an aluminum elliptical reflector and an aluminum parabolic reflector (see Fig. 6.19).

The equation for the elliptical reflector in the (*x*, *y*, *z*) coordinate system is given by:

$$\frac{(x-f_e)^2}{B^2}+\frac{y^2}{A^2}=1 \tag{6.25}$$

where, $A = 100$ mm is the major axis of the ellipse, $B = 86.6$ mm is the minor axis, and the focal length of the ellipse $f_e = (A2^{-}B^{2)1/2} = 50$ mm. The reflector has a width of 150 mm in the *x* direction.

The equation for the parabolic reflector in the (x', *y*, z') coordinate system is:

$$z' = -\frac{1}{4f_p}(x'+x_f')^2 - f_p \tag{6.26}$$

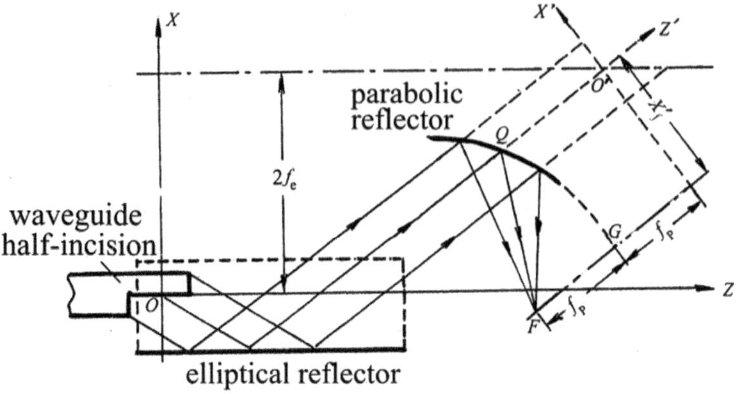

Fig. 6.19 The two-dimensional structure of elliptical-parabolic focusing system

where, the focal length f_p of the parabolic reflector is 74m. The definition of X'_f is shown in Fig. 6.17 and is given as 122.5 mm.

The experiment first examined the purity of the output mode of the gyrotron by measuring the power distribution after the conical waveguide. The absorption plate was placed perpendicular to the z-axis, at a distance of 100 mm from the opening of the conical waveguide. The gyrotron power was set to 70 kW with a pulse width of 1 ms. The measurement result, as shown in Fig. 6.20, was a mode thermal image generated by the microwave absorption plate. This was then recorded using an infrared camera and processed by a computer to obtain a three-dimensional image. The axisymmetric thermal distribution clearly showed correspondence to the TE_{02} mode. Even when the gyrotron output pulse power was changed from a few kilowatts to over 100 kW, the resulting images were almost identical. The gap on the TE_{02} mode electric field image ring was caused by the top of the tapered waveguide being shielded by the thermal imager detector (infrared camera).

Next, with the gyrotron output power fixed at 20 kW and a pulse width of 1 ms, absorption plates were placed at different x positions on the y–z plane behind the elliptical reflector. The thermal images on the absorption plates, processed by a computer, are shown in Fig. 6.21. The top left corner of each image shows a plane image of

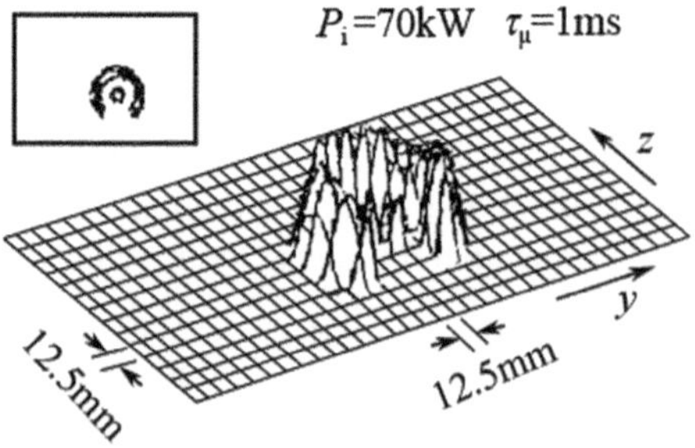

Fig. 6.20 The three-dimensional image of the TE_{02} mode of the 60 GHz gyrotron output, formed after processing the computer-processed infrared thermal imaging (From Ref. [83])

the temperature distribution. It is evident from the images that the microwave beam became a sheet-like beam after being focused by the elliptical reflector. Furthermore, measurements were taken at the focal plane $x = 100$ mm behind the elliptical reflector, with the gyrotron pulse power varying from 20 to 75 kW. The resulting power distribution images showed little change compared to the image displayed in the top right of Fig. 6.21 (the image at $x = 100$ mm), but further power increases could potentially lead to damage of the absorption plate.

To observe the focusing of the microwave beam behind the parabolic reflector, the focus of the parabolic reflector was approximately $x = -70$ mm. We measured the temperature distribution on the absorption plate at different x positions on the y–z plane. In this scenario, it was easy to observe a drastic increase in temperature at

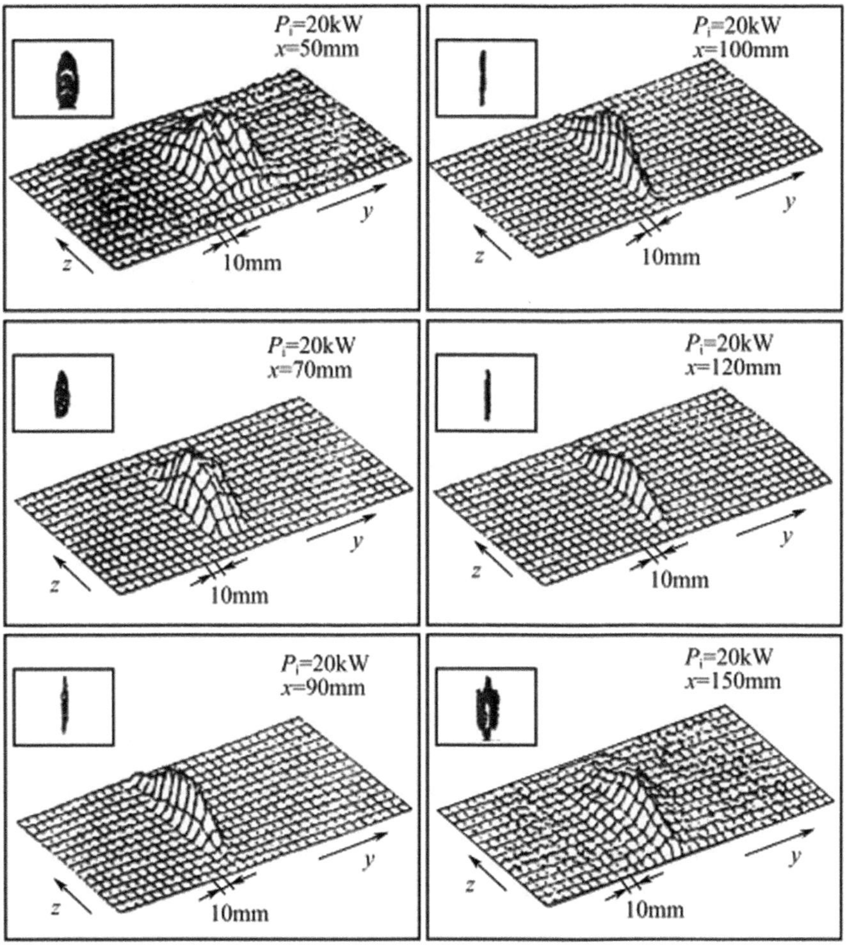

Fig. 6.21 Thermal images of microwave beams measured at different distances behind the elliptical reflector. (From Ref. [83])

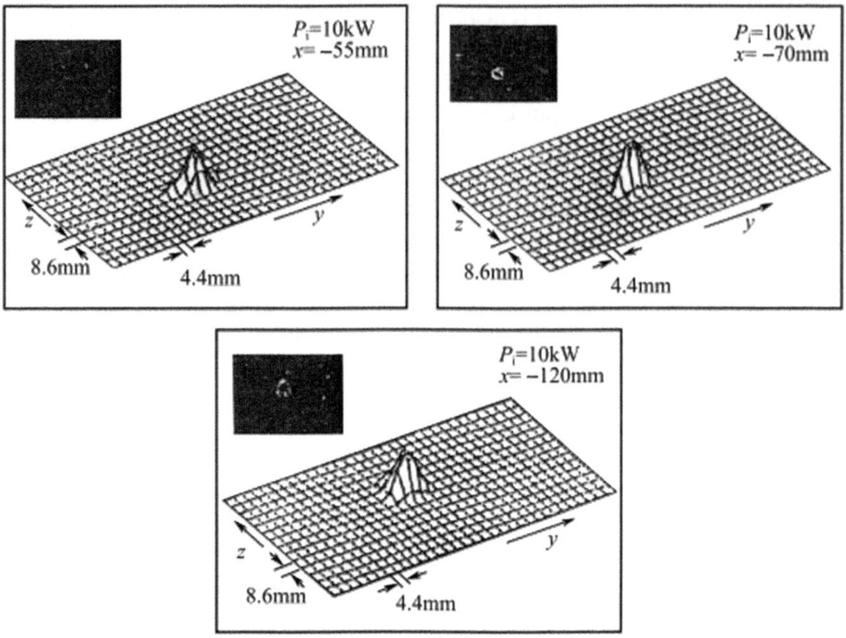

Fig. 6.22 Thermal images of the microwave beam measured at different distances behind the parabolic reflector. (From Ref. [83])

the point of incidence of the microwave beam of the absorption plate, which could lead to the plate being burned or requiring removal. At this point, the microwave power was lowered to 10 kW with a pulse width of only 0.5 ms. Figure 6.22 shows the power distribution at $x = -55$ mm, -70 mm, and -120 mm. At the focus $x = -70$ mm, we obtained a wave beam close to a Gaussian distribution, with a width of approximately 10 mm and a peak temperature of 120 °C. This indicates that when the microwave beam pulse power increased to 100 kW, the pulse energy density easily reaches 100 kW/cm^2.

6.4.2 *Analytical Method of Thermoelectric Detection*

By using a thermoelectric detector array instead of a microwave absorber plate, a microwave mode radiation thermal image can also be obtained. The difference is that the thermoelectric detector array directly converts the thermal image into a current or voltage output, which is recorded and processed by the computer, instead of generating the thermal image through the microwave absorber plate and then having the infrared recorder (infrared camera) record the image for the computer to process.

A thermoelectric detector is composed of thermoelectric materials, such as polar crystals in certain piezoelectric materials like triglycine sulfate (TGS), barium titanate ($BaTiO_3$), and lithium tantalate ($LiTaO_3$), etc. And their spontaneous polarization intensity is a function of temperature and decreases as the temperature rises. The change in temperature causes a change in the polarization state, resulting in a variation of polarized charge on the surface of the crystal perpendicular to the direction of spontaneous polarization, leading to the generation of current. The result is the appearance of an open-circuit voltage that varies with temperature at both ends of the crystal. This type of polar crystal is called a thermoelectric body, and this phenomenon is known as the pyroelectric effect. Infrared detectors (sensors) utilizing the pyroelectric effect have high sensitivity, with a temperature response rate of 4-5uA/°C and a temperature resolution of less than 0.2 °C. Further details about thermoelectric detectors have been discussed in Sect. 2.4.4.

One should place a thermoelectric probe at a certain distance from the microwave radiation port and move it within the plane perpendicular to the radiation waveguide axis to measure the radiation field. Additionally, one should use a fixed thermoelectric probe placed outside the scanning area of the measurement probe in the radiation area as a normalization signal for the measurement results. Input the measured field values and the normalized signal simultaneously into the computer. After computer processing, the distribution image of the radiation field relative power at iso-power lines can be obtained, or data, curves, etc., can be output as needed. If a single thermoelectric probe is replaced with a planar array, the movement of the probe can be reduced or eliminated, and the computer can directly output mode images, thereby improving measurement efficiency.

Figure 6.23 shows radiation images obtained using a thermoelectric detection array at different z distances from the output window of the FUVA gyrotron under different magnetic fields and resonant modes, as measured by Fukui University in Japan. The working magnetic fields and frequencies corresponding to these mode images are shown in Table 6.1 [84].

It's clear that the mode images obtained using the thermoelectric detection method are much clearer than those obtained using image display methods. Based on the same principle, other microwave effect detectors, such as the Kerr effect detector and the resistive sensor detector, can also be used for mode measurement. We've introduced these detectors in Sects. 2.4 and 5.2, but they are applied in different scenarios. Originally, we discussed their application in high-power microwave measurement, but here they are used for microwave mode measurement. The principles are the same, so we won't go into further detail about them. In principle, any electromagnetic effect that can produce changes in voltage, current, resistance, or polarization surface deflection under the action of a microwave field can be used for microwave power measurement and mode field measurement.

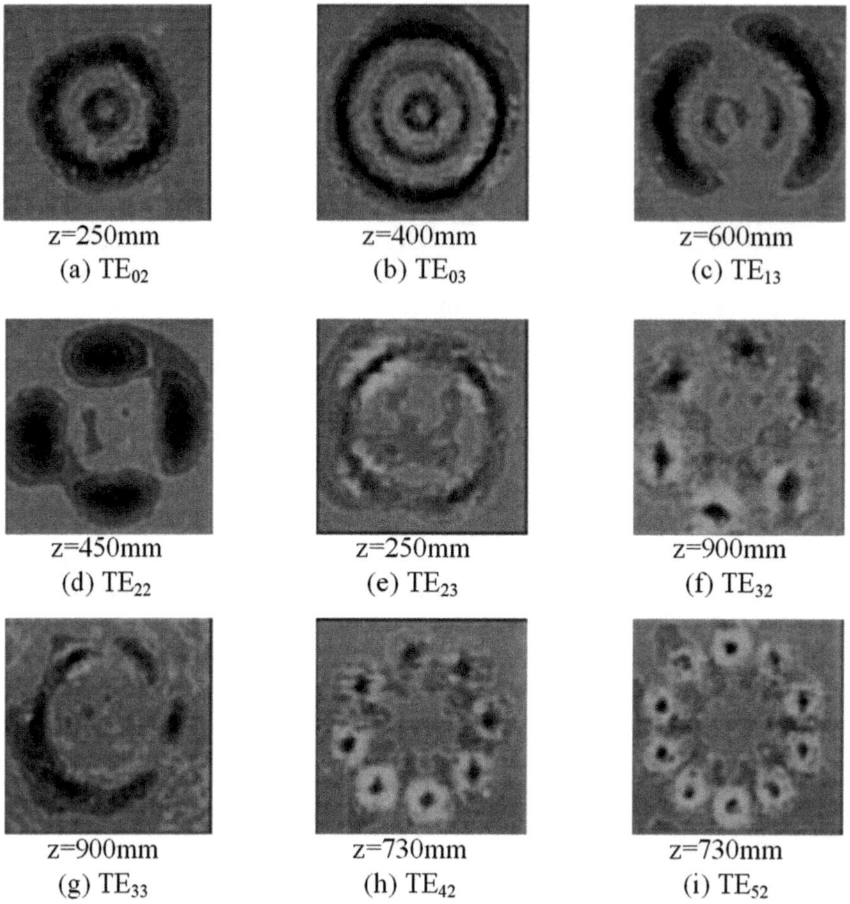

Fig. 6.23 The mode image of the output from the gyrotron measured by a thermoelectric detector array and processed by a computer

Table 6.1 Operating magnetic field and frequency corresponding to the FUVA output mode of the gyrotron

Mode	TE_{22}	TE_{02}	TE_{32}	TE_{13}	TE_{42}	TE_{23}	TE_{03}	TE_{52}	TE_{33}
Magnetic field/T	4.6	4.8	5.5	5.8	6.4	6.8	7.2	7.2	7.8(8.0)
Frequency /GHz	128	132	152	162	175	190	194	199	215(217)

6.5 Measurement Method of the Field in Waveguides [85]

American scholar D. Stone proposed a method for directly measuring the field distribution inside a microwave transmission system, such as a waveguide, or at its port, to determine the waveguide mode. This method is called the waveguide field measurement method or beat wave field measurement method, which avoids errors caused by the radiation of the field. However, in order to measure the complete field structure, measurements must often be made in two-dimensional or even three-dimensional directions, making the measurement setup very complex. Moreover, such a setup will inevitably disturb the original field distribution in the system, leading to measurement errors. Because this method not only involves complex equipment but also tedious measurement processes, it is less commonly used.

Any two mode waves transmitted in an overmoded waveguide will combine to form a beat wave, and, similar to the well-known incident and reflected wave synthesis standing wave in a single-mode waveguide, the ratio of the maximum to minimum value of the standing wave electric field is called the voltage standing wave ratio (VSWR). We define the ratio of the maximum to minimum value of the beat wave electric field as the voltage traveling wave coefficient or voltage traveling wave ratio (VTWR). In general, VTWR is a function of the position on a plane perpendicular to the direction of wave propagation in the waveguide, and the measurement of VTWR can easily be related to the power of each mode or the share of each waveguide mode in the total transmission power. This measurement can be used to:

(a) Reflect the characteristics of the operating mode of a multi-mode output high-power microwave source (such as a gyrotron);
(b) Analyze the characteristics of mode variation in overmoded waveguides;
(c) Determine the optimal position of lossy materials in the transmission line;
(d) Achieve impedance matching (including interference from one or more parasitic modes).

6.5.1 Fundamental Principles

The composite wave of two waves, m mode and n mode, traveling in the z direction, can be expressed as:

$$C(z) = |A_m|e^{-j(k_m z+\varphi_m)} + |A_n|e^{-j(k_n z+\varphi_n)} \tag{6.27}$$

where,

$$\varphi_{m(n)} \equiv \arctan\left\{\frac{\mathrm{Im}(A_{m(n)})}{\mathrm{Re}(A_{m(n)})}\right\} \tag{6.28}$$

where, $k_{m(n)}$ is the propagation constant or wave number for the m or n mode; $\phi_{m(n)}$ is their initial phase and $A_{m(n)}$ is their amplitude. The superposition of two waves will form a beat wave, and when detected using a square law detector, its output will be proportional to

$$|C(z)|^2 = |A_m|^2 + |A_n|^2 + 2|A_m||A_n| \cos[(k_m - k_n)z + (\phi_m - \phi_n)] \tag{6.29}$$

Define $K_{mn} = |k_m - k_n|/2$ as the wave number of the beat wave and $\lambda_{mn} = 2\pi/K_{mn}$ as the wavelength of beat waves. By choosing the position of origin $z = 0$, the beat wave phase (ϕ_m-ϕ_n) can be eliminated. The maximum and minimum values of $|C(z)|$ can then be determined, defining the voltage traveling wave ratio as

$$\text{VTWR} = \frac{|C(z)|_{\max}}{|C(z)|_{\min}} = \frac{||A_m| + |A_n||}{||A_m| - |A_n||} = \frac{|A_m| + |A_n|}{\pm|A_m| \mp |A_n|} \tag{6.30}$$

where, when $|A_m| > |A_n|$, use the upper calculation symbol; when $|A_m| < |A_n|$, use the lower calculation symbol. Thus, we have

$$\frac{|A_n|^2}{|A_m|^2} = \left(\frac{\text{VTWR} \mp 1}{\text{VTWR} \pm 1}\right)^2 \tag{6.31}$$

When $|A_m| > |A_n|$, use the upper calculation symbol; when $|A_m| < |A_n|$ use the lower calculation symbol. Since VTWR is a measurable quantity, the ratio of $|A_n|^2/|A_m|^2$ can be determined from this.

The measurement process of the beat wave field method is: designing and manufacturing a polarized probe that can move in the r, θ, and z directions. This probe is installed at the circular waveguide output port of the microwave source. The polarized probe measures the square value $|E_\theta|^2$ of the angular component of the traveling wave. Figure 6.24 illustrates the structure of such a device. Then, make it measure the relative values of $|E_\theta|^2$ of waveguide field point-by-point along the radial direction $r = -a$ to $r = +a$ of the circular waveguide, and draw the measured distribution of $|E_\theta|2$, as shown in Fig. 6.25. Then, determine the mode composition. Subsequently, the probe is used to measure the value of $|E_\theta|^2$ of the beat field along the z direction at a specific r position, and the VTWR is calculated based on the ratio of its maximum and minimum values, so as to determine the power ratio between the mode components.

6.5.2 Practical Measurement Example

The measurement was carried out on the Varian VGA-800 gyrotron, operating at a frequency of 28 GHz and outputting the TE_{02} mode. The output waveguide of the gyrotron is an overmoded waveguide with a diameter of 6.35 cm. Using a three-dimensional movable probe, the relative values of $|E_\theta|2$ along the radial direction from $r = -a$ to $r = +a$ (where a is the radius of the output circular waveguide) were measured at various z points, as shown in Fig. 6.25. The field distribution in the

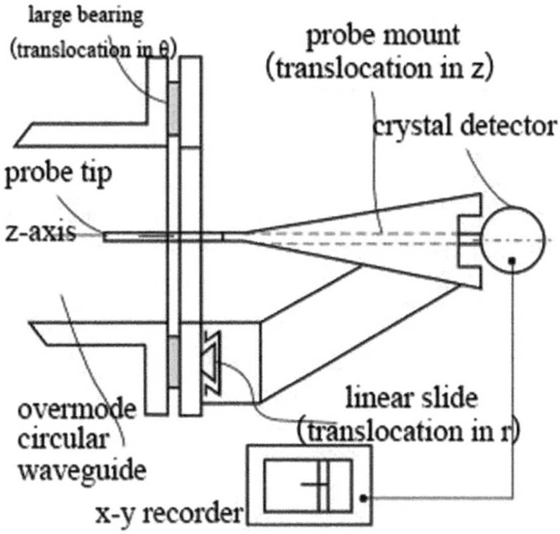

Fig. 6.24 Schematic diagram of the structure of a 3D moving probe for VTWR measurement in a circular waveguide (From Ref. [85])

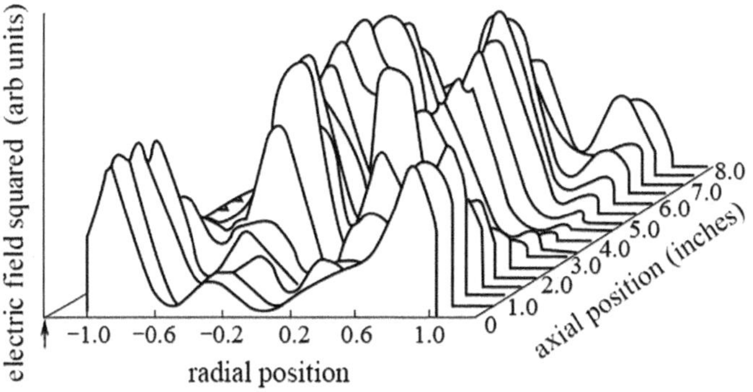

Fig. 6.25 Beat wave field distribution of gyrotron VGA-8000 (From Ref. [85])

figure indicates that the output of the TE_{02} mode has been partially converted into other modes, and is no longer a single TE_{02} mode distribution image.

Assuming the output mode of the gyrotron is now composed of TE_{0m} mode and TE_{0n} mode, with their amplitude coefficients being a_m and a_n, then when the normalization condition

$$\sum_i |a_i|^2 = 1 \tag{6.32}$$

is satisfied, the beat wave field amplitude is obtained:

$$C(r,z)=\frac{\sqrt{2}|a_m|}{|J_0(\mu'_m)|}J_1\left(\frac{\mu'_m r}{a}\right)e^{-j(k_m z+\varphi_m)}+\frac{\sqrt{2}|a_n|}{|J_0(\mu'_n)|}J_1\left(\frac{\mu'_n r}{a}\right)e^{-j(k_n z+\varphi_n)} \quad (6.33)$$

where, for the TE_{0m} mode, μ_m' is the m-th root of J_0' (μ_m'r/a) = 0, and for the TE_{0n} mode, μ_n' is the n-th root of J_0' (μ_n'r/a) = 0. Taking the square of the absolute value $|C(r, z)|$ of Eq. (6.33), according to Eq. (6.29), we can obtain the beat wave wavenumber $K_{mn} = (k_m\text{-}k_n)/2$ for modes m and n, from which the beat wave wavelength $\lambda_{mn} = 2\pi/K_{mn}$ is obtained.

$$\lambda_{mn}\equiv\frac{2\pi}{k_{mn}}=\frac{2\lambda}{\left|\sqrt{1-\frac{\lambda^2\mu_m'^2}{4\pi^2a^2}}-\sqrt{1-\frac{\lambda^2\mu_n'^2}{4\pi^2a^2}}\right|}=\frac{16\pi^2a^2}{\lambda\left|\mu_m'^2-\mu_n'^2\right|}\quad\text{when }\frac{\lambda^2\mu_{m,n}'^2}{4\pi^2a^2}\ll 1 \quad (6.34)$$

where, λ is the free space-free wavelength.

According to the definition of VTWR, comparing Eq. (6.33) with Eq. (6.27), it is not difficult to obtain the relative power ratio of mode m to mode n from Eq. (6.31).

$$\frac{|a_n|^2}{|a_m|^2}=G_{mn}(r)\left[\frac{\text{VTWR}(r)\mp 1}{\text{VTWR}(r)\pm 1}\right]^2 \quad (6.35)$$

where,

$$G_{mn}(r)=\frac{\left|J_0(\mu'_n)\right|^2\left|J_1(\frac{\mu'_m r}{a})\right|^2}{\left|J_0(\mu'_m)\right|^2\left|J_1(\frac{\mu'_n r}{a})\right|^2};\quad G_{mn}(r)=G_{mn}^{-1}(r) \quad (6.36)$$

In order to determine the symbolic choice in Eq. (6.35), we can measure VTWR at the positions r_1 and r_2 and calculate the G_{mn} values: when $G_{mn}(r_1) > G_{mn}(r_2)$ and VTWR(r_1) > VTWR(r_2), or $G_{mn}(r_1) < G_{mn}(r_2)$ and VTWR(r_1) < VTWR(r_2), take the lower symbol in Eq. (6.35), and conversely, when $G_{mn}(r_1) > G_{mn}(r_2)$ and VTWR(r_1) < VTWR(r_2), or $G_{mn}(r_1) < G_{mn}(r_2)$ and VTWR(r_1), > VTWR(r_2), the upper symbol is used.

In order to determine the power ratio of TE_{01} mode and TE_{02} mode, we first try to remove the influence of TE_{03} mode, for this reason, $|E_\theta|2$ was measured along the z-axis at the two positions of $r = \mu_1$'a/μ_3' and $r = \mu_2$'a/μ_3', respectively, where μ_3' is the third root of J_0'(μ_3'r/a) = 0, and the E_θ value of TE_{03} mode is 0 at these two positions, because according to the above selection of r, J_0(μ_3'r/a) = J_0(μ_1') = J_0(μ_2') = 0, therefore, we would only measure the beat wave field of the TE_{01} mode and the TE_{02} mode. The measurement results are shown in Fig. 6.24 and can be obtained using the VTWR definition (6.30) from Fig. 6.24:

$$\mathrm{VTWR}\left(\frac{\mu'_1 a}{\mu'_3}\right) = 3.9$$

$$\mathrm{VTWR}\left(\frac{\mu'_2 a}{\mu'_3}\right) = 6.9$$

and the calculation yields:

$$G_{12}\left(\frac{\mu'_1 a}{\mu'_3}\right) = 0.797$$

$$G_{12}\left(\frac{\mu'_2 a}{\mu'_3}\right) = 1.258$$

In the above symbol of μ' and G, the subscript 1 represents the TE_{01} mode, 2 represents the TE_{02} mode, and 3 represents the TE_{03} mode.

Figure 6.25 shows that the injection mode TE_{02} is partially converted into other circular electric modes, and if we draw the relationship between $|E_\theta|2$ and z in two different radial positions, as pointed out above, when $r = \mu_1'a/\mu_3'$, $r = \mu_2'a/\mu_3'$, the TE_{03} mode will be zero, and the result is shown in Fig. 6.26, which shows the beat wave of TE_{01} mode and TE_{02} mode, and the beat wave wavelength is about 40.6 cm, close to the result obtained by Eq. (6.34). Substituting the VTWR and G values into Eq. (6.35) gives the ratio of the squares of the amplitudes of the beat wave of the TE_{02} and TE_{01} modes, i.e., their power ratios. In exactly the same way, a measurement on $r = \mu_1'a/\mu_2'$ gives the beat wave patterns of the TE_{01} and TE_{03} modes, as well as their power ratios. After normalizing the amplitude utilizing Eq. (6.32), the mode composition of the output power of VGA-800 gyrotron can be obtained as follows: TE_{01} mode (27 ± 5)%, TE_{02} mode (63 ± 5)%, TE_{03} mode (10 ± 5)%, all other modes (0 ± 5)%.

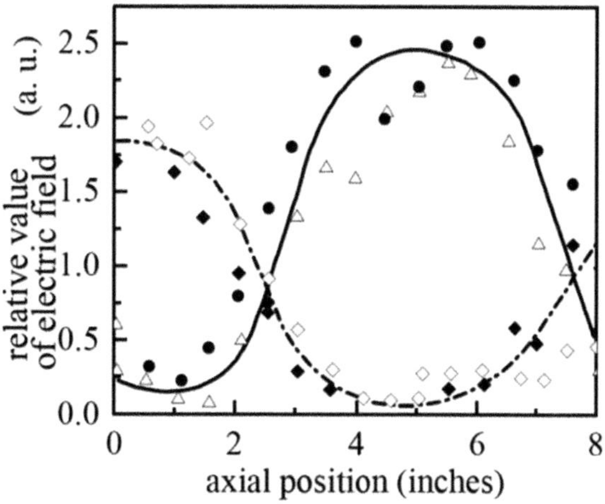

Fig. 6.26 Beat wave distribution of TE_{01} and TE_{02} modes in the output mode of the VGA-8000 gyrotron ●: $r = +\mu'_1 a/\mu'_3$; △: $r = -\mu'_1 a/\mu'_3$;◇: $r = +\mu'_2 a/\mu'_3$; ◆: $r = -\mu'_2 a/\mu'_3$

Chapter 7
Mode Spectrum Measurement of High-Power Microwave Mode

7.1 Introduction

7.1.1 Differences Between Mode Spectrum and Mode Field Measurement Methods

1. **Shortcomings of mode field measurement method**

All types of mode field measurement methods described above first obtain the field structure diagram of the mode field through different methods, and then perform mode recognition and analysis. Among them, the image display method, the direct discrimination method in the radiation field measurement, and the thermal imaging analysis method are relatively simple, convenient, and intuitive, so they have been widely used. Other methods have some common shortcomings.

(1) **Accuracy is limited**

Since they all analyze the mode through the measurement of the radiation field or waveguide field, they will inevitably be limited by the spatial resolution of the receiving antenna or thermal detector used for measurement. And there is the limitation of difficult recognition of pattern image overlap when multiple modes are present. This makes it difficult to obtain accurate field distribution images, and the accuracy of mode recognition and analysis is seriously affected, thereby limiting the number of modes they can identify and their application range. The image display method, direct discrimination method, and thermal imaging analysis method also have serious limitations on the number of identified modes.

(2) **Difficult to recombine modes when multiple modes coexist**

It is precisely because the spatial resolution of the mode field measurement method is not high, and what is measured is a combined field, that the measured mode field is not fine enough, and the structural details of the mode field distribution cannot be

W. Wang et al., *High-Power Microwave Measurement*,
https://doi.org/10.1007/978-981-96-7763-4_7

obtained. Therefore, it is difficult to obtain a reasonable mode composition through mode recombination in the case of multiple modes coexisting, and the recombination process is also very cumbersome and time-consuming.

(3) **Analysis process is relatively cumbersome**

The analysis of the mode in the mode field measurement method, whether it is the radiation field measurement method or the waveguide field measurement method, is very cumbersome in the case of multiple modes, and the actual application is very inconvenient, so it is rarely used.

(4) **Cannot be measured online**

All mode field measurement methods also have a common shortcoming, that is, in order to obtain the mode image or mode field distribution, it is not possible for the microwave system to perform measurements with the workload. Instead, the microwave source must be disconnected from the workload and turned on specifically for the mode measurement. That is to say, these mode field measurement methods cannot achieve online measurement. Moreover, since the measurement process usually takes much longer than the duration of high-power microwave pulses, this method can only be applied to microwave sources with stable continuous output, including continuous pulse output (such as gyrotron), and of course it is impossible to achieve real-time measurement and dynamic measurement. Although the thermal imaging analysis method has a fast response speed and can achieve real-time measurement, it can only achieve real-time imaging of a single pulse of high-power microwaves. Since it is necessary to wait for the image formed on the imaging plane to disappear before it can be imaged again, this time is usually no less than several seconds. Therefore, strictly speaking, it cannot achieve real-time measurement and dynamic measurement in the true sense.

2. **Characteristics of mode spectrum measurement method**

The so-called mode spectral measurement is to distinguish the multiple microwave modes transmitted in the high-power microwave transmission system in the radiation space by some method, that is, the radiation fields of different modes will be distributed in different positions in the space, or coupling the fields of different modes into different sub-transmission systems, so that the mode spectrum is formed at different spatial positions or different sub-systems, so that we only need to perform microwave output measurement at different spatial positions corresponding to different modes or in different sub-systems corresponding to different modes to directly determine the existence and relative size of each mode. Its greatest feature is that it does not need to measure the field structure of the mode field again, which greatly improves the resolution and accuracy of mode recognition.

(1) **High accuracy and fast speed of mode recognition**

Since the mode spectrum measurement method is to separate the original coexisting modes in the transmission system then measure them, the modes can be immediately identified according to the different spatial positions. As long as the different modes are sufficiently separated in space, the measurement accuracy can be greatly improved. Additionally, the mode spectrum measurement method can also couple the different modes to different subsystems, as long as a subsystem has enough isolation between the mode that should be coupled and the mode that should not be coupled, the presence of modes can be determined based on which subsystems have outputs, thereby obtaining accurate mode recognition.

(2) **Can be measured online**

Most mode spectrum measurement methods (except for dispersion reflection antenna measurement methods) only radiate or couple out a small part of the power from the microwave system for measurement, which will not affect the normal operation of the system itself, so it can be measured online, that is, the microwave system can perform mode measurement under the condition of connecting the working load and working normally. At the same time, the mode selection coupling measurement in the mode spectrum measurement method can also achieve dynamic measurement and real-time measurement.

(3) **Relative power measurement of modes can be performed simultaneously**

Since the mode spectral measurement method is measured separately after separating each mode, if the radiation power or coupling power of each mode can be measured and their power in the original transmission waveguide can be obtained after conversion, then their sum is the total power transmitted in the system. At the same time, the power ratio between modes or the proportion of each mode's power in the total transmission power can also be determined.

(4) **The manufacture of measurement equipment is relatively complicated, but the measurement process is relatively simple and fast**

Mode spectrum measurement often involves some special measurement equipment, such as radiation equipment that separates modes in space or special coupling devices that couple different modes into different sub-transmission systems. These types of measurement devices generally have a large size, high machining accuracy requirements, complex design and processing, and relatively high one-time costs, which are their shortcomings. However, after the equipment is manufactured, the measurement only needs to be performed on each single mode, so the mode measurement is much simpler, which greatly simplifies the process of mode recognition and analysis.

7.1.2 Mode Spectrum Measurement Method

Using special antennas such as dispersive reflection antennas, pinhole array antennas, or special coupling methods like mode selective coupling, the different modes in a high-power microwave system are separated to form a mode spectrum. Each mode is then individually measured to achieve the goal of mode recognition and analysis.

This method is known as the mode spectrum measurement method. Currently, there are mainly several specific measurement methods as follows:

1. **Dispersion reflection antenna method**

When microwaves are radiated from a circular waveguide with a stepped-incision or oblique-incision terminal to a dispersion reflection surface, the direction of the beam reflected by the reflection surface will be different due to different modes. Each mode has a unique reflection angle, so different modes will be distinguished in the reflection space. If an infrared receiver board is placed at an appropriate position in the reflection space, a thermal image of each mode will be formed on the receiver board, which can be easily used to identify and analyze the existence and power ratio of each mode by inputting the thermal image into a computer.

2. **Small hole array antenna method**

A series of small holes are opened on the waveguide wall of the circular waveguide high-power microwave transmission line according to a certain rule to form a radiation antenna. The radiation direction of the small hole array antenna depends on the type of transmission mode, and different modes have different radiation direction angles, so the modes are distinguished in different spatial directions, and mode measurement can be performed at different spatial positions. The modes identified by this method will be limited only by the angular resolution of the receiving horn, but the small hole array antenna is relatively large, and the design and processing requirements are relatively high.

3. **Cross-coupling method**

Since the TE_{mn} mode or TM_{mn} mode has symmetry or dissymmetry in the angular direction of the circular waveguide, when coupling holes or coupling probes are set at positions 90° degrees apart in the angular direction of high-power microwave circular waveguide transmission lines, then, the coupled signals from the coupling holes or coupling probes spaced 180° of two apart will either add up or cancel out depending on their phase, i.e. the same phase or opposite phase, so that the symmetric mode and the antisymmetric mode are distinguished.

4. **Mode selective coupling method**

By using the characteristic that the coupling coefficient of the directional coupler varies with the mode, a selective coupler that couples only a specific mode and does not couple other modes can be designed, which is called a mode-selective coupler. As long as a coupler coupling only that mode is manufactured for each possible mode that may exist in the microwave transmission system, the existence and relative power of the mode in the transmission system can be determined based on the presence or absence of output and output magnitude of each mode. This achieves the purpose of mode recognition and analysis.

7.2 Measurement Method Using Dispersive Reflector Antenna [86]

7.2.1 Measurement Principles of Dispersive Reflector Antenna

1. **Measurement device**

If we assume that high-power microwaves are transmitted along a circular waveguide with a radius of a in the z direction. Take the coordinate system shown in Fig. 7.1. A curved reflection surface parallel to the z-axis, perpendicular to the x-axis, and centered at $x = -D$, and $y = 0$ is placed. The mode wave beam radiated from the circular waveguide terminal stepped-incision or oblique incision is projected onto the reflection surface, and the beam reflected from the reflection surface is projected onto the microwave absorption plate of the infrared imaging, due to different reflection directions of different modes from the reflection surface. Therefore, according to the spatial distribution of the temperature rise area recorded by the infrared camera on the absorption plate, the mode image can be quickly analyzed. The structural schematic diagram of the measurement system is shown in Fig. 7.1.

Figure 7.1 also shown the case where the wave beam is projected from the circular waveguide incision to the reflection surface and is reflected. R is the reflection surface, L is the waveguide half-incision antenna, A is the microwave absorption plate, C is the infrared camera for recording the temperature distribution on the absorption plate, and P is the receiving antenna, which is used to directly measure the power distribution of each mode when needed. And the different reflection directions of different modes form a mode spectrum which is recorded.

Since the reflection angle of different modes is different, the difference in reflection angle uniquely determines each mode. The spatial distribution of the TE_{mn} mode on

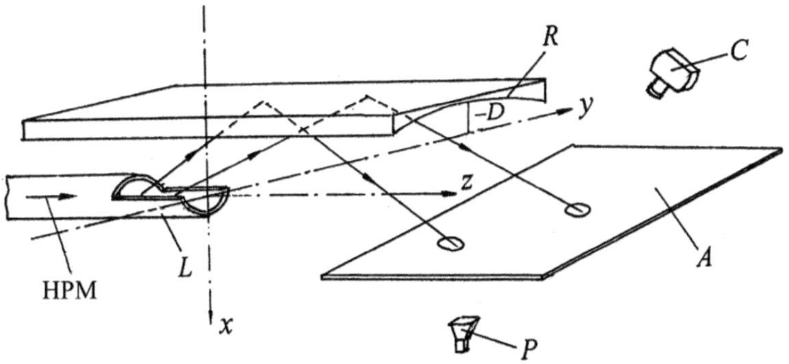

Fig. 7.1 Structural diagram of the measurement device of the dispersion reflection antenna measurement method HPM—high power microwave, R—dispersion reflection surface, L—waveguide half-incision, A—absorption plate, C—infrared camera, P—receiving horn

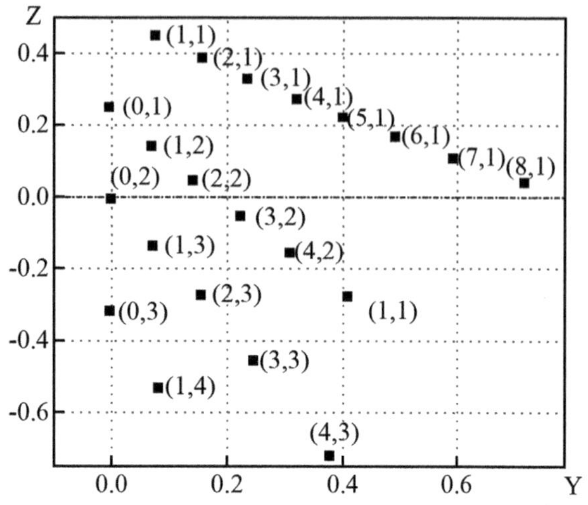

Fig. 7.2 The distribution of the center of the reflected wave beam of each TE_{mn} mode on the microwave absorption plate

the microwave absorption plate is shown in Fig. 7.2, where the numbers in parentheses represent the characteristic values *m* and *n* of the mode, and the normal direction of the absorption plate is consistent with the reflection wave beam direction of the TE_{02} mode, and the position of the TE_{02} mode is basically in the center of the absorption plate. It is precisely because the reflection surface can separate different modes in space that this reflection surface is called a dispersion reflection antenna.

3. **Reflection surface**

Obviously, the key component of the dispersion reflection antenna measurement device is the dispersion reflection surface. This surface distributes the projected wave beam into a mode spectrum in space based on the mode type. And the reflection angle of the wave beam is uniquely determined by the mode. The dispersion reflection surface satisfies the following relationship, that is,

$$\frac{x}{D} = \ln\left\{\frac{1 + \left[1 - (y/D)^2\right]^{1/2}}{2}\right\} - \left[1 - (y/D)^2\right]^{1/2} \tag{7.1}$$

where, *D* is the distance from the reflection surface to the center of the circular waveguide, and the *x* and *y* coordinates are oriented as shown in Fig. 7.1.

The power density distribution, calculated theoretically, of different TE_{mn} modes on the surface of the absorption plate after reflection from such a dispersion reflective surface is shown in Fig. 7.3. The figure shows the calculated image of the microwave absorption plate. The plate is 40 cm away from the radiation port in the *x* direction, 45 cm wide in the *z* direction, and 30 cm high in the *y* direction. The position of the TE_{02} mode radiation beam is used as the coordinate origin, so the spatial positions of each mode in the figure are drawn relative to the position of the TE_{02} mode, and this distribution is arranged neatly as expected in Fig. 7.2.

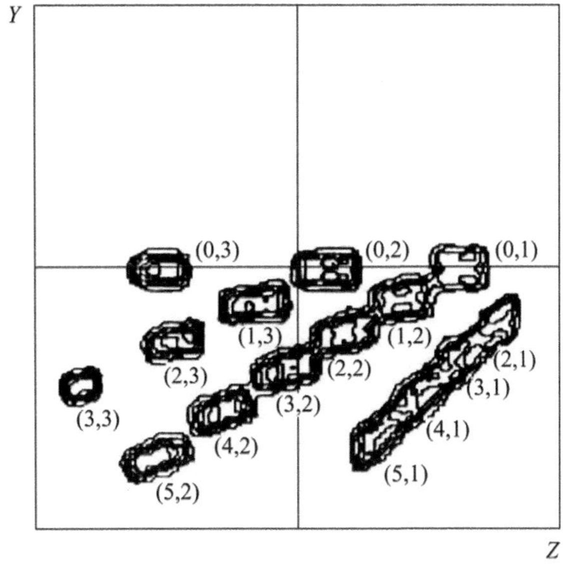

Fig. 7.3 The power distribution of the TE_{mn} mode spectrum calculated on the absorption plate at a distance of $x = 40$ cm from the radiation port (From Ref. [86])

7.2.2 *Experimental Results*

1. **Experimental method**

When using the dispersion reflection antenna measurement method for actual testing, the device shown in Fig. 7.1 is used. The high-power microwaves generated by the microwave source are output from a circular waveguide with a diameter of 37 mm. This circular waveguide is connected to a rising cosine transition waveguide, and the diameter of the waveguide is then reduced to 32 mm. The beam reflected by the reflection antenna is projected onto a microwave absorption plate (Eccosorb AN72, Emerson & Cuming), and the distance between the absorption plate and the center of the waveguide terminal opening in the normal direction can be adjusted. This direction is close to the expected theoretical transmission direction of the TE_{02} mode beam.

The temperature distribution formed on the absorption plate due to absorbing microwaves, i.e., the power density distribution of each mode, is measured by an infrared camera (Japan Avionics TVS-2200). The rise in temperature has a good linear relationship with the absorbed power over a relatively wide power range. In most cases, the relative position of the temperature peaks is sufficient to distinguish the various mode clearly, as shown in Figs. 7.2 and 7.3. In order to evaluate the power components of each mode contained, geometric corrections are performed by a computer to obtain the true two-dimensional temperature distribution.

The small-power experiment was carried out using a 41 GHz klystron. The output mode of the klystron was first transformed into the TE_{01} mode by a mode converter. Then, part or all of the power was further converted into the TE_{02} mode by a six-stage

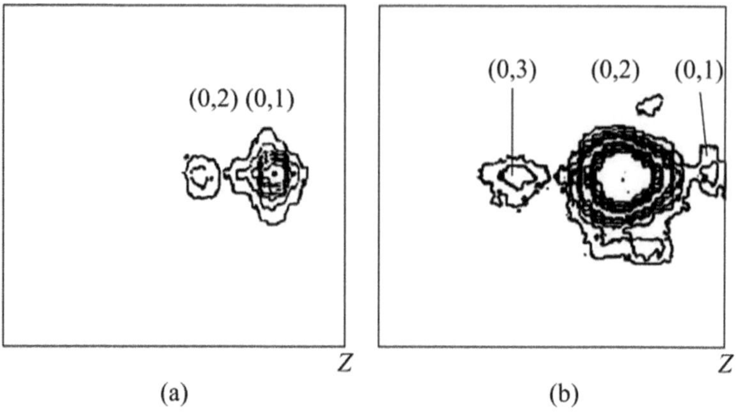

Fig. 7.4 Results of small-power experiment of dispersion reflection antenna measurement method, **a** Infrared image of the output mode spectrum of the klystron, **b** infrared image of the output mode spectrum of the TE_{01}–TE_{02} mode transformer. (From Ref. [86])

stepped converter. The high-power experiment was performed by a Toshiba E3962 gyrotron operating at 41 GHz. It directly outputs the TE_{02} main mode, or converts it into the TE_{01} mode again through a mode converter.

2. **Results**

The radiation image obtained from the low-power experiment of the stepped converter, which used the reflection surface given by Eq. (7.1), is shown in Fig. 7.4. In the experiment, the absorption plate was 20 cm away from the radiation port, and the size of the plate was $30 \times 30\ cm^2$.

The infrared image of the TE_{01} mode output by the klystron from the waveguide terminal radiation is shown in Fig. 7.4a. It consists of a main peak of the TE_{01} mode and a small peak of the TE_{02} mode next to it. The relative power density distribution of each mode on the image can be obtained by integrating over the entire area of each mode in the image, and the results show that the power component of the TE_{02} mode is estimated to account for about 14% of the total power. Figure 7.4b shows the output image of the TE_{02} mode converted from the TE01 mode by a six-stepped converter. From the figure, it can be observed that the power distribution at the position of the TE_{01} mode has obviously decreased. In the experiment, the power density of the TE_{02} mode has been found to be saturated, and some peak values are distributed at the position of the TE_{03} mode. Integrating the distribution area of each mode shows that the output of the mode converter contains 99% of the TE_{02} mode, 0.3% of the TE_{03} mode, and less than 0.5% of the TE_{01} mode component.

7.3 Mode Selective Coupling Method

The method of mode recognition and analysis implemented based on the mode selective coupling function of the mode selective coupler is called the mode selective coupling method. The mode selective coupler only couples a specified single mode or a certain class of modes, and other modes are not coupled and will not have output. Since the mode selective coupler only needs to couple out a small part of the power from the microwave system through small holes or gaps for measurement, it does not affect the normal transmission of microwave energy in the microwave system, and the system can work normally. Moreover, since the measurement is performed on the output of the coupler of each mode, there is no need for any measurement device that needs mechanical movement. Therefore, this measurement method is not only online, but can also be real-time and dynamic, especially suitable for measuring modes of high-power single-pulses or low-repetition-pulse, which overcomes the measurement errors caused by the instability of the microwave output during spatial multi-point measurement in the mode field measurement method, and overcomes the limitation of the spatial resolution of the measurement device. The working band of the mode selective coupling method is relatively narrow, which is obviously the inevitable result of the high correlation between the mode selectivity of the mode selective coupler and the frequency.

7.3.1 Cross-Coupling Method [87]

1. **Principle of cross-coupling method**

The measurement system of the cross-coupling method is shown in Fig. 7.5. On the waveguide wall of the same cross-section of the overmoded circular waveguide microwave transmission system, four completely identical coupling holes are distributed at 90° to each other, and the cross-section ports of the four fundamental mode rectangular waveguides are coupled with each hole, and their wide side (x direction) is consistent with the longitudinal direction (z direction) of the circular waveguide. The coupling signal of two pairs of coupling holes spaced 180° in the circular main waveguide enters the rectangular waveguide for transmission and is combined for output. In this way, coupling occurs between the H_z component of the TE_{mn} mode in the circular waveguide and the H_x component of the TE_{10} mode in the rectangular waveguide. For even values of m in the circular waveguide, $m = 0$, 2, 4, …, the fields in the circular waveguide spacing 180° in angular direction have equal magnitudes and are in-phase. The coupled signals will be in-phase and of equal amplitude, resulting in double the output after passing through an equal length of the rectangular waveguide transmission path. Conversely, for the odd-numbered modes with $m = 1, 3, 5, \ldots$, in the circular waveguide, the fields at relative positions are equal in magnitude and opposite in phase, and the signals coupled out will offset each other to zero after being transmitted through the same length of path. The output

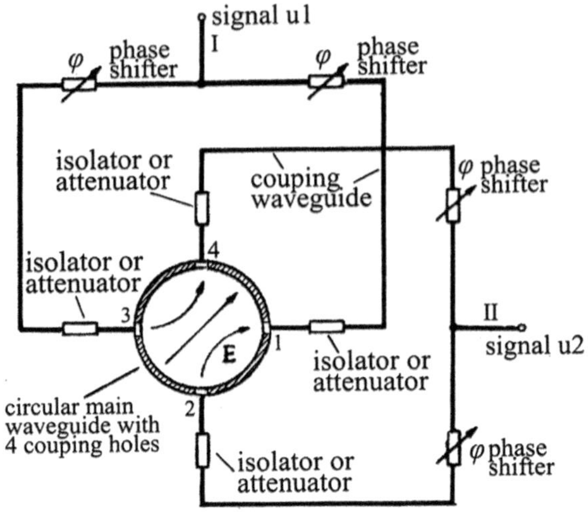

Fig. 7.5 The schematic diagram of the four-hole cross-coupling mode recognition system

results can also be reversed by adjusting the phase shifter on the output waveguide system, that is, the symmetric modes with even-numbered m output zero, and the asymmetric odd-numbered modes output are enhanced by a factor of two. The TM mode in the circular main waveguide has no coupling with the rectangular waveguide because it has no H_z component.

If we take any one radius of the circular waveguide as the reference line of $\varphi = 0$ of the circular waveguide coordinate system, and assume that the phase of the field at the position of the coupling hole 1 or 3 for a symmetric mode (or an asymmetric mode if the output situation is reversed) is φ, then the composite signal $u_{\rm I}$ output from the port I that they couple out will be proportional to $E\sin\varphi$, and the output power P will be proportional to $E^2\sin^2\varphi$, where E is the amplitude of the composite field at port I, and φ is the initial angle of the field. Similarly, the signal coupled out by the coupling holes 2 and 4 relative to the signal coupled out by the coupling holes 1 and 3 changes the phase by 90°, so the signal $u_{\rm II}$ output from the composite output port II will be proportional to $E\cos\varphi$, and the output power P will be proportional to $E^2\cos^2\varphi$. Similarly, E is the amplitude of the composite field at port II.

$$\begin{aligned} P_{\rm I} &\propto E^2 \sin^2 \phi \\ P_{\rm II} &\propto E^2 \cos^2 \phi \end{aligned} \tag{7.2}$$

When the detector is in the square law range, the total output power P is the sum of the two, $P = P_{\rm I} + P_{\rm II}$, which is proportional to E^2. Its phase angle is

$$\phi = \pm \arctan\sqrt{P_I / P_{II}} \tag{7.3}$$

when $\varphi = 0, \pi$, $P_{\rm I} = 0$ and $P_{\rm II} = 1$, when $\varphi = \pi/2, 3\pi/2$, $P_{\rm I} = 1$ and $P_{\rm II} = 0$.

2. **Application of cross-coupling method**

Although the cross-coupling method can distinguish between symmetric and asymmetric modes in the main system, it cannot identify specific modes individually. However, when the output of the microwave source's main mode is already clear, this method can be used to check whether other modes exist and their relative proportions. Since the working mode of the gyrotron often takes the TE_{0n} mode as the main mode, this method can be used to detect whether the TE_{1n}, TE_{3n} and other modes are present in the output of the gyrotron.

7.3.2 Theoretical Foundation of Mode Selective Coupler with Overmoded Waveguide

The basis of the mode selective coupling method is the mode selective coupling of the directional coupler. We know that the coupling coefficient of any directional coupler is different for different modes. If we can consciously enhance the coupling of a certain mode of the overmoded waveguide directional coupler, and there is no coupling or very weak coupling for all other modes, we can make a directional coupler with mode selectivity, called the mode selective coupler [88, 89].

The design of mode selectivity of the mode selective coupler is based on the small hole diffraction theory and the principle of phase superposition. We have discussed this in detail in Sect. 4.4.2, where we derived the expressions (4.80) and (4.82) for the relative total amplitude of the waves coupled to the secondary waveguide during multi-hole coupling.

1. **Coupling strength of small holes in overmoded waveguide**

We discuss the coupling situation of the main waveguide as the overmoded waveguide and the TE_{10} mode rectangular waveguide as the secondary waveguide.

(1) **Normalization of the field during coupling of the overmoded waveguide**

According to the defining expressions (4.64) and (4.65) of the main parameter coupling coefficient of the directional coupler, the coupling coefficient should be calculated based on the power size in the main and secondary waveguides. If the main and secondary waveguides are the same, the mode is also the same, and the coupling holes are located at the identical position on the identical waveguide wall of both waveguides (such as the holes are all on the centerline of the wide side of the main and secondary waveguides), then the mode field components at the position of the coupling hole in the main and secondary waveguides will also be the same. The power ratio of them can be directly replaced by the field amplitude ratio, and Eq. (4.65) can be simplified to

$$C = 10\lg\frac{(A_1)^2}{\left(A_2^+\right)^2} = -20\lg\left|a^+\right| \tag{7.4}$$

where, A_1 is the amplitude of the input signal of the directional coupler, A_2^+ is the amplitude of the forward signal excited in the secondary waveguide, the two modes are the same, the waveguides are the same, and a^+ is the coupling strength of the forward propagating signal. This calculation formula is the most commonly used design basis for directional couplers.

But if the transmission modes of the main and secondary waveguides are different, for example, when the high-order mode in the overmoded waveguide is coupled with the TE_{10} mode in the fundamental mode waveguide, their field expressions are different, and their amplitude coefficients are not comparable. Therefore, Eq. (7.4) is no longer applicable to the overmoded coupling system. For this reason, it is necessary to directly proceed from the definition formula of coupling coefficient, that is, from the power ratio to calculate the coupling coefficient. At this time, the normalization of the field in the Eq. (4.70) of the coupling strength should be power-normalized, so that the mode waves in the main and secondary waveguides have comparability.

Let

$$\left.\begin{aligned} \boldsymbol{E} &= C\boldsymbol{e} \\ \boldsymbol{H} &= C\boldsymbol{h} \end{aligned}\right\} \tag{7.5}$$

where C is the amplitude coefficient of the field; $\boldsymbol{e}$ and $\boldsymbol{h}$ are the other parts in the field expression except the amplitude coefficient, which are called the mode function or eigenfunction.

For the coupling of the same mode, since C is an arbitrary constant, its specific size can be different, but the amplitude coefficient C of the mode field in the main and sub-waveguides is the same. Therefore, the coupling degree calculated by power can be directly replaced by amplitude calculation. However, when two different modes are coupled, C can no longer be arbitrarily given respectively, otherwise the size of the coupling amount cannot be determined. At this time, we should find the relationship between the amplitude coefficients of the two modes with equal power transmission, so that the amplitude can be used instead of power to calculate the coupling degree. For this reason, the most direct method is to use the normalization condition

$$\int_s (\boldsymbol{E} \times \boldsymbol{H}^*)\cdot \mathrm{d}\boldsymbol{S} = C^2 \int_s (\boldsymbol{e} \times \boldsymbol{h}^*) \cdot \mathrm{d}\boldsymbol{S} = C^2 p_s = 1 \tag{7.6}$$

where, $\boldsymbol{h}^*$ is the conjugate value of $\boldsymbol{h}$, p_s is called the normalization constant.

$$p_s = \int_s (\boldsymbol{e} \times \boldsymbol{h}^*) \cdot \mathrm{d}\boldsymbol{S} \tag{7.7}$$

Then Eq. (7.5) can be rewritten as

$$\left.\begin{aligned} \boldsymbol{E} &= \frac{1}{\sqrt{p_s}}\boldsymbol{e} \\ \boldsymbol{H} &= \frac{1}{\sqrt{p_s}}\boldsymbol{h} \end{aligned}\right\} \tag{7.8}$$

After such normalization processing of the fields in the main and secondary waveguides, $\boldsymbol{E}$ and $\boldsymbol{H}$ become the normalized field components for calculating the coupling strength in the over-moded waveguide directional coupler design. Or, they will be the field components obtained under the condition of equal power transmission (power normalization condition), and the coupling of the two will be comparable. Here we give the coupling strength calculation results of several coupling methods commonly used in high-power microwave systems.

(2) **Coupling between circular overmoded waveguide and fundamental mode rectangular waveguide wide side**

As shown in Fig. 7.6, the coupling hole is located at the center of the wide side of the rectangular waveguide, and the center point of the wide side coincides with the diameter of the circular waveguide.

(a) Coupling between the TM_{mn} mode in the circular waveguide and the TE_{10} mode in the rectangular waveguide.

In this case, at the position of the coupling hole, the TE_{10} mode in the rectangular waveguide has both a magnetic field tangential component H_x and an electric field normal component E_y, but on the centerline of the wide side, the other tangential

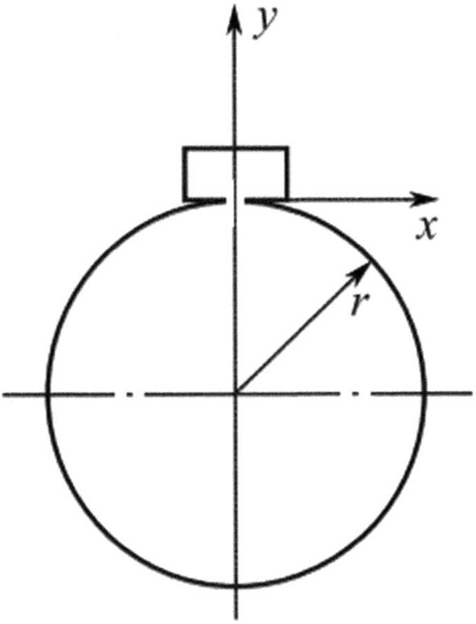

Fig. 7.6 Schematic diagram of the coupling between circular overmoded waveguide and fundamental mode rectangular waveguide wide side

component $H_z = 0$; while in the circular waveguide, the TM_{mn} mode has a magnetic field tangential component H_φ and an electric field normal component E_r at the position of the small hole, which are respectively consistent with the directions of the H_x and E_y components in the rectangular waveguide. Therefore, the main and secondary waveguides can be coupled through H_φ and H_x, E_r and E_y.

Let's take the example of the TM_{01} mode transmitted by the overmoded circular waveguide to derive the expression of the single-hole coupling strength a according to the above theory.

From Eq. (7.6), we obtain the normalization constant of the TE_{10} mode in the rectangular waveguide as

$$\sqrt{p_{s10}} = \frac{a\sqrt{ab\omega\mu\beta_{10}}}{\pi\sqrt{2}} \tag{7.9}$$

where a and b are the dimensions of wide side and narrow side of the rectangular waveguide, respectively, and β_{10} is the phase constant of the TE_{10} mode. From this, the normalized field component that can produce coupling of the TE_{10} mode is obtained as

$$\begin{aligned} E_y^{\mp} &= -\mathrm{j}\frac{\sqrt{2\omega\mu}}{\sqrt{ab\beta_{10}}}\sin\left(\frac{\pi}{a}x\right) \\ H_x^{\mp} &= \mp\mathrm{j}\frac{\sqrt{2\beta_{10}}}{\sqrt{ab\omega\mu}}\sin\left(\frac{\pi}{a}x\right) \end{aligned} \tag{7.10}$$

Similarly, the normalization constant of the TM_{01} mode in the circular waveguide can be obtained as

$$\sqrt{p_{s01}} = -\frac{R^2\sqrt{\pi\omega\varepsilon\beta_{01}}}{2.405\sqrt{2}}J_1(2.405) \tag{7.11}$$

where β_{01} is the phase constant of the TM_{01} mode, and R is the waveguide radius. From this, the normalized field component that can produce coupling of the TM_{01} mode is obtained as

$$\begin{aligned} E_r &= -\mathrm{j}\frac{\sqrt{\beta_{01}}}{R\sqrt{\pi\omega\varepsilon}}\frac{1}{J_1(2.045)}J_1\left(\frac{2.045}{R}r\right) \\ H_\phi &= -\mathrm{j}\frac{\sqrt{\omega\varepsilon}}{R\sqrt{\pi\beta_{01}}}J_1\left(\frac{2.045}{R}r\right) \end{aligned} \tag{7.12}$$

By substituting the field component Eqs. (7.10) and (7.12) into the coupling strength Eq. (4.70), we obtain the single hole coupling strength of the TM_{01} mode of the overmoded circular waveguide and the TE_{10} mode of the rectangular waveguide when coupled, which is

$$a^{\pm} = \frac{\sqrt{2}}{3}\frac{r_0^3}{R\sqrt{\pi ab}}\frac{\omega}{c}\left[\sqrt{\frac{\beta_{01}}{\beta_{10}}} \mp 2\sqrt{\frac{\beta_{10}}{\beta_{01}}}\right] \tag{7.13}$$

where r_0 is the coupling hole radius. In general, similar derivation methods can be used to obtain the coupling strength of the TM_{mn} mode of the overmoded circular waveguide and the TE_{10} mode of the rectangular waveguide when coupled, which is

$$a^{\pm} = \frac{\sqrt{2}}{3}\sqrt{N_m}\frac{r_0^3}{R\sqrt{\pi ab}}\frac{\omega}{c}\left[\sqrt{\frac{\beta_{mn}}{\beta_{10}}} \mp 2\sqrt{\frac{\beta_{10}}{\beta_{mn}}}\right] \tag{7.14}$$

where c is the speed of light, β_{mn} is the phase constant of the TM_{mn} mode in the circular waveguide, and N_{mn} is the Neumann coefficient, that is,

$$N_m = \begin{cases} 1 & \mathrm{m} = 0 \\ 2 & \mathrm{m} \neq 0 \end{cases} \tag{7.15}$$

(b) Coupling between TE_{mn} mode of circular waveguide and TE_{10} mode of rectangular waveguide.

In this case, the field at the position of the small hole is similar to when coupling with the TM_{mn} mode above, that is, the field of the TE_{10} mode in rectangular waveguide has both a tangential magnetic field component H_x and a normal electric field component E_y, while the field components of the TE_{mn} mode in circular waveguide not only have tangential magnetic field component H_φ, but also normal electric field component E_r. Therefore, coupling can occur between E_y and E_r, as well as between H_x and H_φ, but not between H_z, because the TE_{10} mode in rectangular waveguide has $H_z = 0$ on the center line of the wide side. Through a similar derivation to the circular waveguide TM_{mn} mode, the coupling strength is obtained.

$$a^{\pm} = \frac{\sqrt{2}}{3}\sqrt{N_m}\frac{r_0^3}{R\sqrt{\pi ab}}\sqrt{\frac{\beta_{10}}{\beta_{mn}}}\frac{m}{\left[\left(\mu'_{mn}\right)^2 - m^2\right]^{1/2}}\left[\left(\frac{\omega}{c}\right)^2\frac{1}{\beta_{10}} \mp 2\beta_{mn}\right] \tag{7.16}$$

where β_{mn} is the phase constant of the TE_{mn} mode in circular waveguides, μ'_{mn} is the n-th root of the derivative of the m order Bessel function of the first kind, and the remaining symbols are the same as in Eq. (7.14).

(3) **Coupling between circular overmoded waveguide and the narrow side of the fundamental mode rectangular waveguide**

The form and coordinates of this coupling are shown in Fig. 7.7. The coupling hole is located at the center of the narrow side of the rectangular waveguide, and its axis coincides with the diameter of the circular waveguide.

(a) Coupling of the TM_{mn} mode in circular waveguide and the TE_{10} mode in rectangular waveguide.

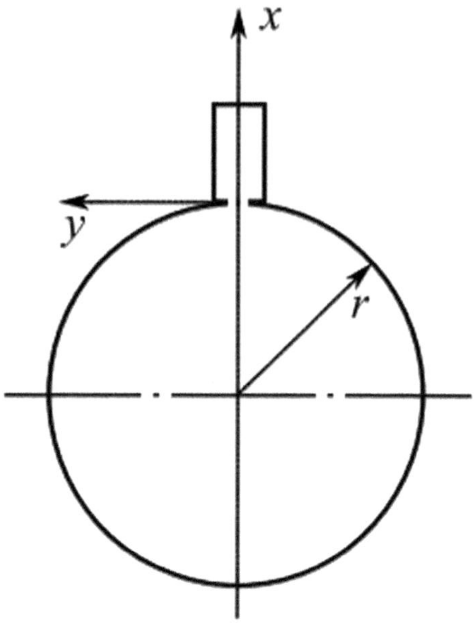

Fig. 7.7 Schematic diagram of coupling between circular overmoded waveguides and fundamental mode rectangular waveguides narrow side

For the TE_{10} mode of the rectangular waveguide, at the position of the small hole with $x = 0$, $y = 0$, the electric field normal component and the magnetic field tangential component are respectively

$$\begin{aligned} E_n &= E_x = 0 \\ H_u &= H_y = 0, \quad \mathrm{H_v} = H_z \neq 0 \end{aligned} \tag{7.17}$$

For the TM_{mn} mode of the circular waveguide, at the position of the small hole with $x = 0$, $y = 0$, there is no normal component of the electric field, and the magnetic field has only one tangential component H_z

$$\begin{aligned} E_n &= E_r \neq 0 \\ H_u &= H_\phi \neq 0\,, \mathrm{H_v} = H_z = 0 \end{aligned} \tag{7.18}$$

That is, there is a normal component of the electric field, but the magnetic field has only one tangential component H_φ.

In the above formulas: the subscript n represents normal field u and v represent tangential fields. According to the coupling rule, there is only a tangential H_z component at the position of the small hole in the secondary waveguide, while in the main waveguide, although there are normal electric field components E_r and tangential magnetic field components H_φ, there is no tangential magnetic field component H_z. As a result, the primary and secondary waveguides have neither common normal electric fields nor common tangential magnetic fields, so the primary and secondary waveguides cannot be coupled, that is

$$a^{\pm} = 0 \tag{7.19}$$

So, the TM_{mn} mode in the circular waveguide cannot excite the TE_{10} mode in the rectangular waveguide when the circular waveguide is coupled with the narrow side of the rectangular waveguide.

(b) Coupling of the TE_{mn} mode in circular waveguide and the TE_{10} mode in rectangular waveguide.

At the position of the small hole, the field of the TE_{10} mode in the rectangular waveguide remains unchanged, and there is only a tangential magnetic field component H_z. At this time, in the circular waveguide TE_{mn} mode, $H_z \neq 0$, so only they can be coupled. The other tangential magnetic field component H_φ and the normal electric field component E_r cannot be coupled because both H_y and E_x in the rectangular waveguide are zero.

Through similar derivation, we obtain the coupling strength of a single hole at this time:

$$a^{\pm} = \frac{2}{3}\left[\frac{2\pi}{\beta_{10}\beta_{mn}} \frac{r_0^6}{a^3bR^4} \frac{\left(\mu'_{mn}\right)^4}{\left(\mu'_{mn}\right)^2 - m^2} N_m\right]^{1/2} \tag{7.20}$$

where,

a and b are the dimensions of wide side and narrow side of the rectangular waveguide,,

R is the radius of the circular waveguide;

r_0 is the radius of the coupling hole,

β_{10} is the phase constant of the TE_{10} mode in the rectangular secondary waveguide,

β_{mn} is the phase constant of the TE_{mn} mode in the circular waveguide,

μ'_{mn} is the nth root at which the derivative of the m order Bessel function of the first kind is zero,

Other symbols are the same as in formula (7.14).

(4) **Coupling between the wide side of the rectangular over-moded waveguide and the wide side of the fundamental mode rectangular waveguide with the TE_{10} mode.**

As shown in Fig. 7.8, the wide side of the rectangular overmoded waveguide is coupled with the wide side of the fundamental mode rectangular waveguide, and the coupling hole is located on the centerline of the common wall of the two waveguides, taking the coordinate system shown in the figure.

(a) Coupling of the TM_{mn} mode in the rectangular overmoded waveguide and the TE_{10} mode in the rectangular fundamental waveguide.

In this case, at the position of the small hole, there is an electric field normal component $E_{y'}$ and a magnetic field tangential component $H_{x'}$ in the secondary waveguide

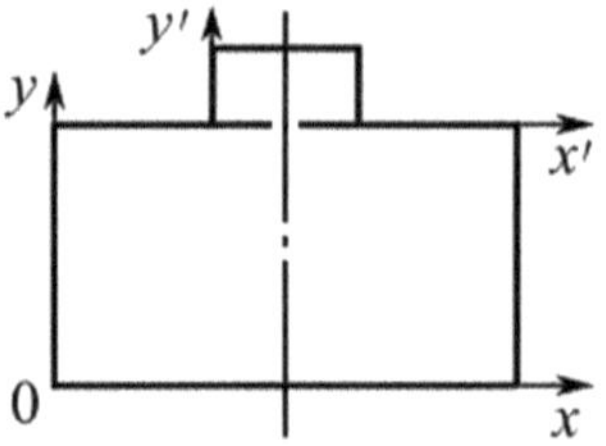

Fig. 7.8 Schematic diagram of coupling between the wide side of the rectangular overmoded waveguide and the wide side of the fundamental mode rectangular waveguide

fundamental mode rectangular waveguide, and another magnetic field tangential component $H_{z'} = 0$. In the main waveguide overmoded rectangular waveguide, the TM_{mn} mode has an electric field normal component E_y, a magnetic field tangential component H_x, and no other tangential magnetic field component H_z. Therefore, the coupling occurs between $E_{y'}$ and E_y and between $H_{x'}$ and H_x. The expression for the coupling strength of a single hole is

$$a^{\pm} = \frac{2\sqrt{2}}{3}\frac{r_0^3}{b\sqrt{aba'b'}}\frac{n\pi}{k_c}\sqrt{\frac{\beta_{10}}{\beta_{mn}}}R_{mn}\left[\frac{\beta_{mn}}{\beta_{10}} \mp 2\right] \tag{7.21}$$

where

a, b, a', and b' are the width and narrow dimensions of the over-moded rectangular waveguide and the fundamental mode rectangular waveguide, respectively,

β_{10} and β_{mn} are the phase constants of the TE_{10} mode in the fundamental mode waveguide and the TM_{mn} mode in the overmoded waveguide, respectively,

k_c is the cutoff wavenumber of the TM_{mn} mode in the overmoded waveguide,

R_m is

$$R_m = \begin{cases} 0 \ (m = 0, 2, 4, \ldots) \\ 1 \ (\mathrm{m} = 1, 3, 5, \ldots) \end{cases} \tag{7.22}$$

From Eq. (7.22), it can be seen that when m is even, since $E_y = H_x = 0$ in the center of the waveguide's wide side, $a^{\pm} = 0$ i.e. there is no coupling. Only when m is odd, the TM_{mn} mode of the rectangular overmoded waveguide can couple with the TE_{10} mode of the rectangular fundamental mode waveguide. To achieve coupling between the TM_{mn} mode and the TE_{10} mode.

When m is even, in order for the coupling to occur, the position of the coupling hole only needs to be deviated from the center of the overmoded waveguide's wide side, that is, away from the position where $E_y = H_x = 0$.

(b) Coupling of rectangular overmoded waveguide TE_{mn} mode and rectangular fundamental mode waveguide TE_{10} mode

In this case, similar to the coupling between the TM_{mn} mode and the TE_{10} mode analyzed above, coupling can only occur between $E_{y'}$ and E_y as well as $H_{x'}$ and H_x. The expression for the coupling strength is

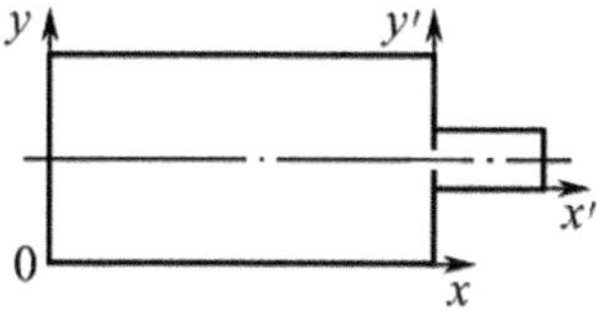

Fig. 7.9 Schematic diagram of coupling between the narrow side of the rectangular over-moded waveguide and the narrow side of the fundamental mode rectangular waveguide

$$a^{\pm} = \frac{\sqrt{2}}{3}\frac{r_0^3}{a\sqrt{aba'b'}}\frac{1}{\sqrt{\beta_{10}\beta_{mn}}}\frac{m\pi}{k_c}R_m\left[(\frac{\omega}{c})^2 \mp 2\beta_{10}\beta_{mn}\right] \tag{7.23}$$

where, the symbols have the same meaning as in Eq. (7.21). It can be seen that the coupling between the TE_{mn} mode and the TE_{10} mode is completely similar to that of the TM_{mn} mode mentioned above. Only when m is odd, can the TE_{mn} mode couple with the TE_{10} mode. When m is even, $E_y = H_x = 0$, although $H_z \neq 0$ in the over-moded waveguide at this time, $H_{z'} = 0$ in the fundamental rectangular waveguide, so coupling cannot exist. Only by making the coupling hole deviate from the center of the overmoded waveguide's wide side, can the TE_{mn} mode and the TE_{10} mode be coupled.

(5) **Coupling between the narrow side of the rectangular overmoded waveguide and the narrow side of the fundamental mode rectangular waveguide.**

The structure of the coupling between the narrow side of the rectangular overmoded waveguide and the narrow side of the rectangular fundamental mode waveguide is shown in Fig. 7.9, and the coupling hole is located at the center of the narrow side of the corresponding waveguide.

(a) Coupling between the TM_{mn} mode of the rectangular overmoded waveguide and the TE_{10} mode of the fundamental mode rectangular waveguide.

At this time, at the position of the coupling hole ($y' = b'/2$, $y = b/2$), the TE_{10} mode has neither the electric field normal component E_x nor the tangential magnetic field component H_y, only the longitudinal tangential magnetic field component H_z. However, in the overmoded waveguide, the TM_{mn} mode has no magnetic field longitudinal component, and $H_z = 0$. Therefore, there is no common direction field component between the two waveguides at the position of the small hole, and coupling cannot occur, so

$$a^{\pm} = 0 \tag{7.24}$$

(b) Coupling between the rectangular overmoded waveguide TE_{mn} mode and the fundamental mode rectangular waveguide TE_{10} mode

In this case, the TE_{10} mode still has no electric field normal component E_x, but at this time, the TE_{mn} mode already has a tangential magnetic field component H_z. Therefore, the two rectangular waveguides can be coupled through H_z, and the coupling strength can be derived as

$$a^{\pm} = \pm\frac{2}{3}\sqrt{2}\frac{\pi r_0^3}{a'\sqrt{aba'b'}}\frac{k_c}{\sqrt{\beta_{10}\beta_{mn}}}R_n \tag{7.25}$$

where

$$R_n = \begin{cases} 0 \ (\mathrm{n} = 1, 3, 5, \ldots) \\ 1 \ (\mathrm{n} = 0, 2, 4, \ldots) \end{cases} \tag{7.26}$$

It can be seen that only when n is even, the H_z component of the TE_{mn} mode exists at the position of the coupling hole at $b/2$, and coupling can occur. When n is odd, $H_z = 0$, and only by letting the small hole position deviate from the $b/2$ position, can coupling occur.

The coupling between the overmoded rectangular waveguide and the fundamental mode rectangular waveguide can also occur when the wide side of the overmoded rectangular waveguide is coupled with the narrow side of the fundamental mode rectangular waveguide, or when the narrow side of the overmoded rectangular waveguide is coupled with the wide side of the fundamental mode rectangular waveguide. We will no longer list them one by one here.

2. **Specific application of phase superposition condition**

In Sect. 4.4.2, we introduced the principle of phase superposition between the coupled waves through small holes and obtained the total coupling strength expressions after superposition when the total number of coupling holes is $N = 2n$ and $N = 2n + 1$ respectively in Eqs. (4.80) and (4.82). Now we will further discuss their applications in special cases.

(1) **Equidistant but unequal strength coupling**

In multi-hole coupling, if the spacing S_k between the coupling holes is identical and equal to S, it is called equidistant coupling. Then we have

$$S_1 = S_2 = \cdots = S_k = \cdots = S_n = S \tag{7.27}$$

In this way, Eqs. (4.81) and (4.83) can be rewritten as

$$d_k = \begin{cases} (2k-1)S \ (\text{when } N = 2n) \\ 2kS \ (\text{when } N = 2n+1) \end{cases} (k = 1, 2, \ldots) \tag{7.28}$$

and

$$\theta_k = \begin{cases} (2k-1)\varphi^{\pm} \text{ (when } N = 2n) \\ 2k\varphi^{\pm} \quad\quad\quad \text{(when } N = 2n+1) \end{cases} \quad (k = 1, 2, \ldots) \tag{7.29}$$

where

$$\varphi_{\pm} = (\beta_1 \mp \beta_2)\frac{S}{2} \tag{7.30}$$

As long as

$$\phi^{\pm} = i^{\pm}\pi \ (i^{\pm} = 1, 2, \ldots) \tag{7.31}$$

the requirement for in-phase superposition can be satisfied. When i is even, it is equivalent to Eq. (4.84), and when i is odd, it is equivalent to Eq. (4.85). At this time, the condition for anti-phase cancellation becomes

$$\phi^{\pm} = \left(i^{\pm} - \frac{1}{2}\right)\pi \quad (i^{\pm} = 1, 2, \ldots) \tag{7.32}$$

In the case where the hole spacing S is the same, but the size of each pair of coupling holes is different (the coupling strength is different), this is the case of equidistant but unequal strength coupling. The coupling strength is determined by Eqs. (4.80) or (4.82) which can be expressed as

$$A^{\pm} = \begin{cases} 2\left|\sum\limits_{k=1}^{n} a_k^{\pm}\cos(2k-1)\varphi^{\pm}\right| & (N = 2n) \\ \left|a_0^{\pm} + 2\sum\limits_{k=1}^{n} a_k^{\pm}\cos 2k\varphi^{\pm}\right| & (N = 2n+1) \end{cases} \tag{7.33}$$

(1) **Equal strength but unequal spacing coupling**

When all the coupling holes have the same size and shape, assuming the coupling is very weak, the single-hole coupling strength remains the same as long as the incident wave amplitude of the main waveguide is unchanged throughout the coupling process. This is the so-called equal-strength multi-hole coupling. This is the so-called equal-strength multi-hole coupling. At this time,

$$a_0^{\pm} = a_1^{\pm} = a_2^{\pm} = \cdots = a_k^{\pm} = \cdots = a_n^{\pm} = a^{\pm} \tag{7.34}$$

Therefore, Eqs. (4.80) or (4.82) become

$$A^{\pm} = \begin{cases} \left|2a^{\pm}\sum\limits_{k=1}^{n}\cos\theta_k^{\pm}\right| & (N = 2n) \\ a^{\pm}\left(1 + 2\left|\sum\limits_{k=1}^{n}\cos\theta_k^{\pm}\right)\right| & (N = 2n+1) \end{cases} \tag{7.35}$$

Since the hole spacing is not equal at this time, that is, the case of equal strength but unequal spacing coupling, the size of θ_k should be given by Eqs. (4.81) or (4.83). The phase superposition conditions under this condition are still the same as Eqs. (4.84), (4.85), and (4.86).

(3) **Equal spacing and equal strength coupling**

In the actual design of directional couplers, we often encounter situations where the spacing between the coupling holes is equal. Moreover, the size and shape of the holes are also the same. This is the case of equal spacing and equal strength coupling. At this time, Eqs.(4.80) or (4.82) are simplified to the following form:

$$A^{\pm} = \begin{cases} \left| 2a^{\pm} \sum\limits_{k=1}^{n} \cos(2k-1)\varphi^{\pm} \right| & (N = 2n) \\ a^{\pm}\left(1 + 2\left| \sum\limits_{k=1}^{n} \cos 2k\varphi^{\pm} \right|\right) & (N = 2n+1) \end{cases}$$
$$= \left| a^{\pm} \frac{\sin N\varphi^{\pm}}{\sin \varphi^{\pm}} \right| \tag{7.36}$$

Clearly, the phase superposition conditions (7.31) and (7.32) for equal spacing coupling are also applicable to equal spacing and equal strength coupling.

7.3.3 *Design of Mode Selective Coupler [90]*

1. **Introduction to mode selective coupler**

Mode selectivity can be achieved by utilizing the principle of phase superposition. That is, we select the hole spacing of the directional coupler's coupling holes. This selection ensures that the microwave signals coupled out of all holes to a specified mode (referred to as the coupled mode) in the secondary waveguide are in-phase when forward output. As a result, the coupling of this mode will be enhanced, and there will be maximum output in the forward direction of the directional coupler. At the same time, the hole spacing should also ensure that the signals coupled out of each hole for one or more other modes (referred to as non-coupled modes or modes that need to be suppressed) are out of phase and cancel each other when forward output. Thus, these modes will not have output in the forward direction of the directional coupler. Since such a mode selective coupler can only couple the microwave energy of one mode, when the microwave system has multiple modes in the main waveguide, it is necessary to set a mode selective coupler for each mode. Then, based on whether each coupler has coupling output or the size of the output of each coupler, we can determine the existence and relative size of these modes when the microwave system works. The actual practice is to set a secondary waveguide coupling arm for each mode on the same main waveguide, and multiple coupling arms are set for multiple modes at the same time, which can greatly save the processing cost and the volume

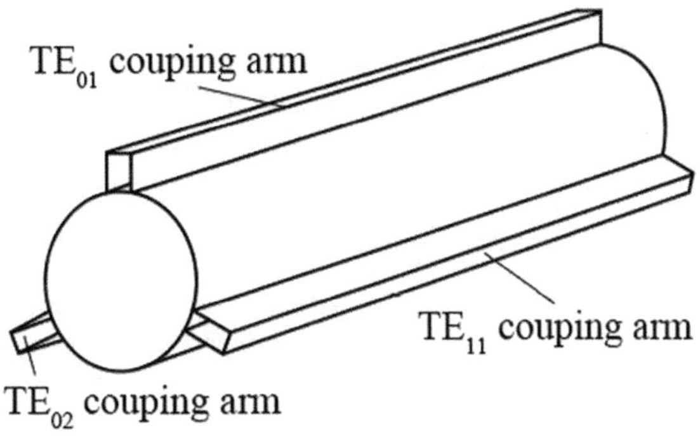

Fig. 7.10 Mode selective coupler with three coupling arms

and weight of the mode selector. Figure 7.10 shows a schematic diagram of a mode selective coupler that identifies three modes TE_{01}, TE_{02}, and TE_{11}.

From the above analysis, it is easy to see that to identify the mode, it is sufficient to focus on the forward output of the coupler, and there is little relationship with the backward output. Therefore, when designing the mode selective coupler, its directional property can be ignored, that is, its backward output can be ignored, which will greatly simplify the design of the mode selective coupler. Even if it is necessary to determine the relative power of each mode, it is only necessary to calibrate the coupling coefficient of its coupled mode for each mode selective coupler. Therefore, when calling the mode selective coupler, we intentionally omit the words "directional", because it is actually no longer a mode selective directional coupler with directional coupling, but only a mode selective coupler.

In the following discussion, we will only consider the forward wave of the mode selective coupler, but the design principles and methods are also applicable to the case of considering the backward wave. In the following discussion, it is assumed that the main waveguide of the mode selective coupler is an overmoded circular waveguide, and the secondary waveguide is a fundamental mode rectangular waveguide operating in a TE_{10} mode, but the specific methods in these discussions are also applicable to other waveguide types.

It is obvious that the mode recognition method of the mode selective coupler is very simple, easy to use, and the recognition process is simple and does not affect the normal operation of the system. It is not only an online measurement method, but also has real-time and dynamic characteristics because there is no need to adjust or move any elements during the measurement process, and the output of each coupling arm is indicated by a crystal detector and an oscilloscope. At the same time of identifying the mode, power measurement can also be performed. The main disadvantage is that a large one-time investment is required to make a coupling arm for each possible mode; in addition, because it is difficult to make a single-mode generator for high-order modes, calibration of the coupling coefficient and suppression coefficient is relatively time-consuming. The author of this book has proposed a method that does not require a special single-mode generator, and uses back-to-back measurement to calibrate the coupling coefficient of the mode selective coupler with multiple coupling arms. These contents will be introduced in Sect. 7.4. The bandwidth of the mode selective coupler is narrow, generally only a few percent, and the author has also proposed a method to use the Chebyshev distribution or binomial distribution of the coupling hole and other auxiliary measures to increase the bandwidth to 10–20%. Interested readers can refer to relevant literature [91–94].

2. **Determination of hole spacing**

From the above discussion of the principle of phase superposition, it can be seen that the in-phase superposition and out-of-phase cancellation of waves depend on the phase shift of the electromagnetic wave between the coupling holes in the mode selective coupler. In other words, for a given mode selective coupler and a given mode, it depends on the hole spacing of the coupling holes. We stipulate that for the coupled mode in the mode selective coupler, all its parameters are represented by the subscript "*w*", and for the non-coupled mode or the mode that needs to be suppressed, it is represented by the subscript "*u*".

Using the in-phase superposition conditions (4.84) and (4.85) obtained in Sect. 4.4.2 for multi-hole coupling, and considering Eq. (4.81) or (4.83) of θ_k, and noticing $\beta = 2\pi/\lambda_g$, where λ_g is the waveguide wavelength, we can obtain:

$$d_{k,w} = 2i_{k,w}\left|\frac{\lambda_w\lambda_{10}}{\lambda_w - \lambda_{10}}\right| \quad (i_{k,w} = 1, 2, \ldots) \tag{7.37}$$

or

$$d_{k,w} = (2i_{k,w} - 1)\left|\frac{\lambda_w\lambda_{10}}{\lambda_w - \lambda_{10}}\right| \quad (i_{k,w} = 1, 2, \ldots) \tag{7.38}$$

where, λ_w is the waveguide wavelength of the coupled mode in the circular main waveguide, and λ_{10} is the waveguide wavelength of the TE_{10} mode in the rectangular secondary waveguide.

Similarly, the out-of-phase cancellation condition (4.86) can be written as:

$$d_{k,u} = \left(i_{k,u} - \frac{1}{2}\right)\left|\frac{\lambda_u\lambda_{10}}{\lambda_u - \lambda_{10}}\right| \quad (i_{k,u} = 1, 2, \ldots) \tag{7.39}$$

where, λ_u is the waveguide wavelength of the non-coupled mode (mode that needs to be suppressed) in the main waveguide.

In the case of equal spacing coupling, the in-phase superposition condition (7.31) and the out-of-phase cancellation condition (7.32) can be simplified to:

$$S_w = i_w\left|\frac{\lambda_w\lambda_{10}}{\lambda_w - \lambda_{10}}\right| \quad (i_w = 1, 2, \ldots) \tag{7.40}$$

$$S_u = \left(i_u - \frac{1}{2}\right)\left|\frac{\lambda_u \lambda_{10}}{\lambda_u - \lambda_{10}}\right| \quad (i_u = 1, 2, \ldots) \tag{7.41}$$

3. **Setting of multiple groups of coupling holes**

(1) **The hole spacing simultaneously satisfies the limitation of in-phase superposition and out-of-phase cancellation**

In general, for the case of equal spacing coupling, when the hole spacing S is already determined (it means that this S is both S_w and S_u, meaning it can only be one unified S), if we want both the coupled mode w to undergo in-phase superposition and a specific non-coupled mode u_1 to be suppressed (out-of-phase cancellation), S should satisfy both Eqs.(7.40) and (7.41) simultaneously.

$$S = \begin{cases} i_w \left|\frac{\lambda_w \lambda_{10}}{\lambda_w - \lambda_{10}}\right| & (i_w = 1, 2, \ldots) \\ \left(i_{u1} - \frac{1}{2}\right)\left|\frac{\lambda_{u1} \lambda_{10}}{\lambda_{u1} - \lambda_{10}}\right| & (i_{u1} = 1, 2, \ldots) \end{cases} \tag{7.42}$$

In the case where the modes and dimensions of the main and secondary waveguides in the mode selective coupler coupling arm have been determined. That is, λ_w, λ_{u1}, and λ_{10} have been determined. In this situation, it will be somewhat difficult to make S satisfy both equations at the same time. The only adjustable variables are i_w and i_{u1}, which are two independent variables that can be artificially adjusted separately. In principle, such an S can still basically be found, but it may cause S to be quite large (i_w and i_{u1} selected are relatively large). To overcome this obstacle, the size of the main and secondary waveguides can be adjusted repeatedly, which means the values of λ_w, λ_{u1}, and λ_{10} can be changed. By combining this with the selectionof i_w and i_{u1} to make S not too large. Meanwhile, we can allow S not to strictly satisfy the requirements of Eq. (7.42), permitting a certain reasonable tolerance range.

Obviously, if the coupling arm of the mode selective coupler not only needs to make the coupled mode w in the in-phase superposition, but also needs to suppress two different non-coupled modes u_1 and u_2 undergo in-phase superposition but also needs to suppress two different non-coupled modes u1 and u2 out of phase, the same S must satisfy three equations at the same time. Moreover, S cannot be too large to avoid making the mode selective coupler very large. Such an S will be difficult to find, and can only be found by chance. In general, a determined S value can only couple one coupled mode and suppress one non-coupled mode at the same time.

In Eq. (7.42), the non-coupled mode u_1 should be selected as the mode that we want to suppress first in the mode selective coupler. Often, the non-coupled mode with the waveguide wavelength closest to the waveguide wavelength of the coupled mode in the main waveguide is the easiest to excite. Therefore, it is generally selected as the mode u_1 that should be suppressed first. Of course, this is not absolute, and the selection of the u_1 mode should be determined according to the actual situation.

As for the number of holes N, we can temporarily choose an N value, and then combine it with the calculation of the coupling hole radius after all the dimensions of the mode selective coupler have been basically determined, and then make necessary adjustments.

(2) **Setting of multiple groups of coupling holes**

When more non-coupled modes need to be suppressed, such as non-coupled modes u_2, u_3, etc. in addition to u_1, this means that the hole spacing S needs to satisfy three, four, or even more equations at the same time (including one equation satisfied by the coupled mode). Obviously, it is completely impossible to find such a unified S. In this case, multiple groups of coupling holes can be set up to solve this problem, and each additional group of coupling holes can increase the suppression of one or two non-coupled modes.

The S calculated according to Eq. (7.42) and the selected number of holes N form the first group of coupling holes. At this time, the obtained S can suppress a non-coupled mode u_1 while ensuring that the coupled mode w is in phase. If we need to suppress other non-coupled modes u2 and u3, we can use the method of setting up the second group of coupling holes. The arrangement, hole spacing, and hole radius of the second group of coupling holes correspond one-to-one to those of the first group of coupling holes. That is, the second group of coupling holes completely duplicates the first group of coupling holes. However, the second group of coupling holes is shifted by a distance Sp1 relative to the first group of coupling holes as a whole. And Sp1 must satisfy the suppression condition (7.41) for a non-coupled mode u_2 or two non-coupled modes u_2 and u_3, that is,

$$S_{p1} = \left(i_{u2} - \frac{1}{2}\right)\left|\frac{\lambda_{u2}\lambda_{10}}{\lambda_{u2} - \lambda_{10}}\right| \quad (i_{\mathrm{u}2} = 1, 2, \ldots) \tag{7.43}$$

or

$$S_{p1} = \begin{cases} \left(i_{u2} - \frac{1}{2}\right)\left|\frac{\lambda_{u2}\lambda_{10}}{\lambda_{u2}-\lambda_{10}}\right| & (i_{\mathrm{u}2} = 1, 2, \ldots) \\ \left(i_{u3} - \frac{1}{2}\right)\left|\frac{\lambda_{u3}\lambda_{10}}{\lambda_{u3}-\lambda_{10}}\right| & (i_{\mathrm{u}3} = 1, 2, \ldots) \end{cases} \tag{7.44}$$

Whether to use Eqs. (7.43) or (7.44) to calculate S_{p1} depends on how many non-coupled modes need to be suppressed, and when suppressing two non-coupled modes at the same time, it also depends on the size of S_{p1} that can be tolerated, that is, the length of the mode selective coupler that can be allowed.

When it is necessary to set up the third group of coupling holes to suppress more non-coupled modes, the arrangement, hole spacing, and hole radius of the third group of coupling holes should correspond one-to-one to those of all existing coupling holes (that is, the total coupling holes of the first and second groups together), and the overall displacement S_{p2} should also satisfy Eqs. (7.43) or (7.44) for the new non-coupled modes. This process can be continued to set up the fourth group and more groups of coupling holes, but in practical design, it is rare to have more than three groups of coupling holes, because three groups of coupling holes can already suppress at least three non-coupled modes, and even up to five non-coupled modes.

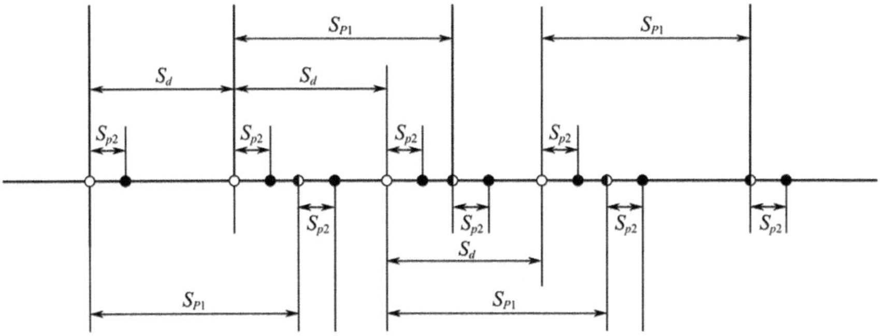

Fig. 7.11 Hole distribution of coupling arm with three groups of coupling holes (○—first group of holes, ◐—second group of holes, ●—third group of holes)

In general, the higher-order or lower-order modes with characteristic values m and n far away from the characteristic value of the coupled mode often have very small amplitudes and can be ignored.

Figure 7.11 shows a schematic diagram of the distribution of coupling holes on a coupling arm with three groups of coupling holes, the first group has four coupling holes, the second group duplicates the first group and also has four coupling holes, and the third group should duplicate the sum of the first and second groups, with eight coupling holes. In this figure, S_d is the hole distance of the first group of coupling holes, S_{p1} is the displacement of the second group of coupling holes relative to the first group of coupling holes, and S_{p2} is the displacement of the third group of coupling holes relative to the first and second groups of coupling holes.

In the case of multiple sets of coupling holes, the total number of holes N_T will increase exponentially with the number of groups.

$$N_T = 2^{(m-1)} \times N_1 \tag{7.45}$$

where, m is the number of groups of coupling holes, and N_1 is the number of coupling holes in the first group. At this time, the length L of the coupling region of the mode selective coupler is

$$L = (N_1 - 1)S + \sum_{l=1}^{\mathrm{m}-1} S_{p,\,l} \tag{7.46}$$

The $S_{p,\,l}$ of each group is determined by Eqs. (7.43) or (7.44).

The coupling holes of the first group are regarded as a coupling whole unit, according to the previous analysis, after adding the coupling holes of the second group, the coupling holes of the second group are also regarded as a coupling whole unit. Therefore, according to the double-hole coupling theory analyzed in Sect. 4.4.2, the same theory can be completely applied to the case of coupling holes of two groups. The total coupling strength can be obtained by multiplying the coupling strength of the coupling holes of a single group by the cosine function determined by the spacing S_p between the coupling holes of two groups, just like the expression given in Eq. (4.78). Similarly, when there are coupling holes of multiple groups, Eq. (7.33) should be rewritten as follows:

$$A^{\pm} = \begin{cases} 2\left|\sum_{k=1}^{n} a_k^{\pm}\cos(2k-1)\varphi^{\pm}\right|\left|2\prod_{l=1}^{m-1}\cos\theta_l^{\pm}\right| & (N=2n) \\ \left|a_0^{\pm}+2\sum_{k=1}^{n} a_k^{\pm}\cos 2k\varphi^{\pm}\right|\left|2\prod_{l=1}^{m-1}\cos\theta_l^{\pm}\right| & (N=2n+1) \end{cases} \tag{7.47}$$

$$\theta_l^{\pm} = (\beta_{10} \mp \beta_2)S_{p,l}/2 \quad (\mathrm{p} = 1, 2, \ldots, \mathrm{m}) \tag{7.48}$$

where, $\varphi^{\pm}$ should be calculated according to Eq. (7.30). Since β_2 is the phase constant of the wave in the secondary waveguide, when the modes coupled by the same coupling arm (i.e., the secondary waveguide) are different, β_2 is different; when the sizes of different coupling arms are different, β_2 is also different. Therefore, β_2 should be divided into the following two cases, namely

For coupled modes

$$\beta_2 = \beta_{w,q}, \quad \theta_{\mathrm{q}}^{\pm} = \theta_{w,q}^{\pm} \tag{7.49}$$

where, $q = 1,2, \ldots, n$ is the coupling arm serial number,

For non-coupled modes

$$\beta_2 = \beta_{u,q}, \quad \theta_{\mathrm{q}}^{\pm} = \theta_{u,q}^{\pm} \tag{7.50}$$

Similarly, the subscript $q = 1, 2, \ldots, n$ is the coupling arm serial number.

At this time, if the lth group of coupling holes need to simultaneously suppress two non-coupled modes, the phase changes $\theta_{u,q}$ caused by the two modes on the $S_{p,l}$ displacement should be close to the condition of anti-phase cancellation, so the size of A^- should be calculated separately for the two non-coupled modes.

(3) **Precautions for the design of mode selective couplers**

(a) The number of the groups of coupling holes depends on how many non-coupled modes need to be suppressed, but it should not be too much, because if the number of groups m is too large, not only will the number of coupling holes increase exponentially, but the length of the mode selective coupler will increase rapidly, and the volume and weight will increase accordingly, and overlapping of coupling holes is also likely to occur,, so the number of groups of coupling holes is generally not more than three.

(b) The determination of $S_{p,1}$ should make it meet the requirements of simultaneously suppressing two non-coupled modes as much as possible, that is, to satisfy Eq. (7.44). For this purpose, we can carefully adjust the size of the main circular waveguide and the secondary rectangular waveguide, that is, change the size of λ_{10} and $\lambda_{u,1}$ to change $S_{p,1}$, so as to satisfy Eq. (7.44) as much as possible, so as to reduce the number of coupling hole groups. However, it should be noted that after the size of the main waveguide and the secondary waveguide is changed, S itself should be recalculated accordingly, and the condition of the same-phase superposition of the coupled modes cannot be destroyed.

(c) The selection of non-coupled modes should consider the relative component size of the non-coupled modes in the main waveguide and their influence on the coupling coefficient of the coupled modes. Generally speaking, the closer the waveguide wavelength is to the waveguide wavelength of the lowest mode in the system or the mode with the waveguide wavelength closest to the waveguide wavelength of the coupled mode, the easier it is to excite the mode, and the greater the impact on the coupling coefficient of the coupled mode. Therefore, when designing multiple groups of coupling holes, this type of mode should be given priority for suppression.

(d) It should be noted in particular that the size of $S_{p,l}$ should avoid being close to $(i_w–1/2)S_w$ $(i_w = 1,2,\ldots)$, because the forward waves of the modes that need to be coupled will cancel each other out at $S_{p,l} = (i_w–1/2)S_w$, which will reduce the coupling of the coupled mode. If $S_{p,l}$ is closer to an integer multiple of S_w, this situation of cancelling each other is less likely to occur, and conversely, it can also enhance the forward coupling of the coupled mode.

(e) Whether it is the calculation of S or $S_{p,l}$, when two equations are required to be satisfied at the same time, that is, when Eqs. (7.42) or (7.44) is required to be satisfied, it is not necessary to be strictly accurate, because this will make the calculation of S and $S_{p,l}$ quite difficult. Generally speaking, as long as the two equations are satisfied within an acceptable tolerance range, it suffices. For example, if the equation is satisfied within an accuracy of one-tenth to one percent of a millimeter, it is sufficient. Higher accuracy has little effect on the performance of the mode selection coupler and will also bring difficulties or increase processing costs to the processing. Of course, with the increase of microwave frequency, the overall size of the mode selection coupler will decrease, and the precision requirements for all sizes should also be increased.

4. **Determination of waveguide size**

(1) **Determination of main waveguide size**

The size of the main waveguide of the mode selection coupler is mainly determined according to the power capacity and the transmission conditions of the coupled mode. The transmission conditions in the overmoded waveguide can be written as:

For overmoded circular waveguide,

$$\lambda_1 \leq 2\pi R / \mu_{mn} \tag{7.51}$$

For overmoded rectangular waveguide,

$$\lambda_1 \leq 2 \Big/ \sqrt{\left(\frac{m}{A}\right)^2 + \left(\frac{n}{B}\right)^2} \tag{7.52}$$

where

λ_1 is the wavelength corresponding to the lowest frequency of the coupled mode in the working frequency range of the mode selective coupler, or the maximum working wavelength,

For TM modes, μ_{mn} is the n-th root of the first kind of m-th order Bessel function. For TE modes, μ_{mn} is the n-th root of the derivative of the Bessel function of m-th order of the first kind,

R is the radius of the main circular waveguide,

m, n are the characteristic values of the highest order mode (generally the coupled mode) allowed to be transmitted in the main waveguide,

A, B are the wide and narrow side sizes of the rectangular waveguide as the main waveguide.

Increasing the size of the main circular waveguide radius R or the dimensions A and B of the rectangular main waveguide will increase the higher order modes, but appropriately increasing the size of the waveguide (radius R of the circular waveguide or dimensions A and B of the rectangular waveguide) will have certain benefits for expanding the working frequency range of the mode selection coupler. R (radius of the circular main waveguide) or A and **B** (dimensions of the rectangular main waveguide) should primarily be selected according to the precautions (b) and (c) in the mode selection coupler design.

(2) **Determination of secondary waveguide size**.

When the main waveguide is an overmoded circular waveguide and the secondary waveguide is a fundamental mode rectangular waveguide, after determining the radius R of the main circular waveguide, the width a of the secondary rectangular waveguide can be determined by Eqs. (7.40) and (7.41) under the equidistant coupling condition, so that the forward wave of the coupled mode can be maximally superimposed in phase and the forward wave of the non-coupled mode can be canceled out as much as possible. From Eqs.(7.40) and (7.41), we can get

$$\lambda_{10,0} = \begin{cases} \frac{[i_w - (i_u - 1/2)]\lambda_w \lambda_u}{i_w \lambda_w - (i_u - 1/2)\lambda_u} & (\lambda_w - \lambda_{10})(\lambda_u - \lambda_{10}) > 0 \\ \frac{[i_w + (i_u - 1/2)]\lambda_w \lambda_u}{i_w \lambda_w + (i_u - 1/2)\lambda_u} & (\lambda_w - \lambda_{10})(\lambda_u - \lambda_{10}) < 0 \end{cases} \tag{7.53}$$

where, the subscript "0" represents the parameter at the center frequency of the working band of the mode selective coupler. In this way, the size of a of the secondary waveguide rectangular waveguide can be obtained according to the calculated

$$\lambda_{10,0} = \frac{\lambda_0}{\sqrt{1 - (\lambda_0 / 2a)^2}} \tag{7.54}$$

where, λ_0 is the wavelength corresponding to the center frequency, and a must also satisfy the single-mode transmission condition of the TE_{10} mode. When $a = 2b$, this condition is

$$\lambda_2 > a > \lambda_1/2 \tag{7.55}$$

where, λ_2 is the minimum wavelength in the working band of the mode selective coupler, and λ_1 is the maximum working wavelength in the working frequency range of the mode selective coupler.

As for the size of the narrow side b of the rectangular waveguide, it can be selected according to the actual situation of the mode selective coupler structure design.

5. **Calculation of coupling hole radius and selective of hole number**

According to Eq. (7.36) under equal strength coupling, Eq. (7.47) under equal spacing coupling can be further simplified as

$$A^{\pm} = \left|a^{\pm}\frac{\sin N\phi^{\pm}}{\sin\phi^{\pm}}\right|\left|2\prod_{l=1}^{m-1}\cos\theta_l^{\pm}\right| \tag{7.56}$$

To calculate the size of $A^{\pm}$, we first need to know the value of $a^{\pm}$. To do this, we should set a required coupling coefficient for the coupled mode at the center frequency of the working band of the mode selective coupler, denoted as C_0 (in dB). Then, we can obtain

$$C_0 = -20\lg\left|A_w^+\right|,\ \ A_w^+ = 10^{-C_0/20} \tag{7.57}$$

From this, we can obtain

$$a_{w,0}^+ = 10^{C_0/20} \Big/ \left|\frac{\sin N\phi_{w,0}^+}{\sin\phi_{w,0}^+}\right|\left|2\prod_{l=1}^{m-1}\cos\theta_{l,0}^+\right| \tag{(7.58)}$$

where, the subscript "0" represents the parameter value at the center frequency.

Once $a_{w,0}^+$ is determined, we can use Eqs. (4.70) or (4.75) derived from the theory of pinhole diffraction to calculate the radius r_0 of the coupling hole in the case of equal spacing and equal strength coupling under the condition of power normalization. Substituting the obtained r_0 back into Eq. (4.70) or (4.75), we can then calculate the coupling strength at other frequencies. Finally, by using Eq. (7.56), we can determine the total coupling strength and coupling coefficient at each frequency.

When calculating the total coupling strength $A^{\pm}$, the number of coupling holes N must also be given. The determination of N is mainly based on the desired coupling coefficient C_0 for the coupled mode. That is, when determining the radius r_0 of the coupling hole using the obtained $a_{w,0}^+$ from Eq. (7.58), it can be mutually adjusted in conjunction with the value of N. Increasing N can decrease r_0, while decreasing N can increase r_0. Generally, r_0 needs to satisfy the condition for small holes, meaning

$r_0 \ll \lambda$, yet it should not be too small, which will bring difficulties to processing. Adjusting the size of N in this case can ensure that r_0 is of an appropriate size.

By adopting a completely similar calculation process, the coupling coefficient of the backward wave can also be calculated, and then the directionality of the mode selective coupler can be obtained. As we have already mentioned, the directionality can be ignored, so no further discussion will be made here.

7.3.4 Design Examples of Mode Selection Coupler [90]

1. **Calculation of the parameters of the mode selective coupler**

On the basis of the above discussion, we can calculate the parameters of the mode selective coupler with equal spacing and equal strength using Eq. (7.56).

Couple coefficient C of the coupled mode

$$\begin{aligned} C &= -20\lg A_w^+ \\ &= -20\lg a_w^+\left|\frac{\sin N\phi_w^+}{\sin\phi_w^+}\right| - 20\lg\left|2\prod_{l=1}^{m-1}\cos\theta_{w,l}^+\right| \end{aligned} \quad (7.59)$$

The definition of the suppression coefficient for the non-coupled mode is given by

$$\begin{aligned} p &= 20\ \lg\frac{A_w^+}{A_u^+} \\ &= 20\ \lg\frac{a_w^+\left|\frac{\sin N\phi_w^+}{\sin\phi_w^+}\right|}{a_u^+\left|\frac{\sin N\phi_u^+}{\sin\phi_u^+}\right|} + 20\ \lg\frac{\left|2\prod_{l=1}^{m-1}\cos\theta_{w,l}^+\right|}{\left|2\prod_{l=1}^{m-1}\cos\theta_{u,l}^+\right|} \end{aligned} \quad (7.60)$$

From this, we can also obtain the coupling coefficient of the non-coupled mode as

$$C_u = -20\lg A_u^+ = +20\lg\left(\frac{1}{A_w^+}\frac{A_w^+}{A_u^+}\right) = p + C \quad (7.61)$$

If necessary, the directionality of the mode selective coupler can also be calculated, i.e.,

$$\begin{aligned} d &= 20\ \lg\frac{A_w^+}{A_w^-} \\ &= 20\ \lg\frac{a_w^+\left|\frac{\sin N\phi_w^+}{\sin\phi_w^+}\right|}{a_w^-\left|\frac{\sin N\phi_w^-}{\sin\phi_w^-}\right|} + 20\ \lg\frac{\left|2\prod_{l=1}^{m-1}\cos\theta_{w,l}^+\right|}{\left|2\prod_{l=1}^{m-1}\cos\theta_{w,l}^-\right|} \end{aligned} \quad (7.62)$$

By replacing the subscript "w" with the subscript "u" in Eq. (7.62), the directionality of the non-coupled mode, d_u, can be obtained.

2. **Design example of selected mode selective couplers**

We have designed two mode selective couplers for overmoded circular waveguides, with center operating frequencies of 35 GHz and 70 GHz, respectively. Assuming that the main modes transmitted in the main waveguide of the 35 GHz coupler are TE_{01}, TE_{02}, and TE_{11} modes, this mode selective coupler will be configured with coupling arms for these three modes. Similarly, for the 70 GHz coupler, the main modes transmitted in the primary waveguide are TE_{01}, TE_{02}, and TE_{03}, so the coupler will be designed with coupling arms for these three modes. The coupling coefficient requirement for each coupling arm at the center frequency is 30 dB. The radius R of the primary waveguide overmoded circular waveguide is 10 mm and 13.9 mm, respectively. The narrow side dimension b of the secondary waveguide rectangular waveguide has been adjusted multiple times in the design process and is finally determined to be 3.56 mm and 1.88 mm, respectively, while its wide side dimension should be determined during the design process. Two groups of coupling holes are set for each coupling arm. After repeated adjustments, we have designed the structural dimensions of these two mode selective couplers, including the wide side a of the rectangular waveguide, the hole spacing S_w of the coupling hole, the displacement S_p between the two groups of coupling holes, the determination of six coupling holes in each group, and the calculation of the radius r_0 of the coupling hole and the total length L of the coupling region. Then, the suppression coefficient of the non-coupled mode for each coupling arm is calculated.

The two mode selective couplers are denoted as No.1 and No.2, and their design and calculation results are provided in Table 7.1.

The coupling coefficient and suppression coefficient as a function of frequency are respectively shown in Figs. 7.12 and 7.13. From the curves in the figures, it can be observed that the coupling coefficient generally meets the requirements over a fairly wide frequency range. However, the band of suppression coefficient is relatively narrow, achieving high suppression only within a limited frequency range. According to the definition formula (7.60) of the suppression coefficient, a high suppression coefficient indicates that the amplitude of the coupled mode w, $A_w{}^+$, in the secondary waveguide is large, while the amplitude of the non-coupled mode to be suppressed, $A_u{}^+$, is small. This clearly indicates a high selectivity of the mode, and thus the level of the suppression coefficient directly reflects the quality of the mode selectivity of the mode selective coupler. The suppression coefficient generally only reaches a relatively ideal level within a narrow frequency range, which is a limitation of the mode selective coupler. However, it is evident that at the center frequency, the suppression coefficient of each arm's coupled mode to the non-coupled mode is very high, all exceeding 40 dB and even reaching 90 dB, indicating a relatively successful design. As previously mentioned, methods to expand the frequency band include using unequal spacing or unequal strength coupling hole distributions, such as binomial distributions and Chebyshev distributions, while also implementing other measures such as adjusting the spacing between the two outermost holes on either

Table 7.1 Design and calculation results of 35 GHz and 70 GHz mode selective couplers

	Discriminator						
Parameters		No. 1			No. 2		
		TE_{01} coupling arm	TE_{02} coupling arm	TE_{11} coupling arm	TE_{01} coupling arm	TE_{02} coupling arm	TE_{03} coupling arm
Given parameters	f_0 (GHz)	35.0			70.0		
	C_0 (dB)	30			30		
	R (mm)	10.0			13.9		
	b (mm)	3.56			1.88		
Designed dimensions	a (mm)	5.51	5.74	6.34	3.10	2.92	2.77
	S_w (mm)	38.18	22.91	37.15	16.51	16.51	18.54
	S_p (mm)	12.61	17.17	9.61	14.88	11.44	7.06
	r_0 (mm)	1.55	1.37	1.61	1.01	1.0	0.99
	n	3	3	3	3	3	3
	N_T	12	12	12	12	12	12
	L (mm)	203.51	129.25	195.36	97.43	93.99	99.76
Calculated suppressions in (dB) at f_0	TE_{01}		77.23	63.38		45.12	47.57
	TE_{02}	62.68		60.90	40.33		63.36
	TE_{03}				57.16	57.00	
	TE_{11}	78.10	97.21		5.98	19.76	31.45
	TE_{21}	5.67	8.57	2.98	0.90	15.46	29.87
	TE_{31}	11.26	15.52	9.64	5.76	2.58	28.20
	TE_{41}	0.80	17.89	5.98	4.01	1.53	36.07
	TE_{51}	5.83	20.36	1.77	8.41	5.93	29.70

(continued)

Table 7.1 (continued)

	Discriminator						
Parameters		No. 1			No. 2		
		TE_{01} coupling arm	TE_{02} coupling arm	TE_{11} coupling arm	TE_{01} coupling arm	TE_{02} coupling arm	TE_{03} coupling arm
	TE_{12}	3.53	8.90	0.68	3.16	19.23	31.27
	TE_{22}	0.94	25.37	7.69	4.39	− 0.85	30.19

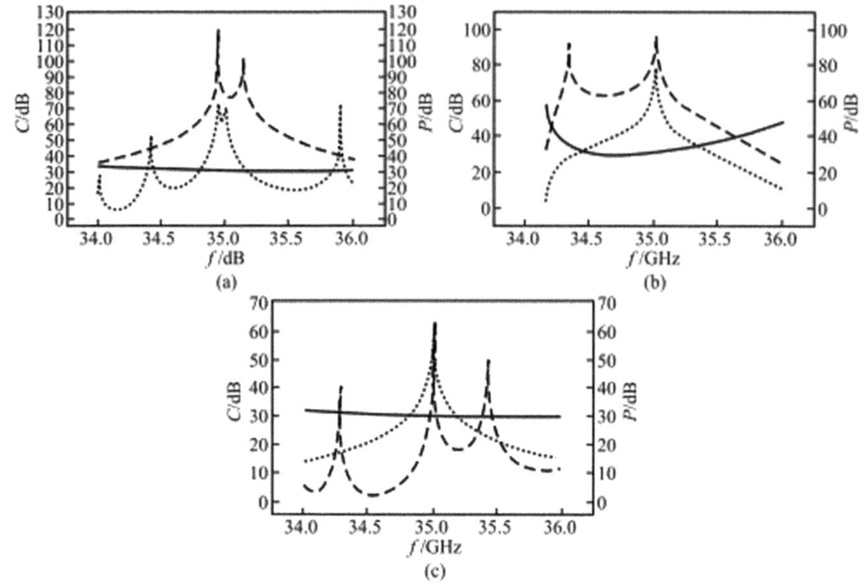

Fig. 7.12 The relationship curves between coupling coefficient and suppression coefficient as a function of frequency for the 35 GHz mode selective coupler, **a** TE_{01} coupling arm: –TE_{01} mode coupling coefficient, - - -suppression coefficient to TE_{02} mode, - - -suppression coefficient to TE_{11} mode, **b** TE_{02} coupling arm: –TE_{02} mode coupling coefficient, - - -suppression coefficient to TE_{01} mode, - - -suppression coefficient to TE_{11} mode, **c** TE_{11} coupling arm: –TE_{11} mode coupling coefficient, - - -suppression coefficient to TE_{01} mode, - - -suppression coefficient to TE_{02} mode

side of the same group of coupling holes. Interested readers can refer to the relevant papers listed in the references.

7.4 Calibration and Power Measurement of Mode Selective Coupler

7.4.1 Direct Calibration Method

After the design and manufacture of the mode selective coupler, it is apparent that we need to calibrate the coupling coefficient of the coupled mode and the suppression coefficient of the non-coupled mode for each coupling arm of the mode selective coupler. Then, we can directly measure the output power of each coupling arm using a small power meter, or use a calibrated crystal detector in combination with an oscilloscope to measure the output power of each coupling arm. In this way, the power levels of each coupled mode and their relative proportions can be accurately

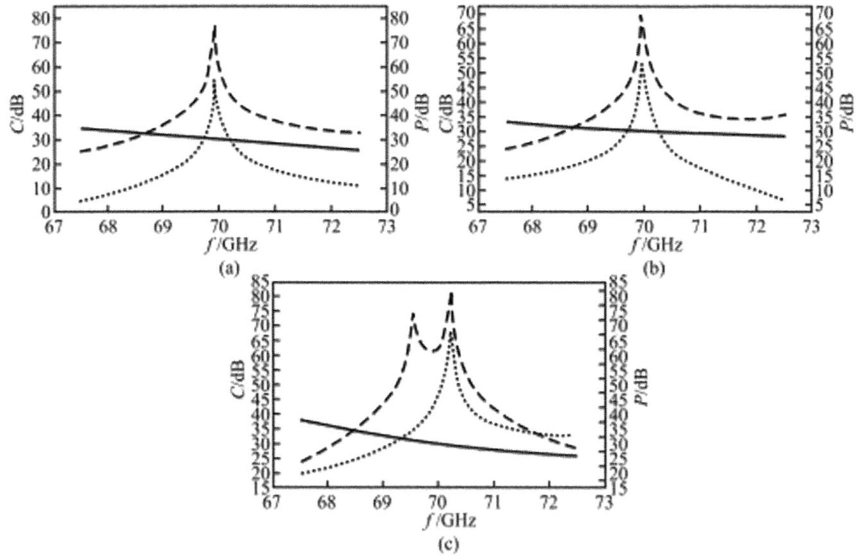

Fig. 7.13 The relationship curves between coupling coefficient and suppression coefficient as a function of frequency for the 70 GHz mode selective coupler, **a** TE_{01} coupling arm: –TE_{01} mode coupling coefficient, ---suppression coefficient to TE_{02} mode, ---suppression coefficient to TE_{03} mode, **b** TE_{02} coupling arm: –TE_{02} mode coupling coefficient, ---suppression coefficient to TE_{01} mode, ---suppression coefficient to TE_{03} mode, **c** TE_{03} coupling arm: –TE_{03} mode coupling coefficient, ---suppression coefficient to TE_{01} mode, ---suppression coefficient to TE_{02} mode

determined. Therefore, the calibration of the coupling coefficient and suppression coefficient is crucial for power measurement using the mode selective coupler.

1. **Measurement method of direct calibration**

In practice, if the suppression coefficient of each coupled mode to the non-coupled mode is sufficiently high, such as exceeding 30 dB, the power of the non-coupled mode in each coupling arm can be neglected, and it can be assumed that only the coupled mode exists in the coupling arm during calibration. In this case, it is only necessary to measure the coupling coefficient of the coupled mode during calibration, and there is no need to measure the suppression coefficient of the non-coupled mode. By measuring the power of the coupled mode output from each coupling arm and using the coupling coefficient of that mode to calculate the power of that mode in the main waveguide, the sum of the powers of each mode is the total power in the system.

Considering only the calibration of the coupling coefficient of each coupling arm for the coupled mode, this constitutes the fundamental principle of the direct calibration method.

The coupling coefficient can be directly calibrated using a scalar network analyzer, but for calibration, a single mode generator generating the signal of the mode to

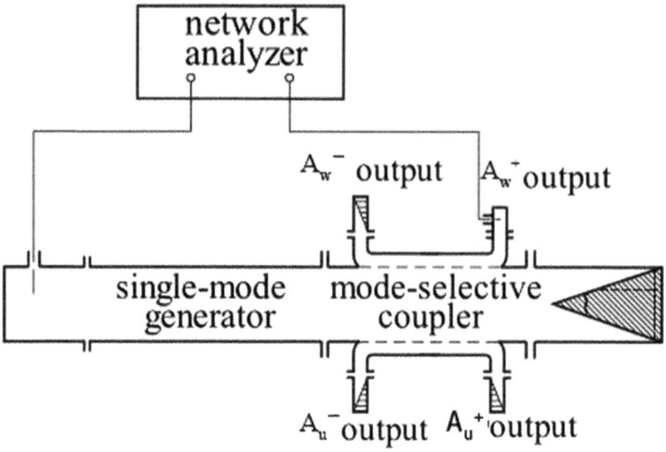

Fig. 7.14 Schematic diagram of a measurement system used to calibrate the coupling coefficient using a network analyzer

be calibrated is required. The measurement system for this calibration method is relatively simple, as shown in Fig. 7.14.

When the presence of the non-coupled mode in each coupling arm is ignored, in order to identify the presence and measure the power of each coupled mode, it is only necessary to calibrate the coupling coefficient of the coupling arm of each coupled mode of the mode selective coupler. Therefore, during calibration, the output signal of the network analyzer should first be transformed into the mode to be calibrated using a single-mode generator, then input to the forward end of the mode selective coupler, and the output signal of the coupling arm for that mode should be fed back to the network analyzer (note that the signal output from the secondary waveguide is already the TE_{10} mode, so it can be directly input, or input to the network analyzer after waveguide-coaxial conversion), and S_{21} should be measured. The coupling coefficient of that mode can then be directly read from the network analyzer. By simply changing the single mode generator and the coupling arm corresponding to that mode output, the coupling coefficient of each coupled mode of the mode selective coupler can be measured using the same method. When using the calibrated coupling coefficient of the mode selective coupler for high-power microwave measurements, a small-power meter (including a crystal detector small-power meter) can be connected to the output end of the coupling arm of a certain mode, and based on the reading from the small-power meter, the input power of that mode into the main waveguide can be calculated using the coupling coefficient. The sum of the powers calculated in this way for each coupling arm is the total power in the main waveguide. Since the definition of the coupling coefficient is already the logarithm of the ratio of the power input to the mode selective coupler to the power output from the forward direction of the coupling arm of that mode selective coupler, the power calculated using the coupling coefficient is the input power of that mode, and is independent of the directivity of the coupling arm, meaning it is not affected by the presence of the backward wave in the coupling arm, as its impact is directly reflected in the magnitude of the coupling coefficient of the forward wave.

However, if the suppression coefficient of the mode selective coupler for the non-coupled mode is not sufficiently high, then the non-coupled mode will also have some power coupled into the coupling arm and output in the forward direction. In this case, the measured coupling coefficient actually no longer solely depends on the output of the coupled mode, but includes contributions from the non-coupled mode. This issue is difficult to resolve in the direct calibration method unless the single-mode generator can produce a completely pure output without any spurious modes (non-coupled modes).

During measurement, except for the input and output ports of the signal, all other ports should be connected to matching loads.

2. **Difficulties of direct calibration method**

Obviously, using a high-power microwave source directly as the calibration signal source is not feasible because the high-power microwave source itself outputs multiple modes, which are the objects we want to identify and measure. Therefore, using it as a calibration signal source cannot accurately measure the coupling coefficient of a single mode. In addition, calibration can only be performed at small power levels, as the output power of the high-power microwave source far exceeds the power capacity of the calibration measuring instruments and components. Even with multiple attenuations, it is difficult to achieve the required small power, and multiple attenuations can easily introduce more sources of error. Therefore, it is more reasonable to use a small-power signal source, such as the signal source of the network analyzer itself, to generate the required mode through a mode converter for calibration. For example, when calibrating the coupling coefficient of the TE_{01} mode, a mode generator or mode converter capable of generating the pure TE_{01} mode should be used in the main waveguide of the mode selective coupler.

When performing mode recognition and mode power measurement using a calibrated mode selective coupler for a high-power microwave source, the coupling coefficient of each coupling arm of the mode selective coupler must have a high level of attenuation, generally at least 40 to 60 dB, or even higher, which will also bring about some difficulties.

The main difficulty of the direct calibration method is the requirement to create corresponding single-mode generators for each coupling arm of the mode selective coupler. As mentioned, only relatively low-order modes, i.e., modes with relatively simple field structures, are easily obtained with single-mode generators. For higher-order modes, either there is no ideal transformation method, or the mode must be obtained through a mode converter, which has a complex structure, large volume, and may require multiple transformations to obtain the desired mode. This becomes a major obstacle for the calibration of the mode selective coupler and is the primary difficulty of the direct calibration method. Additionally, even if single-mode generators are manufactured, it is extremely difficult to assess their performance, primarily in terms of the purity of the output mode.

Given the difficulties of the direct calibration method for the mode selective coupler, the authors and colleagues proposed a back-to-back calibration method that does not require single-mode generators.

7.4.2 *Back-to-Back Calibration Method [95]*

1. **Back-to-back measurement system and basic equations for mode selective coupler**

(1) **Physical foundation**

The physical foundation of the back-to-back calibration method is that passive linear microwave components have reciprocity, meaning that the input and output ports can be interchanged without changing the performance of the component. Specifically, when a multi-mode microwave signal is input at one end of the main waveguide of the mode selective coupler, Due to the mode selectivity, the forward direction of each of its coupling arms will output only the microwave signal of the TE_{10} mode excited by the coupling mode w of that arm. The principle of reciprocity tells us that, in turn, when we input the forward output port of the coupling arm as the input port to the microwave signal of TE_{10} mode, also due to mode selectivity, it will only stimulate the signal of w mode in the main waveguide. Moreover, at this time, the original input port of the main waveguide becomes the output port, and the single-mode w mode will output at that port. In fact, according to the principle of reciprocity, the forward and reverse directions of the coupling arm are arbitrary. Similarly, the forward and reverse directions of the main waveguide are also arbitrary, with the direction that is the same as the direction of the input microwave signal transmission being the forward direction, and the opposite being the reverse direction. Therefore, such a mode selective coupler can itself serve as a single-mode generator, and regardless of which port the microwave signal is input into the coupling arm, it will only excite the same mode as the coupling mode in the main waveguide. Using such a mode selective coupler to calibrate the coupling coefficient of the mode selective coupler under test eliminates the trouble of manufacturing single-mode generators. Since the mode selective coupler serving as a single-mode generator and the mode selective coupler under test are identical, with only the input and output directions of the two couplers being reversed when connected, it is called back-to-back measurement, a method that is quite commonly used in microwave measurements.

(2) **Back-to-back calibration system for mode selective couplers**

The schematic diagram of the measurement system for back-to-back calibration is shown in Fig. 7.15. Except for the input and output ports, all other ports should be terminated with matching loads. The mode selective coupler used to generate a single mode is referred to as Coupler I, while the mode selective coupler under test is referred to as Coupler II.

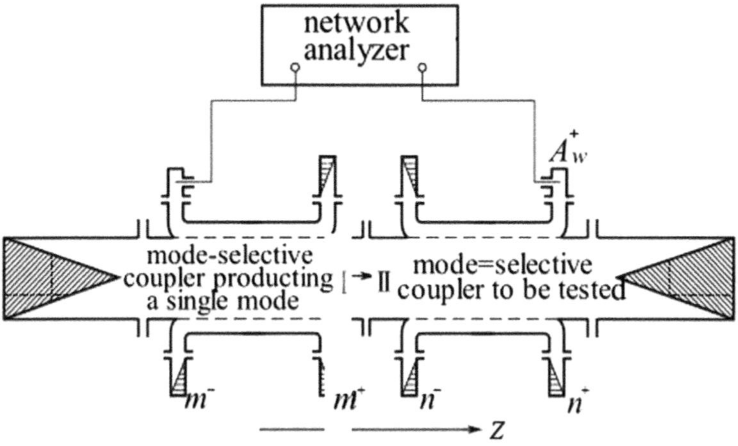

Fig. 7.15 Schematic diagram of the measurement system used to calibrate the coupling coefficient of a mode-selective coupler using the back-to-back calibration method

The back-to-back calibration method simplifies the calibration system by using the mode selective coupler as a single-mode generator. However, this method imposes higher requirements on the dynamic range of the network analyzer, as the input signal needs to go through two couplings before it can be output from the output port. Thus, the coupling coefficient of the output signal relative to the input signal will be twice the value of the coupling coefficient of the mode selective coupler. For example, if a mode's coupling arm has a coupling of 50 dB, the signal excited in the main waveguide by the input signal from the mode selective coupler acting as a single-mode generator will be 50 dB lower than the input signal. When this mode signal is output from the secondary waveguide of the mode selective coupler under test, it undergoes an additional 50 dB attenuation. This means the output signal will be 100 dB lower than the input signal, requiring a network analyzer with a dynamic range of at least 100 dB.

To simplify the subsequent equation derivation, let's assume there are only three modes transmitted in the main waveguide of the mode selective coupler, denoted as 1, 2, and 3. Correspondingly, there are three coupling arms set up for those three modes on the main waveguide, denoted as arms 1, 2, and 3, with the arm numbers corresponding to the mode numbers.

(3) **Basic Equations and Symbol Definitions**

According to the definition of coupling coefficient, we can write:

$$C_w^+ = -20\lg A_w^+ = C_w$$

$$C_u^+ = -20\lg A_u^+ = -20\lg\left(A_w^+ \frac{A_u^+}{A_w^+}\right) = C_w + p_u$$

$$
\begin{aligned}
C_w^- &= -20\lg A_w^- = -20\lg\left(A_w^+\frac{A_w^-}{A_w^+}\right) = C_w + d_w \\
C_u^- &= -20\lg A_u^- = -20\lg\left(A_u^+\frac{A_u^-}{A_u^+}\right) = C_u + d_u \\
&= C_w + p_u + d_u
\end{aligned}
\tag{7.63}
$$

where

the subscripts w and u represent the coupled and uncoupled modes, while the superscripts $+$ and $-$ represent the forward and backward waves,

C represents the coupling coefficient,

p represents the suppression coefficient of the coupled mode to the uncoupled mode,

d represents the directivity.

$A_w{}^{\pm}$ and $A_u{}^{\pm}$ respectively represent the relative amplitudes of the forward and backward waves excited in the secondary waveguide for the coupled and uncoupled modes, which is the coupling strength. We have already provided their calculation formulas above. From Eq. (7.63), it can be seen that the coupling coefficients of the coupled or uncoupled modes can be obtained by measuring C_w, d_w, P_u, and d_u, all of which are represented by positive values.

The reason for writing out the coupling coefficient of the uncoupled mode in Eq. (7.63), as we have pointed out above, is that the presence of the uncoupled mode affects the accuracy of the coupling coefficient calibration of the coupled mode. In practice, the suppression coefficient of the uncoupled mode in the manufactured mode selective coupler cannot reach the ideal level, especially the suppression coefficient decreases rapidly after the frequency deviates from the center frequency. Therefore, to accurately calibrate the coupling coefficient, the uncoupled mode must be considered.

For convenience, we use k to denote the mode serial number 1, 2, 3, m to denote the serial number of the coupling arm of mode selective coupler I, and n to denote the serial number of the coupling arm of mode selective coupler II. In addition, it has been stipulated earlier that the subscript w represents the coupled mode and the subscript u represents the uncoupled mode, which is the mode that needs to be suppressed. In this way, for the convenience of future writing, we have defined the following series of symbols.

$C_{m,k}$: The coupling coefficient of mode k in the m coupling arm of mode selective coupler I.

$C_{m,k=m}$: The coupling coefficient of the coupled mode in the m coupling arm of coupler I. As mentioned earlier, the coupling arm serial number corresponds to the coupled mode serial number, so if $k = m$, then mode k represents the coupled mode of arm m, equivalent to $C_{m,k} = C_{m,w}$.

$C_{m,k\neq m}$: The coupling coefficient of the uncoupled mode in the m coupling arm, because $k \neq m$ implies that mode k is not the coupled mode, equivalent to $C_{m,,k} = C_{m,u}$.

$C_{n,k}$: The coupling coefficient of mode k in the n coupling arm of mode selective coupler II.

$C_{n,k=n}$: The coupling coefficient of the coupled mode in the n coupling arm of coupler II, because the coupling arm serial number corresponds to the coupled mode serial number, so n represents the coupled mode in the n coupling arm, i.e., $C_{n,k} = C_{n,w}$. Also, because the two mode selective couplers are exactly the same, if the n coupling arm and the m coupling arm are the same coupled mode, i.e., $n = m$, then their coupling coefficients should be equal, so we should have $C_{n,k} = C_{m,k}$ $(n = m)$.

$C_{n,k\neq n}$: The coupling coefficient of the uncoupled mode in the n coupling arm of coupler II, equivalent to $C_{n,k} = C_{n,u}$.

$p_{m,k\neq m}$: The suppression coefficient of the coupled mode to the uncoupled mode in the m coupling arm of mode selective coupler I, because the $k \neq m$ modes are the uncoupled mode, so only $k \neq m$ modes have suppression.

$p_{n,k\neq n}$: The suppression coefficient of the coupled mode to the uncoupled mode in the n coupling arm of mode selective coupler II. Similarly, because $k = n$ is the coupled mode, $k \neq n$ modes are the uncoupled modes. Also, based on the fact that the two mode-selective couplers are exactly the same, so, if $n = m$ and k is the same mode, then $p_{n,k\neq n} = p_{m,k\neq m}(n = m)$.

$d_{m,k}$: The directivity of mode k in the m coupling arm of mode selective coupler I. When $k = m$, $d_{m,k}$ is the directivity of the coupled mode in the m coupling arm, and when $k \neq m$, it represents the directivity of the uncoupled mode in the m coupling arm.

$d_{n,k}$: The directivity of mode k in the n coupling arm of mode selective coupler II. When $k = n$, $d_{n,k}$ is the directivity of the coupled mode in the n coupling arm, and when $k \neq n$, it represents the directivity of the uncoupled mode in the n coupling arm. Also, based on the fact that the two mode selective couplers are exactly the same, if $n = m$ and k is the same mode, $d_{n,k} = d_{m,k}$.

$P_k(m^{\pm})$: The power of mode k excited in the main waveguide when the signal is input from the $+ z$ direction port m^+ or-z direction port m^- of the m coupling arm of mode selective coupler I.

$P_k(m^{\pm}, n^{\pm})$: When the signal is input from port $m^{\pm}$ in the ±z direction of the m coupling arm of the mode-selective coupler I, the k mode excited in the main waveguide outputs power through port $n^{\pm}$ in the ±z direction of the n coupling arm of the mode-selective coupler II.

2. **Establishment of the calibration equations**

From Eq. (7.63), to obtain the coupling coefficients of the coupled and uncoupled modes, for a mode-selective coupler with three coupling arms, there will be a total of 18 quantities that need to be measured, namely the coupling coefficients $C_{1,1}$, $C_{2,2}$, $C_{3,3}$ of the three coupled modes; the suppression coefficients $p1,2, p1,3, p2,1, p2,3, p3,1, p3,2$ of each coupled mode to the two uncoupled modes in each coupling arm; the directivities $d1,1, d2,2, d3,3$ of the three coupled modes and the directivities $d1,2, d1,3, d2,1, d2,3, d3,1, d3,2$ of the uncoupled modes in each coupling arm.

(1) **Signal P_{in} is Input from M^- Port**

When the signal is input from the-z end of the m coupling arm of coupler I, the power of mode k excited in the main waveguide in the $+z$ direction is given by

$$\begin{aligned} P_k(m^-) &= 10^{-[(C_{m,k=m}+p_{m,k\neq m})/10]}P_{in} = K_{m,k}P_{in} \\ K_{m,k} &= 10^{-[(C_{m,k=m}+p_{m,k\neq m})/10]} \end{aligned} \tag{7.64}$$

where, $m = 1,2,3,k = 1,2,3$; when $k = m$, $p_{m,k} = 0$. Obviously, for different combinations of m and k, Eq. (7.64) should be able to be written as 9 specific equations, i.e., $P_1(1^-)$, $P_2(1^-)$, $P_3(1^-)$, $P_1(2^-)$, $P_2(2^-)$, $P_3(2^-)$, $P_1(3^-)$, $P_2(3^-)$ and $P_3(3^-)$. Since there are two non-coupled modes for $k \neq m$, the term $p_{m,k\neq m}$ should be the sum of two terms.

(a) $P_k(m^-)$ is output from n^+ port

When the power of mode k excited in the main waveguide of coupler I is output from the $+z$ end n^+ of the n coupling arm of coupler II, since k can be $k = 1,2,3$, the total output power is given by

$$\begin{aligned} P(m^-, n^+) &= \sum_{k=1,2,3} 10^{-[(C_{n,k=n}+p_{n,k\neq n})/10]}P_k(m^-) = \sum_{k=1,2,3} K_{n,k}K_{m,k}P_{in} \\ K_{n,k} &= 10^{-[(C_{n,k=n}+p_{n,k\neq n})/10]} \end{aligned} \tag{7.65}$$

where, $m = 1,2,3$, $n = 1,2,3$. when $k = n$, $p_{n,k} = 0$. For different combinations of m and n, Eq. (7.65) can also be written as 9 specific equations, i.e., $P(1^-, 1^+)$, $P(1^-, 2^+)$, $P(1^-, 3^+)$, $P(2^-, 1^+)$, $P(2^-, 2^+)$, $P(2^-, 3^+)$, $P(3^-, 1^+)$, $P(3^-, 2^+)$ and $P(3^-, 3^+)$. Since coupler I and II are exactly the same, according to the reciprocity principle, when $m \neq n$, there exists the relationship:

$$\sum_{k=1,2,3} K_{n',k}K_{m',k}P_{in} = \sum_{k=1,2,3} K_{n,k}K_{m,k}P_{in} \quad (n' = m, m' = n) \tag{7.66}$$

Equation (7.66) indicates that

$$P(m^-, n^+) = P(n^-, m^+) \tag{7.67}$$

The significance of Eqs. (7.66) or (7.67) is that when the coupling arm serial numbers of coupler I are exchanged with the coupling arm serial numbers of coupler II, the output in the coupling arm of coupler II remains unchanged. This is not difficult to understand, as exchanging the coupling arms of the two couplers is equivalent to exchanging the entire couplers, and since the two couplers are exactly the same, such an exchange does not change their performance. However, this means that among the 9 equations originally contained in Eq. (7.65), the 6 equations that satisfy Eq. (7.66) are equivalent in pairs, i.e., $P(1^-, 2^+) = P(2^-, 1^+)$, $P(1^-, 3^+) = P(3^-, 1^+)$, $P(2^-, 3^+) = P(3^-, 2^+)$, so Eq. (7.65) actually contains only 6 independent equations.

(b) $P_k(m^-)$ is output from n^- port

If the power of mode k ($k = 1,2,3$) excited in the main waveguide of coupler I is output from the n^-port of the n coupling arm of coupler II along $-z$ direction, its output power is:

$$P(m^-, n^-) = \sum_{k=1,2,3} 10^{-[(C_{n,k=n}+p_{n,k\neq n}+d_{n,k})/10]} P_k(m^-) = \sum_{k=1,2,3} R_{n,k} K_{m,k} P_{in}$$
$$R_{n,k} = 10^{-[(C_{n,k=n}+p_{n,k\neq n}+d_{n,k})/10]} \quad (7.68)$$

where, $m = 1,2,3$, $n = 1,2,3$. when $k = n$, $p_{n,k} = 0$. Equation (7.68) also contains 9 specific equations with different combinations of m and n, and these 9 equations are all independent equations.

(2) **Signal P_{in} is Input from M^+ port**

When the signal is input into the $+z$ terminal of the m coupling arm of coupler I, only the backward wave excited by P_{in} in the main waveguide is transmitted in the $+z$ direction, and according to the definition of the backward wave coupling coefficient given in Eq. (7.63), we can write that the power of the k mode excited in the $+z$ direction is

$$P_k(m^+) = 10^{-[(C_{m,k=m}+p_{m,k\neq m}+d_{m,k})/10]} P_{in} = R_{m,k} P_{in}$$
$$R_{m,k} = 10^{-[(C_{m,k=m}+p_{m,k\neq m}+d_{m,k})/10]} \quad (7.69)$$

In the equation: $m = 1,2,3$, $k = 1,2,3$; when $k = m$, $p_{m,k} = 0$. Obviously, similar to Eqs. (7.64), and (7.69) should also contain 9 specific equations for different combinations of m and k, and $p_{m,k\neq m}$ should be the sum of two terms.

(a) $P_k(m^+)$ is output from n^+ port

When the power of mode k is output from the port n^+ of the n coupling arm of the coupler II along the $+z$ direction, since $k = 1, 2, 3$, the total output power is:

$$P(m^+, n^+) = \sum_{k=1,2,3} 10^{-[(C_{n,k=n}+p_{n,k\neq n})/10]} P_k(m^+)$$

$$= \sum_{k=1,2,3} K_{n,k} R_{m,k} P_{in} \tag{7.70}$$

where, $K_{n,k}$ and $R_{m,k}$ are given in Eqs. (7.65) and (7.69). For the same reasons as Eq. (7.66), it can also be written as:

$$\sum_{k=1,2,3} K_{n',k} R_{m',k} P_{in} = \sum_{k=1,2,3} R_{n,k} K_{m,k} P_{in} \quad (n' = m,\ m' = n) \tag{7.71}$$

which is:

$$P(m^+, n^+) = P(n^+, m^+) \tag{7.72}$$

It is easy to see that Eq. (7.70) is actually completely equivalent to Eq. (7.68), they just swap the positions of K and R, which does not affect the calculation result of power. It can be seen that Eq. (7.70) does not provide new independent equations, but it provides another way to measure power.

(b) $P_k(m^+)$ is output from n^- port

If the power of mode k in the main waveguide of coupler I, $P_k(m^+)$ is output from the port n^- of the n coupling arm of coupler II along –z direction, then the output power is given by:

$$\begin{aligned} P(m^+, n^-) &= \sum_{k=1,2,3} 10^{-[(C_{n,k=n} + p_{n,k\neq n} + d_{n,k})/10]} P_k(m^+) \\ &= \sum_{k=1,2,3} R_{n,k} R_{m,k} P_{in} \end{aligned} \tag{7.73}$$

where, $R_{n,k}$ and $R_{m,k}$ are already given in Eqs.(7.68) and (7.69). $m = 1,2,3$, $n = 1,2,3$, when $k = n$, $R_{n,k} = 0$. Equation (7.73) also contains 9 specific equations for different combinations of m and n, but it can be found that, similar to Eqs.(7.66) and (7.67), when $m \neq n$, the following relationship exists:

$$\sum_{k=1,2,3} R_{n',k} R_{m',k} P_{in} = \sum_{k=1,2,3} R_{n,k} R_{m,k} P_{in} \quad (n' = m,\ m' = n) \tag{7.74}$$

which is:

$$P(n^+, m^-) = P(m^+, n^-) \tag{7.75}$$

Equations (7.74) and (7.75) show that $P(1^+, 2^-) = P(2^+, 1^-)$, $P(1^+, 3^-) = P(3^+, 1^-)$, $P(2^+, 3^-) = P(3^+, 2^-)$, which means that among the 9 equations in Eq. (7.73), 6 are equivalent in pairs, so it actually only contains 6 independent equations.

7.4.3 Coupling Coefficient Calibration, Mode Discrimination, and Power Measurement

1. **Calibration of coupling coefficient**

   (1) **Establishment of the equation group**

Equations (7.65), (7.68) or (7.70), and (7.73) together provide a total of 21 independent equations, while the unknown coupling coefficient, suppression coefficient, and directivity that we need to solve for are only 18. Therefore, it is completely possible to measure the output power of the $\pm z$ terminals of the three coupling arms of coupler II in order to determine these 18 parameters. In reality, the parameters of interest are the coupling coefficients C_{11}, C_{22}, C_{33} of each coupling arm, and the suppression coefficients p_{12}, p_{13}, p_{21}, p_{23}, p_{31}, p_{32} for each coupling mode to two non-coupled modes, totaling 9 parameters. As we have pointed out before, the parameters relevant to the power of a mode at the input of the mode selective coupler are only these 9, and are independent of the power of the backward direction, and therefore independent of directivity. However, the problem lies in the fact that without considering directivity and relying solely on coupling coefficients and suppression coefficients, it would be impossible to establish 9 independent equations, and thus solve for these 9 parameters. The same situation occurs even if only considering the coupling coefficients of the 3 coupling modes and ignoring the influence of suppression coefficients on power calculation, where the number of independent equations that can be established is less than the number of parameters to be solved for. For this reason, we have to establish 21 equations to solve for these 18 parameters, and in the end, when calculating power, only the aforementioned 9 parameters are actually used.

Each coupling arm of coupler I has two input ports ($+z$ end and $-z$ end, respectively), and with three coupling arms, there are a total of 6 ports that can be used as input ports ($m^{\pm}$, $m = 1, 2, 3$). Each input signal input into the main waveguide will excite the same three mode waves ($k = 1, 2, 3$). When these modes transmit to coupler II, there will be 6 output ports ($n^{\pm}$, $n = 1, 2, 3$) from which power can be measured, or their ratios can be directly measured using a network analyzer, resulting in a total of 36 absolute or relative power values that can be measured. However, as mentioned earlier, only 21 of these measurements are truly independent, and the rest are theoretically repeated. By using 18 of the 21 independent measurements to establish independent equations, the required coupling coefficients, suppression coefficients, and directivity can be calibrated, and the remaining three measurements can be used for validation.

(2) **Measurement methods**

When using the measurement system shown in Fig. 7.16 for measurements, there are two methods: method one involves using a network analyzer to measure the ratio of output power to input power, i.e., the value of $|S_{21}|^2$, and method two involves directly using a standard high-power source (output mode only needs to be TE_{10} mode) as

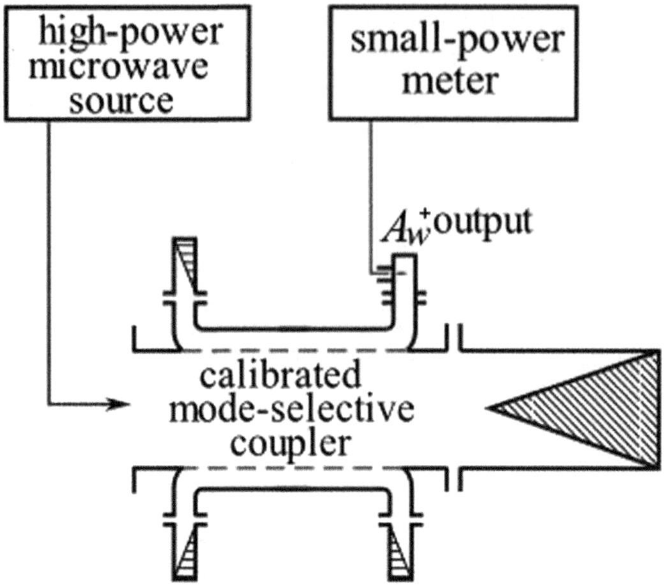

Fig. 7.16 Schematic diagram of the mode discrimination and power measurement system of the mode selective coupler

the input signal and measuring its power P_{in}, while simultaneously using a small-power meter to measure the output power $P(m^{\pm}, n^{\pm})$ at each port. By substituting the obtained $|S_{21}|^2$ value or P_{in} and the sizes of $P(m^{\pm}, n^{\pm})$ into Eqs. (7.65), (7.68) or (7.70), and (7.73), the sizes of all C, d, and p can be determined.

Since the input signal is input from the secondary waveguide (coupling arm) of the mode selective coupler I, and the secondary waveguides all operate in the fundamental mode TE_{10}, there is no need for single-mode generators for each mode in the main waveguide. The waves input into the main waveguide from the secondary waveguide can excite the required modes, thereby achieving calibration of the coupling coefficients, suppression coefficients, and directivity for each mode, and this is the greatest advantage of the back-to-back calibration method.

The main drawback of the back-to-back calibration method is the requirement to produce two completely identical mode selective couplers. Any dimensional errors between them can lead to differences in parameters of couplers, thereby affecting the accuracy of the calibration results. Additionally, the back-to-back calibration method imposes very high demands on the dynamic range of the network analyzer, as the signal will experience significant attenuation from input to output during measurement (maximum up to the sum of positive coupling coefficient, suppression coefficient, and directivity), meaning the output signal will be very weak, thus requiring a high measurable range for the instrument. The back-to-back calibration method involves many measurement parameters and a relatively complex calculation process, which is another drawback of the method.

2. **Mode discrimination and power measurement**
(1) **Measurement system and mode discrimination**

After the mode selective coupler is calibrated, mode discrimination and power measurement of high-power microwaves can be carried out. During measurement, high-power microwaves are input into the main waveguide of the mode selective coupler, and the output power $P(n^{\pm})$ in the $+z$ and- z directions of each coupling arm is measured. Depending on whether each coupling arm has an output and the size of the output, it is determined whether the corresponding mode coupled to that arm exists in the main waveguide. Furthermore, based on the calibrated coupling coefficient and suppression coefficient, the power of that mode in the main waveguide can be calculated. By sequentially measuring each coupling arm, the mode composition in the main system, the power of each mode, and their respective proportions can be identified. The test system is shown in Fig. 7.16, where a calibrated mode selective coupler is connected to the high-power microwave transmission system. High-power microwaves generated by the high-power microwave source are fed into the left end of the coupler, and the right end of the coupler is connected to the system's operating load. The power output from the coupling arm of the coupler is indicated by a calibrated power indicator, and a crystal detector and a memory oscilloscope are commonly used as the calibrated power indicator. To protect the crystal detector, an attenuator can be connected to its front stage when necessary.

(2) **Power Calculation**

Let the input power of high-power microwaves from the left end of the mode selective coupler be P_t, which contains three modes: 1, 2, and 3, with corresponding powers P_1, P_2, and P_3.

The power measured at the $+z$ end of coupling arm 1 of the mode coupler is

$$P(1^{+}) = K_{1,1}P_1 + K_{1,2}P_2 + K_{1,3}P_3 \tag{7.76}$$

The power measured at the $+z$ end of coupling arm 2 of the mode coupler is

$$P(2^{+}) = K_{2,1}P_1 + K_{2,2}P_2 + K_{2,3}P_3 \tag{7.77}$$

The power measured at the + z end of coupling arm 3 of the mode coupler is

$$P(3^{+}) = K_{3,1}P_1 + K_{3,2}P_2 + K_{3,3}P_3 \tag{7.78}$$

Then the total power in the system is given by

$$P_t = \sum_{k=1,2,3} K_{n,k}P_k \tag{7.79}$$

where, $K_{n,k}$ is given by Eq. (7.65). According to the calibration method given above, we have obtained the coupling coefficient and suppression coefficient of each mode, from which we can calculate $K_{n,k}$. It is immediately apparent from Eq. (7.65) that $K_{n,k}$ depends only on the coupling coefficient of the forward wave of each coupling arm and the forward suppression coefficient of the coupled mode to the uncoupled mode. This once again proves our previous conclusion: power measurement is only related to the coupling coefficient and suppression coefficient of the forward wave, and is independent of the reverse direction, i.e., independent of directivity.

Using the measured values of $P(1^+)$, $P(2^+)$, and $P(3^+)$, we can calculate P_1, P_2, and P_3 using Eqs. (7.76), (7.77), and (7.78). Then, using Eq. (7.79), we can obtain the total power in the system.

7.5 Wavenumber Spectrum Measurement Method-Small-Hole Array Antenna Method

When using a mode selective coupler for mode measurement, if there are many possible modes in the main waveguide, that is, in other words, if the waveguide is more severely overmoded, the mode spectrum will be very dense, making it quite difficult to suppress uncoupled modes. This also means that there is a particularly high demand for mode selectivity, which places stringent requirements on the machining precision of the coupling structure. Additionally, the operating frequency band of the mode selective coupler will become narrower with the increase in uncoupled modes. The required coupling arms will also increase with the increase in coupled modes. Therefore, the application of the mode selective coupler method is subject to certain limitations.

The wavenumber spectrum measurement method is a mode measurement method proposed based on the radiation direction of the small hole array antenna depends on the type of transmission mode, with different modes having different radiation direction angles. This method can identify almost an unlimited number of modes, limited only by the spatial angular resolution of the receiving horn. Similar to the mode selective coupler method, it is also a method that can be used for online measurements. However, due to the need to move the receiving horn during the measurement process, it is generally difficult to achieve dynamic and real-time measurements. The measurement device is large in size, requires high precision in processing, and the design and processing are quite complex. The one-time cost is relatively high. These are its drawbacks.

7.5.1 Fundamental Principles and Measurement Instruments of Wavenumber Spectrum Method [96]

1. **Measurement Principle**

A series of small holes can be directly opened on the circular waveguide wall to form a small hole array antenna, or a continuous gap can be opened to form a slot antenna, both of which can be collectively referred to as leaky wave antennas. If a microwave with a wavenumber k_{mn} is propagating in the circular waveguide, a small portion of the wave will radiate into space from the antenna structure. The direction angle θ_{mn} of the main lobe of the radiation pattern is in a one-to-one correspondence with the wavenumber k_{mn}, i.e.,

$$\cos\theta_{mn} = \frac{k_{mn}}{k_0} \tag{7.80}$$

where, θ_{mn} is the angle relative to the waveguide axis at which the main lobe of the mn mode radiates through the leaky wave antenna, i.e., the radiation direction angle; k_{mn} is the wavenumber of that mode in the waveguide; k_0 is the free space wavenumber. It can be seen that each mode that may propagate in the waveguide has a different radiation direction angle, thus separating the radiation of each mode in space to form a spatial mode spectrum. If the microwave frequency ω, i.e., k_0 is known or measured in advance, then by measuring in the direction of θ_{mn} corresponding to each mode, the presence or absence of the mode as well as its relative power of that mode can be determined.

The following discussion will specifically address the case of the small hole array antenna, and the slot antenna has similar performance.

Actual measurements of the small hole array antenna show that the radiation pattern is in good agreement with the radiation characteristics of a magnetic dipole. The normal electric field of the corresponding small hole in the waveguide wall can be considered as the action of an electric dipole, and its contribution is relatively weak, at least 10 dB lower than the radiation of the magnetic dipole. Therefore, from the perspective of radiation field measurement, the magnetic field component on the waveguide wall plays a decisive role.

For the TM_{mn} mode, the magnetic field on the waveguide wall has only an angular component:

$$H_\phi = \sqrt{\frac{4}{\varepsilon_m \pi}}\sqrt{\frac{P}{Z_0}}\sqrt{\frac{k_0}{k_{mn}}\frac{1}{a}}\sin m\varphi \, \mathrm{e}^{-\mathrm{j}k_{mn}z} \tag{7.81}$$

where,

P is the mode power,

Z_0 is the wave impedance in free space,

$k_0 = \omega/c$ is the wavenumber in free space,

m is the angular direction characteristic value of the mode,

a is the radius of the circular waveguide.

If absolute phase is ignored, when $m = 0$, $\varepsilon_m = 2$; when $m \neq 0$, $\varepsilon_m = 1$. The wavenumber k_{mn} of the TM_{mn} mode is given by Eq. (7.82).

$$k_{mn} = \frac{\omega}{c}\left[1 - \left(\frac{\mu_{mn}}{k_0 a}\right)^2\right]^{1/2} \tag{7.82}$$

where, μ_{mn} is the nth root of the m order Bessel function of the first kind, i.e., $J_m(\mu_{mn}) = 0$.

For the TE_{mn} mode, the magnetic field on the waveguide wall has both angular and longitudinal components, i.e.:

$$\begin{aligned} H_\varphi &= \sqrt{\frac{4}{\varepsilon_m \pi}}\sqrt{\frac{P}{Z_0}}\sqrt{\frac{k_{mn}}{k_0}}\frac{1}{a}\frac{m}{(\mu'^2_{mn} - m^2)^{1/2}} \sin m\varphi\, e^{-jk_{mn}z} \\ H_z &= \sqrt{\frac{4}{\varepsilon_m \pi}}\sqrt{\frac{P}{Z_0}}\sqrt{\frac{k_{mn}}{k_0}}\frac{1}{k_{mn}a^2}\frac{\mu'^2_{mn}}{(\mu'^2_{mn} - m^2)^{1/2}} \cos m\varphi\, e^{-jk_{mn}z} \end{aligned} \tag{7.83}$$

Here, the wavenumber of the TE_{mn} mode is given by:

$$k_{mn} = \frac{\omega}{c}\left[1 - \left(\frac{\mu'_{mn}}{k_0 a}\right)^2\right]^{1/2} \tag{7.84}$$

where, μ'_{mn} is the nth root of the derivative of the m order Bessel function of the first kind, i.e., $J'_m(\mu'_{mn}) = 0$.

2. **Measurement device**

(1) **Basic structure**

The wavenumber spectrum mode measurement method was proposed by scholars W. Kasparek and others from the University of Stuttgart, Germany. They named the wavenumber spectrum measurement device wavenumber spectrometer. It is a type of mode spectrum measurement method. The schematic diagram of the experimental device structure is shown in Fig. 7.17. The device consists of a section of circular overmoded waveguide for transmitting high power microwave as the main body. The waveguide wall has a certain number of small holes arranged in a linear pattern with a regular variation in hole radius, forming a small hole array. A receiving horn is positioned at a distance R above the symmetrical center of the small hole array and can move 90° to the left and right in the θ direction on a sliding track. The sliding track, along with the receiving horn, can rotate within a certain range in the φ direction. The measuring device can also be installed using a long support rod of length R, placed at a specific location outside the waveguide, aligning with the symmetrical center of a small hole array in the z-direction and extending along the horizontal radius of the circular waveguide in the r-direction. A receiving horn is fixed at the top of the support rod, protruding from it so that the horn mouth faces directly above the small hole array. The support rod and the receiving horn can rotate horizontally in the θ direction and can also adjust to a definite angle in the φ direction. The picture of real products of the wave number spectrum measurement device shown in Fig. 7.23 is based on this structure. Adjusting the hole size r, the number of holes, and the waveguide wall thickness t can change the radiation intensity of the small hole array.

The coordinate system used by the wave number spectral measurement device is also shown in Fig. 7.17. From this figure, it can be seen that for each hole in the small hole array on the waveguide wall, if R is always much larger than r_0^2/λ (where r_0 is the hole radius), and also much larger than λ, the radiation field of the receiving horn position relative to a single hole can be considered as far-field radiation. However, for the entire radiating small hole array, it is evident that the radiation fields of each

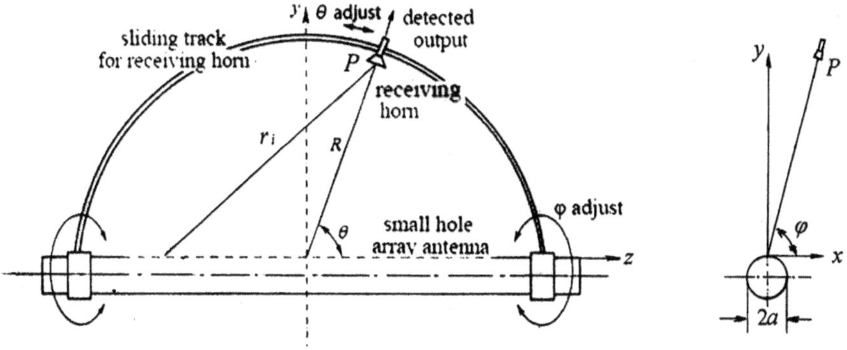

Fig. 7.17 Schematic diagram of the wavenumber spectrum measurement device

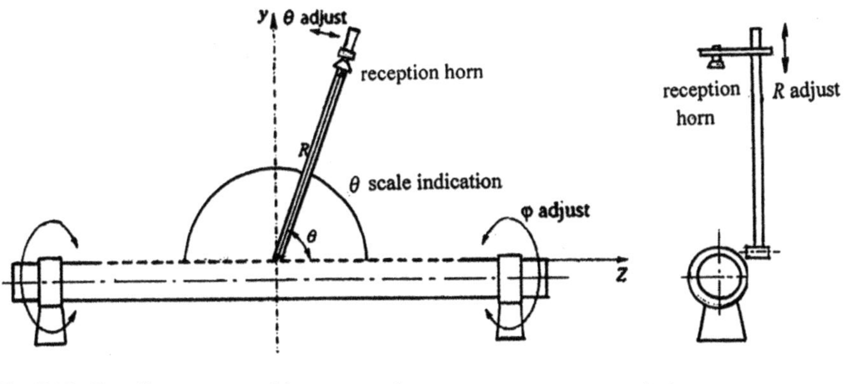

Fig. 7.18 Coordinate system of the wavenumber spectrum measurement device

hole do not reach the receiving horn in parallel, and thus must be treated as near-field radiation. By using the theory of far-field radiation from small apertures and the principle of superposition of fields, it is not difficult to calculate the near-field radiation field of the small hole array at the position of the receiving horn, i.e., at the position of distance *R*. In theory, the radiation pattern and the radiation intensity at the main lobe angle θ_{mn} of the small hole array can be obtained.

(2) **Main dimensions of structure**

The wavenumber spectrum measurement device constructed by W. Kasparek et al. uses a C76 standard circular waveguide as the main waveguide (American EIA standard WC109, Chinese standard BY76) with an inner radius of 13.9 mm for low-power measurements, operating in the V band (50–75 GHz) with a corresponding wavelength of λ0 = 4.28 mm at the center frequency. For high-power measurements, the diameter of the main waveguide will be increased to 63.4 mm.

The waveguide is 2 m long, and 600 small holes are drilled in a straight line on the waveguide to form a radiation array. The diameter of the small holes increases from 0.6 to 1.0 mm in steps of 0.025 mm from the edge to the center of the small hole array, following a Gaussian function distribution exp(- z^2/340 mm^2) and symmetrically distributed on the left and right sides of the center. The spacing between the holes is 2 mm, and the waveguide wall thickness is 0.6 mm. Therefore, the total length of the small hole array is 1200 mm.

To improve measurement sensitivity, a parabolic antenna and a rectangular horn are commonly used as receiving antennas. The parabolic antenna is linearly distributed in the *y*–*z* plane and curved in the *x*–*y* plane. After the radiation field is concentrated by the parabolic reflector antenna, it is then received and detected by the horn antenna for output. The diameter of the parabolic reflector is 170 mm, and the angular resolution of the receiving antenna reaches 2°. The receiving antenna is installed at the top of a support arm, which extends from the center of the corresponding small hole array and can be scanned above the small hole array under the drive of the motor. For high-power applications, an additional receiving antenna is installed to estimate the measurement values and the magnitude of reflections. This antenna can be manually moved or fixed.

The entire measurement system is shielded on both sides of the circular waveguide with two microwave absorbing plates to avoid stray radiation.

7.5.2 *Main Factors to Be Considered in Wavenumber Spectrum Method [96, 97]*

1. **Spacing *d* between small holes**

In order to minimize the impact of side lobes of the radiation from the small holes, the spacing *d* between the small holes must be less than half of the wavelength λ_0 of the microwaves in free space, $d < \lambda_0/2$. At the same time, small holes with a certain wall thickness will form a short section of circular waveguide, which will be cutoff waveguide for all modes, including the fundamental mode TE_{11}, i.e.

$$\lambda_0 > (\lambda_c)_{TE_{11}} \quad (7.85)$$

where, λ_c is the cutoff wavelength of the TE_{11} mode in the circular waveguide formed by the radiation from the small holes. Equations (4.71) to (4.73) provide a correction for the coupling strength of the small holes considering the wall thickness.

2. **Coupling hole radius r_0**

In order to minimize the impact of the coupled small holes on the power transmission in the waveguide, the microwave power coupled out must be only a small fraction of the power transmitted in the waveguide. This requires that the coupling hole meets the conditions for small hole coupling, i.e., the radius r_0 of the small hole must be much smaller than the free-space wavelength λ_0 of the microwaves, $r_0 \ll \lambda_0$.

3. **Side lobe radiation should be below − 30 dB**

When measuring at a certain angle θ_{mn} in space, the radiation side lobes corresponding to the mode of the radiation angle must be sufficiently small, so that the side lobe radiation intensity is at least 30 dB lower than the main lobe radiation intensity in the θ_{mn} direction, in order to avoid the side lobe's impact on the measurement of adjacent modes at the radiation angle.

4. **Gaussian distribution of small hole array**

Different small hole array distributions have different radiation characteristics. After comparing the theoretical calculation results of the radiation field of linear distribution, binomial distribution, Chebyshev polynomial distribution, and Gaussian distribution of the small hole array, considering factors such as the radiation main lobe width, directionality, and suppression of side lobes, the small hole array with a Gaussian distribution is superior to other distributions of small hole arrays [98].

The Gaussian distribution can be represented as:

$$f = e^{-z^2/2\sigma^2} \quad (7.86)$$

where, σ is a constant, which can be repeatedly adjusted and determined during the theoretical calculation of the radiation field. The hole radius of the small hole array should be distributed according to the requirement in Eq. (7.86). In practice, strictly determining the hole radius of each hole according to the Gaussian distribution will cause great difficulty in the processing of the small holes. Generally, from the perspective of manufacturability, a stepped Gaussian distribution of the hole radius is more realistic. However, it should be noted that the variation in aperture is symmetrically distributed from the center position of the small hole array, with the hole radius increasing from both sides to the center.

5. **Reference receiving antenna**

In order to monitor the size of the reflected waves, an additional receiving antenna can be set up. This antenna can be manually adjusted or fixed in position.

6. **Differentiation between TE mode and TM mode**

Due to the fact that the radiation is mainly generated by the magnetic field component at the waveguide wall, the TM mode has only the H_φ component at the waveguide wall, i.e., the direction perpendicular to the main waveguide axis. Since the magnetic field and electric field of the electromagnetic field are perpendicular to each other, the electric field of its radiation field will be in the direction parallel to the waveguide axis, i.e., ***E//K***, where ***K*** is the wavenumber vector of the electromagnetic wave in the waveguide. In space, this means that the radiation magnetic field is in the φ direction (H_φ), and the electric field is in the θ direction (E_θ). Obviously, in order for the horn to receive the E_θ field component, the wide side of the horn must be parallel to the φ direction. The electric field excited in the fundamental mode rectangular waveguide of the receiving antenna is perpendicular to the direction of the wide side, i.e., in the θ direction in the y–z plane (E_θ, i.e., E_y in the rectangular waveguide). It can be seen that when the rectangular horn receives the radiation field in the ***E//K*** manner, it receives the radiation of the TM mode.

TE mode has both H_z component and H_φ component at the waveguide wall. The H_z component is parallel to the main waveguide axis, so the electric field of its radiation field will be in the direction perpendicular to the waveguide axis, i.e., $\boldsymbol{E}\perp\boldsymbol{K}$. In space, this means it should be in the φ direction. When the narrow side of the receiving horn is parallel to the φ direction and the wide side is placed parallel to the θ direction, the electric field direction excited in the fundamental basic mode rectangular waveguide is also the E_φ component (E_y in the rectangular waveguide). It can be seen that when the electric field excited in the receiving antenna is parallel to the φ direction, i.e., when the rectangular horn receives the radiation field in the $\boldsymbol{E}\perp\boldsymbol{K}$ manner, we are receiving radiation of the TE mode. Although the H_φ component of the TE mode can also excite the E_θ electric field in space, like the TM mode, but the radiation direction angles θ_{mn} are different from the TM mode, so measuring at different θ angles, they can be distinguished.

7. **Differentiation of modes with close $\boldsymbol{\theta_{mn}}$**

For modes with very close radiation direction angles θ_{mn} that cannot be distinguished by the receiving horn, they can be measured separately in different φ directions, for example, on the *y–z* plane at $\varphi = 90°$ (Fig. 7.19), measurements are taken at the radiation direction angle of one mode, and then the entire measurement support can be rotated around the waveguide axis, i.e., changing the value of φ, at a position in the *y*-plane where φ is less than 90°, measurements are taken at the radiation direction angle of another mode, thus distinguishing the two modes. Since the two measurements are taken at different angles φ, this avoids the problem of the radiation direction angles being too close for the receiving horn to be placed.

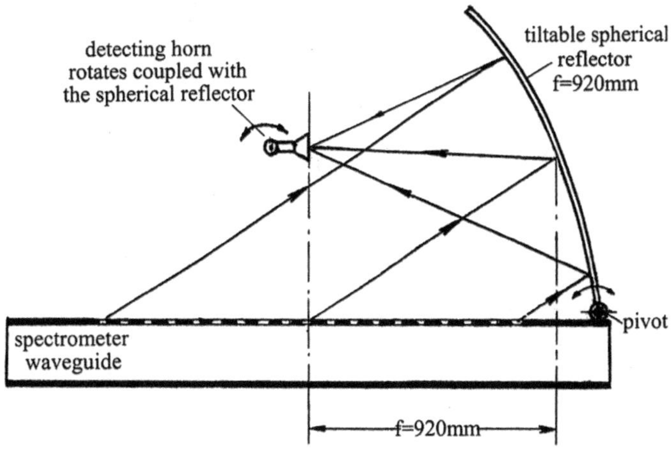

Fig. 7.19 Schematic diagram of the structure of a wavenumber spectrum measurement device for measuring small-angle θ_{mn} radiation

For degenerate modes TE_{0n} and TM_{1n} with the same radiation direction angle θ_{mn}, this method can also be used to distinguish them.

8. **Measurement at small angle θ_{mn}**

Due to the receiving horn aperture has a certain size, when the radiation angle θ_{mn} of the mode is very small, the size of the horn will limit its placement at the corresponding position of θ_{mn}, thus preventing the reception of the radiation signal. This is particularly evident for low-order modes. For example, in the experiments of W. Kasparek et al., the radiation angle of the 70 GHz TE_{01} mode is only 4.7°, and the receiving horn may not be able to be placed at such a small angle for measurement. Therefore, the resolution of the wave number spectrum measurement device will be limited by the size of the receiving antenna.

The measurement of small radiation angles can be solved by using a reflective mirror system, which consists of a tilted spherical reflective surface with a long focal length and a receiving horn. The reflective surface is placed at the end of the wave number spectrum measurement device, with a focal length equal to half the length of the wavenumber spectrum measurement device. The receiving horn is positioned above the center of the wavenumber spectrum measurement device at a certain height (Fig. 7.19). Both the spherical reflective surface and the receiving horn can be rotated around their respective axes to ensure that the spherical reflective surface can converge and reflect the beam onto the aperture of the receiving horn, allowing the radiation field of the small hole array antenna to be received by the receiving horn.

Although this measurement device effectively extends the lower end of the measurement angle range, theoretically it can even measure down to 0°, but the limited size of the reflector restricts the measurement of the upper end of the radiation angle. At a frequency of 70 GHz, a reflector 460 mm long and 100 mm wide can measure a maximum radiation angle of approximately 25°.

7.5.3 Measurement Examples of Wavenumber Spectrum Method

The following describes the calibration and measurement results of the wavenumber spectrum measurement device actually produced by the University of Stuttgart in Germany.

1. **Calibration of the device**

The calibration of the small hole array antenna can be done through calculation, as mentioned above. The ratio of the radiation power at any point in space to the waveguide transmission power can be theoretically determined using the small hole radiation theory and the principle of field superposition. It can also be obtained through experimental methods, i.e., directly measuring the radiation power of different modes and the input waveguide transmission power. Theoretical calculations require the calculation of the waveguide field, the coupling coefficient of the small holes, and the radiation direction angle of each mode, and can only yield the relative value of the radiation power. The existence of processing tolerances makes it difficult to obtain precise values for many structural dimensions, leading to inevitable errors in the calculations. Experimental calibration is simpler, but it requires the use of a single-mode generator for each mode. For example, the calibration signal source output can be first transformed into the TE_{01} mode and then a certain single mode of the TE_{0n} mode can be generated through a stepped circular waveguide, which is then input into the wavenumber spectrum measurement device. The TE_{0n} mode corresponding to the single-mode generator is excited in the wavenumber spectrum measurement device, and the receiving horn receives the signal in the corresponding radiation direction of that mode. The coupling coefficient of this mode can be directly measured through the network analyzer.

A calibrated wavenumber spectrum measurement device can be used for power measurements. The power of each mode can be measured using a small-power meter, and then the total power transmitted in the waveguide can be calculated using the coupling coefficients of each mode.

In practical applications, it is generally not necessary to calibrate the coupling coefficients because the wavenumber spectrum measurement device is mainly used for mode recognition rather than power measurement. Therefore, as long as the radiation direction angle of each mode is theoretically calculated and compared with experimental measurements, the presence of each mode can be determined. After calibrating the receiving system including the detector and oscilloscope, the relative power levels between modes can also be obtained.

2. **Measurement examples** [97]
(1) **Output mode of the gyrotron VGE8070**

Figure 7.20 shows the results of the measurement of the output mode of the gyrotron VGE8070 by the Institute for Plasma Research at the University of Stuttgart and the Max Planck Institute for Plasma Physics using the wave number spectrum measurement method. The output waveguide diameter of the gyrotron is 27.8 mm, the operating frequency is 70 GHz, and the output mode is the TE_{02} mode. From the measurement results, it can be seen that in addition to the main mode TE_{02}, the TE_{03}, TE_{13}, as well as the TM_{11} and TM_{12} modes were also measured, with their relative power levels being TE_{03}: 3.5%, TE_{13}: 6%, and TM_{11} and $TM_{12} < 1\%$. Analysis suggests that the TE_{03} mode is caused by the non-ideal linear variation of the part of conical waveguide, and the asymmetric TE_{13}, TM_{11}, and TM_{12} modes may be due to slight tilting inside the gyrotron (slightly poorer collimation). The mode spectrum in Fig. 7.20 also shows higher-order parasitic modes at $\theta = 53°$ and 62°, corresponding to the TE_{05} and TE_{16} modes, and their power can be minimized by optimizing the gyrotron's working magnetic field to the extent that they can be ignored.

(2) **Output mode of the gyrotron VGE8007**

Figures 7.21 and 7.22 show the wavenumber spectrum measurement results of the output modes of the continuous wave gyrotron VGE8007. The output waveguide diameter is 63.4 mm, and the main mode is still the TE_{02} mode. The total power content of the parasitic modes TE_{01}, TE_{03}, …, TE_{08} is approximately 4–5%, while the total content of the TM_{11}, TM_{12}, TM_{13}, and TM_{14} modes is less than 1%, meaning that the purity of the TE_{02} mode power reaches 94%. This measurement result was

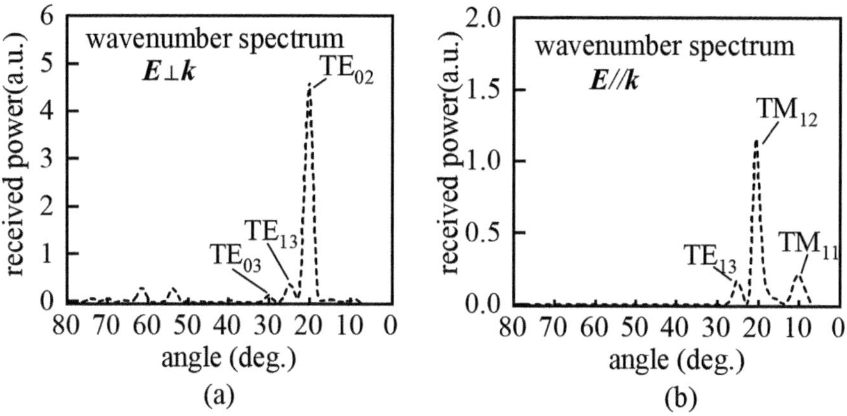

Fig. 7.20 Measurement results of the wavenumber spectrum for the output mode of gyrotron VGE8070

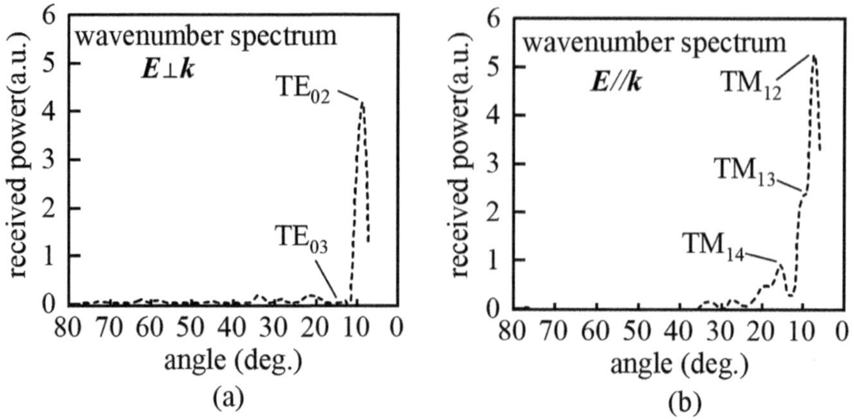

Fig. 7.21 Measurement results of continuous wave gyrotron VGE8007 using a wavenumber spectrum measurement

obtained using a reflector antenna, ensuring the measurement of low-order modes (such as the TE_{01} mode). Figure 7.21 shows the results obtained using a wavenumber spectrum measurement device with a parabolic reflector antenna with a diameter of 170 mm, while Fig. 7.22 shows the results obtained using a 460 mm high, tilted spherical surface reflector and a rectangular horn receiver.

3. **Advantages and disadvantages of wavenumber spectrum measurement method**

Figure 7.23 shows a physical photo of a relatively simple wavenumber spectrum measurement device developed by the author's research group.

Wavenumber spectrum measurement has the following advantages and disadvantages:

The main advantage of the small aperture array antenna method is that it can theoretically measure almost unlimited modes and can perform online measurements. Since the radiation field measured through the small aperture array is only a small part of the power transmitted in the waveguide, the majority of the microwave energy can be transmitted to the working load without affecting the normal operation of the microwave system.

The drawback of this method is that the receiving horn needs to be moved in the θ and even φ directions during measurement, which takes time and makes dynamic and real-time measurements difficult. This brings difficulties to the measurement of high-power microwave single pulses or low-repetition-rate pulses. Although we can cover the receiving horn along the measurement track in advance, or arrange the receiving horns at θ_{mn} values corresponding to the possible modes that may exist in the waveguide, so as to realize dynamic measurement and real-time measurement, it is not very convenient after all. One solution is to use a power normalized measurement method, similar to the one described in Sect. 3.4, where a fixed reference horn

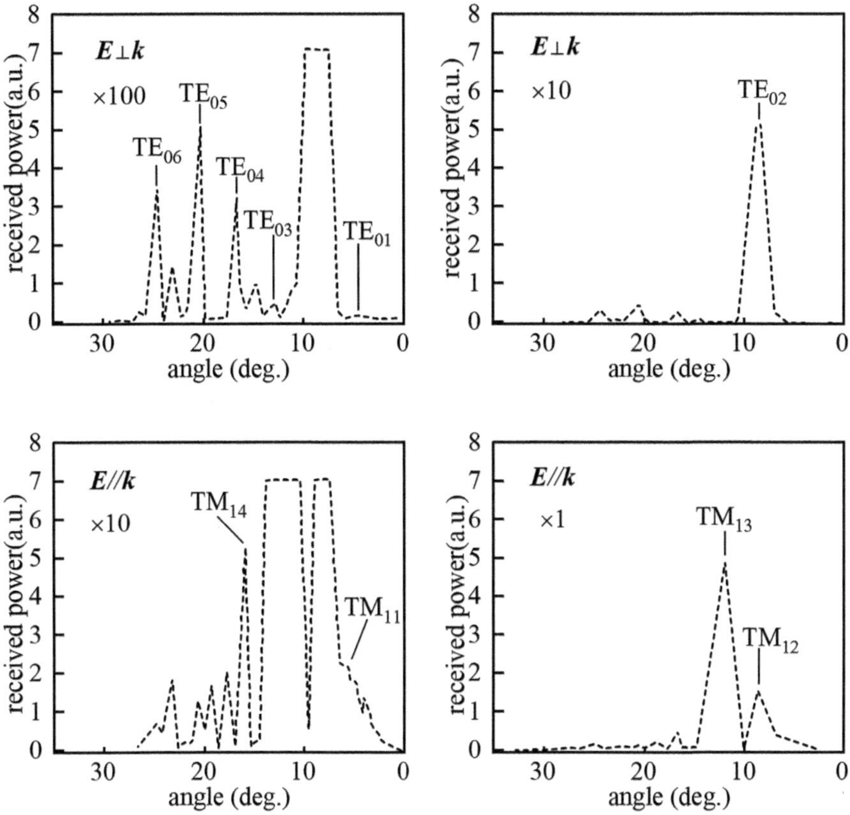

Fig. 7.22 Measurement results of continuous wave gyrotron VGE8007 using a wavenumber spectrum measurement device shown in Fig. 7.21

is used for simultaneous measurements, and the measuring results of single pulses in different radiation directions are normalized to the reference horn's measurements. This avoids uncertainties in relative power measurements between different single pulses and the inconvenience of placing receiving horns for each mode on the measurement track.

Additionally, the aperture size of the receiving horn limits the spatial angular resolution of measurements and makes it difficult to measure low-order modes with small radiation angles θ_{mn}. A rotatable spherical reflector antenna installed at the waveguide terminal is necessary to address this.

The large size and high processing precision requirements are also disadvantage of the wave number spectrum measurement method.

In the various mode discrimination and measurement methods mentioned above, each method has its own advantages and disadvantages, and there is no perfect method. Comparatively, image display method, analytical method of thermal image and direct discrimination methods in the Radiation field measurement method are

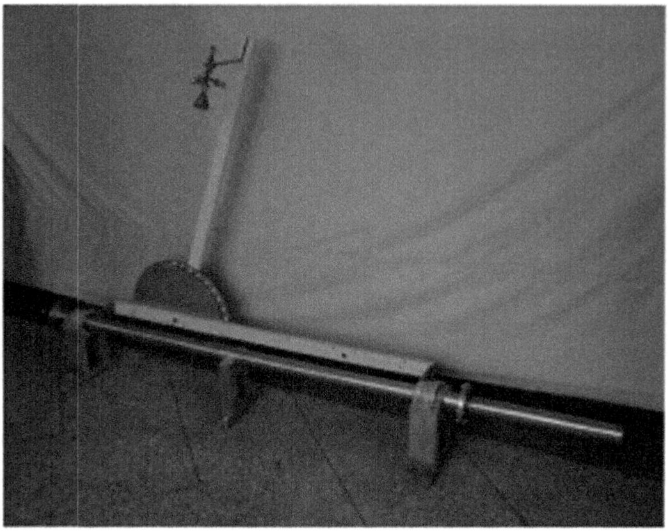

Fig. 7.23 Actual photograph of the wavenumber spectrum measurement device developed by the University of Electronic Science and Technology of China

relatively simple and have sufficient accuracy for mode recognition when a single mode or a small number of modes are present, making them the most widely used mode discrimination methods. On the other hand, wave number spectrum measurement and mode selection coupling measurement have the advantages of online measurement and multi-mode measurement, with the mode selection coupling method being particularly suitable for instantaneous and continuous measurements. The main drawback is the initial high cost and high processing precision requirements of the measurement device, but once the measurement device is processed, the measurement process is relatively simple.

There are also some more complex mode measurement methods, such as measuring field amplitude values at several different cross-sections in periodic inhomogeneous waveguides or radiation fields, and then using iterative method to determine their phases, and finally recombining the field distribution at any cross-section to discriminate the mode. These methods have almost no practical application in mode measurement and are mainly encountered in research in other fields, so they are not further discussed here.

Chapter 8
Frequency Measurement of High-Power Microwave

Due to the pulse width of high-power microwaves being only tens of nanoseconds, conventional frequency measurement methods such as digital frequency meters and resonant wavelength meters are no longer applicable to high-power microwave frequency measurement. However, frequency measurement is one of the most basic parameter measurements in high-power microwave measurement. The frequency parameter is the basis for adjusting the state of the high-power microwave source and calibrating other high-power microwave measurement devices for measuring power, mode, etc. Therefore, high-power microwave frequency measurement has always been a key technology of concern to people, and efforts have been made to study its measurement methods and develop accurate and reliable measurement devices.

8.1 Waveguide Dispersion Method

8.1.1 Conventional Waveguide Dispersion Method

The dispersion line method is the most widely used method in high-power microwave frequency measurement. It is simple and reliable, and uses the relationship between the transmission time and frequency of short-pulse microwaves in waveguide systems with dispersion characteristics to determine the frequency by measuring the delay time of the microwave envelope transmitted in the dispersion line.

1. **Fundamental principle**

(1) **Theoretical basis**

The fundamental principle of the dispersion line method is that when microwaves of different frequencies are transmitted in a waveguide, dispersion occurs, and the phase velocity v_p and group velocity v_g of the microwaves change with frequency.

W. Wang et al., *High-Power Microwave Measurement*,
https://doi.org/10.1007/978-981-96-7763-4_8

$$v_p = \frac{c}{\sqrt{1 - (f_c/f)^2}} \tag{8.1}$$

$$v_g = c\sqrt{1 - (f_c/f)^2} \tag{8.2}$$

where, c is the speed of light in free space, f_c is the cutoff frequency of the transmission mode in the waveguide, f is the microwave frequency. The waveguide used as the dispersion line is generally the transmission waveguide of the microwave signal received by the receiving horn, and the standard waveguide of the fundamental mode is used, so f_c is easy to determine.

After the microwave pulse is detected by the detector, the pulse envelope, that is, the wave packet, is displayed on the oscilloscope. The transmission speed of the wave packet is the group velocity. Therefore, after the microwave is transmitted a certain distance in the waveguide, the signal detected by the detector, that is, the wave packet, will be delayed by a certain time T compared to the time it enters the waveguide.

$$T = \frac{L}{v_g} = \frac{L}{c\sqrt{1 - (f_c/f)^2}} \tag{8.3}$$

where, L is the distance of electromagnetic waves transmitted in the waveguide.

Since the group velocity v_g is related to the microwave frequency, the frequency expression can be obtained from Eq. (8.3) as

$$f = \frac{f_c}{\sqrt{1 - (L/cT)^2}} \tag{8.4}$$

Since the correlation between group velocity v_g and microwave frequency is caused by the dispersion of the waveguide, we call the waveguide with a length of L a dispersion line. In this way, the microwave pulse output by the high-power microwave source is attenuated and divided into two paths by a directional coupler or a tee branch waveguide, magic tee, etc. One path directly detects the pulse envelope, and the other path inputs a certain length of dispersion line, and detects the output. The two paths detect the pulse envelope and input them together into the oscilloscope for display. The delay time T of the pulse envelope after passing through the dispersion line relative to the pulse envelope obtained by direct detection can be read on the oscilloscope. The microwave frequency can be obtained by using Eq. (8.4). This method of measuring microwave frequency is called the dispersion line method.

Obviously, the longer the dispersion line, the greater the time delay of the microwave pulse generated by the dispersion line, and the smaller the reading error of T, the more accurate the frequency measurement. In addition, for different frequency components, the longer the dispersion line, the greater the difference in time delay they produce after passing through the same dispersion line, that is, the higher the frequency resolution.

(2) **Shortcomings of the dispersion line method**

(a) Microwave pulses, especially ultra-short pulses, contain rich spectral components. In general, waveguide dispersion lines can only measure the carrier frequency of microwave signals and cannot measure the frequency spectrum of pulses.

(b) For microwave pulses of different frequencies or different microwave frequency components within the same pulse (referring to different carrier frequencies rather than harmonics in the pulse spectrum), the frequency resolution is proportional to the length of the waveguide dispersion line. To improve the resolution, the length of the waveguide should be increased, which leads to an increase in cost and volume, especially for low-frequency high-power microwaves. The size of the waveguide is relatively large, and the manufacturing cost and volume of the dispersion line that is too long are often quite large.

(c) Microwave pulses are composed of time harmonics of different frequencies. These different frequency components will produce different delays when passing through the dispersion line, which will cause distortion in the shape of the pulse envelope after passing through the dispersion line, making the pulse envelope wider and causing uncertainty errors in accurately determining the time measurement point due to the wider envelope.

(d) The microwave pulse will be severely attenuated after passing through a dispersion line that is tens of meters long, especially for single-pulse signals with weak power, and the signal may not even reach the sensitivity limit of the detector at the end of the dispersion line, and the oscilloscope cannot detect the signal.

(e) The measurable range of the frequency measured by the dispersion line method will be limited by the working frequency range of the waveguide. In addition, at the low end of the frequency, there is a significant difference in group velocity between different harmonic frequencies., and dispersion may cause pulse distortion, which will limit the accuracy of frequency measurement.

2. **Measurement method**

(1) **Measurement system** [99]

The system diagram of the dispersion line method for high-power microwave frequency measurement is shown in Fig. 8.1.

In the figure: HPM is a high-power microwave source, the receiving attenuation refers to the device that attenuates the microwave power output by the high-power source to a level that the measurement system can accept. In general, the microwave mode entering the dispersion line is the TE_{10} mode, and the dispersion line uses the fundamental mode rectangular waveguide. Therefore, the receiving attenuation in the figure usually refers to the radiation horn-receiving horn device, because the TE_{10} mode is excited in the receiving horn. Of course, in order to further reduce the signal power level, the receiving attenuation should also include the attenuation introduced after the receiving horn, such as directional couplers, probes, fixed attenuators, etc.

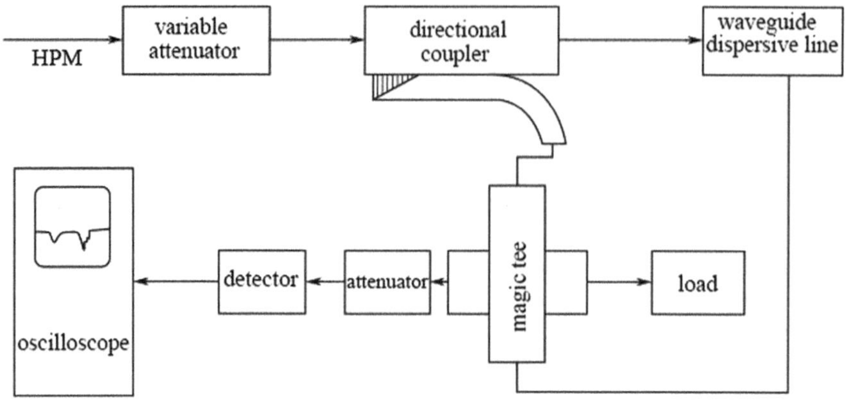

Fig. 8.1 System diagram of high-power microwave frequency measurement using dispersion line method

The directional coupler in Fig. 8.1 is used to divide the measured microwave pulse signal into two paths. The signal output from the attenuator (called the starting signal) enters the directional coupler. Since this signal output from the coupled port will directly enter the detector through the magic tee, the power may be relatively high., the power may be relatively high. Therefore, an attenuator should be introduced if necessary to further ensure that the signal power level is within the allowable range of the crystal detector. Then, after detection, it will be input into the oscilloscope to indicate the starting time. Another signal output from the direct pass port of the directional coupler enters the other arm of the magic tee through the waveguide dispersion line, and is output from the same port of the magic tee together with the starting signal. After being detected by the same detector, the two signals are compared and the delay time is measured to calculate the frequency. Since the two signals are separated in time, they generally do not overlap.

The function of the magic tee lies in its ability to output two microwave signals, derived from a directional coupler, through a single port. These signals are inputted via the E-T and H-T branch ports of the magic tee and subsequently exit through the ports of the main waveguide. One output port directs the signal into a detector, which displays the envelope on an oscilloscope, while the other port outputs a signal absorbed by a matched load. It is important to note that signals entering the H-T branch produce in-phase outputs at both ports of the main waveguide, whereas signals from the E-T branch yield out-of-phase outputs. Consequently, when these two signals are output through the same port, the oscilloscope may display an inverted envelope waveform. In such instances, switching the output port of the main waveguide can output the waveform in the same direction.

(2) **Measurement example** [100]

We take the frequency measurement results of the C-band magnetically insulated transmission line oscillator developed by Chinese scientific researchers using the

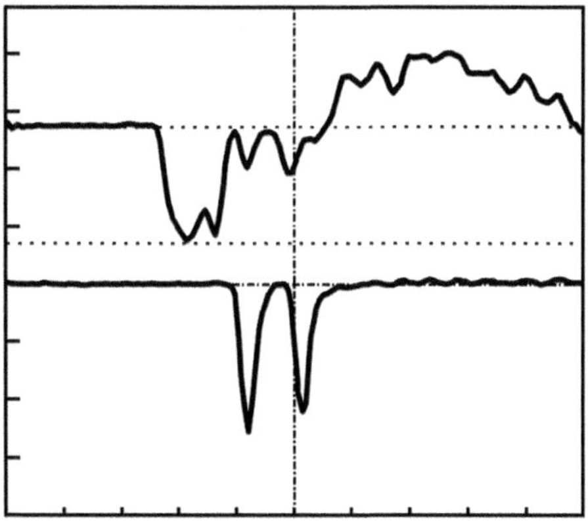

Fig. 8.2 The frequency measurement waveform of the CD-MILO magnetic insulation line oscillator

dispersion line method as an example to illustrate the practical application of this method in high-power microwave frequency measurement. Figure 8.2 shows the measurement results of the magnetically insulated transmission line oscillator CD-MILO. The upper part of the figure is the current waveform curve output by the pulse generator, and the lower part is the frequency measurement curve (pulse envelope waveform). The time difference between the two peaks of the envelope waveform is 96 ns, the length of the dispersion line used is 22.24 m, the WR187 standard waveguide is used, and the cutoff frequency fc of the TE_{10} mode is 3.152 GHz. According to Eq. (8.4), the operating frequency of the magnetically insulated transmission line oscillator can be calculated to be 4.96 GHz.

Figure 8.3 shows the frequency measurement waveform of the CV-MILO magnetically insulated transmission line oscillator. The WR187 standard waveguide was used as the dispersion line, with a length of 20.17 m. The time delay of the two envelope waveforms was measured to be 87.94 ns, and the calculated frequency was 4.89 GHz. The upper curve in the figure is the voltage waveform, and the lower part is the pulse envelope waveform of the frequency measurement.

The time displayed on the oscilloscope has an error of ± 2 ns. Under the above measurement conditions, the corresponding frequency error is ± 0.15 GHz.

8.1.2 Circulating Waveguide Dispersion Method [101]

1. **Basic idea of circulating waveguide dispersion line method**

(1) **Measurement system**

From the above analysis, it can be concluded that in order to improve the frequency resolution of the dispersion line method, it is necessary to increase the length of the waveguide dispersion line. However, with the increase of the length of the dispersion line, the loss will inevitably increase, and the waveform distortion due to the different phase velocities of different frequency components in the microwave pulse will occur,

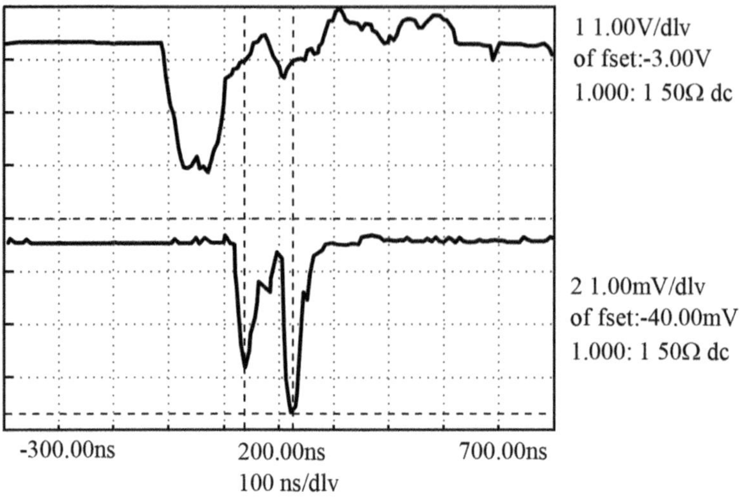

Fig. 8.3 The frequency measured waveform of the magnetically insulated transmission line oscillator CV-MILO

which will affect the accuracy of frequency measurement. If the dispersion line is too long, the cost and volume will also increase rapidly. To overcome this contradiction between frequency resolution and waveguide dispersion line length, Chinese scholar has proposed the circulating waveguide dispersion line measurement method. The measurement system of this method is shown in Fig. 8.4.

The microwave pulse signal is input into the circulating waveguide measurement system after necessary receiving attenuation (total attenuation caused by radiation horn-receiving horn and attenuators, etc.). Like the conventional dispersion line method, when the high-power microwave is input into the measurement system, a

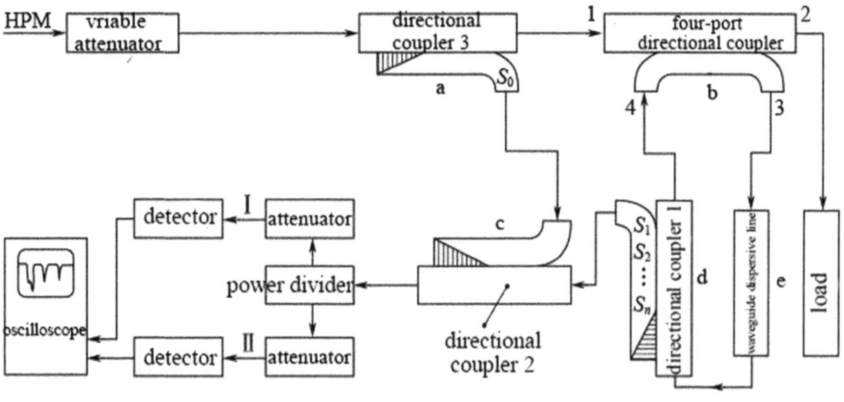

Fig. 8.4 Schematic diagram of the circulating waveguide dispersion line method measurement system

directional coupler 3 is used to obtain an initial reference signal S_0 at the coupling output end. The main pulse signal is output from its direct pass port and input into the 1 port of the main waveguide of the four-port directional coupler. Part of the power coupled out by the four-port directional coupler is output from the 3 port of the secondary waveguide. This part of the signal power enters the waveguide dispersion line, and then enters the main waveguide of the directional coupler 1, and is output through its direct pass port and then enters again the 4 port of the four-port directional coupler, and is directly transmitted to the 3 port in the secondary waveguide. Of course, a small portion of the signal entering port 4 will also couple into the main waveguide, and then couple to the secondary waveguide through the main waveguide for output from port 3, Obviously, this part of the signal will be much smaller than the signal transmitted directly to port 3 in the secondary waveguide. The signal is coupled through the main waveguide to reach port 3 and directly passes through the secondary waveguide to reach port 3, the transmission length in the two paths can be considered basically equal, because the main waveguide and the secondary waveguide are both working in TE_{10} mode, and their waveguide sizes are generally the same, so the signal phase shift through two paths can also be considered the same, the output signal will be superimposed on each other. The signal output from port 3 re-enters the waveguide dispersion line, thus completing one cycle of the pulse signal. The signal passing through the dispersion line then returns to the four-port directional coupler via the main waveguide of directional coupler 1 and is output from port 3 to enter the dispersion line again… This process can continue indefinitely. With each cycle, directional coupler 1 will couple out a portion of the signal, noted as S_1, S_2, …, S_n in sequence. Of course, due to losses in the feedback system itself, the signal will gradually weaken during the cycling process. In this process, the signal output from the direct port of the four-port directional coupler is absorbed by the absorbing load.

(2) **Signal relationship**

The microwave pulse power entering the main waveguide of the four-port directional coupler is mostly output from the direct pass port 2 and absorbed by the load except for a small part that is coupled out from the secondary waveguide port 3. Every cycle of the microwave pulse in the loop, the directional coupler 1 outputs the signal once. After n cycles, it will output n pulse signals S_1, S_2, …, S_n. The initial signal S_0 coupled out by the directional coupler 3 and the signals S_1, S_2, …, S_n coupled out by the directional coupler 1 all enter the directional coupler 2. In theory, the output end of the directional coupler 2 can be directly connected to the detector. After detection, the signal is displayed on the oscilloscope. Since the amplitude of the reference signal S_0 that did not pass through the dispersion line and the signal S_1, S_2, …, S_n that passed through the dispersion line will differ greatly, and the signal will become weaker and weaker during the cycle, the amplitude of $S_1, S_2, \ldots, S_n$ each other will also differ greatly. If the same detector is used, the detector will be difficult to withstand the power of the signal when it is large. In addition, the linear range of the detector is relatively small, and it is difficult for the same detector to detect signals with large differences in amplitude, which will inevitably cause distortion when the signal is large. Therefore, the signal output by the directional coupler 2 must be divided into several paths, and a microwave limiter should be added at the same time to suppress the amplitude of the reference signal and the signal with fewer circulation times to avoid damaging the detector and ensure that the signal is not distorted.

To accurately compare the time of multiple signals detected, it should be required that the lengths of the lines through which each signal passes, excluding the waveguide dispersion line (i.e., the corresponding time length), are equal or the difference in line lengths between them is equal, to ensure that the time difference displayed on the oscilloscope is purely caused by the dispersion line, and does not include the time difference caused by the different transmission line lengths other than the dispersion line. Specifically, the route that S_0 passes from directional coupler 3 to directional coupler 2 only has L_{ac}, the route that S_1 passes from directional coupler 3 to directional coupler 2 excluding the waveguide dispersion line is $L_{ab} + L_{dc}$ (since the directional coupler is generally directly connected to both ends of the waveguide dispersion line, $L_{be} = L_{ed} = 0$ is assumed), and the route that the circulating signal passes from directional coupler 1 through the four-port directional coupler to directional coupler 2 to obtain S_2, excluding the waveguide dispersion line, is $L_{db} + L_{dc}$. Therefore, the difference in the route (time length) between S_0 and S_1 is L_{ac}-L_{ab}-L_{dc}, and the difference in the route between S_1 and S_2 is L_{ab}-L_{db}. The two should be equal, that is, L_{ac}-$L_{dc} = 2L_{ab}$-L_{db}. In this way, the time delay of S_1, S_2, and S_0 displayed on the oscilloscope will depend solely on the time the signal passes through the waveguide dispersion line, thereby ensuring the time relationship between signals S_1, S_2, …, and S_n is consistent.

(3) **Measurement of multi-frequency pulse signals**

In a microwave pulse with a pulse width of τ, if it contains microwave components at different frequencies (carrier frequencies), The different frequency components it contains are not necessarily distributed at the same moment, i.e., microwaves at different frequencies may exist at different moments in a pulse, such as two different frequencies f_1 and f_2 at times t_1 and t_2 within the pulse width. The delay amount T measured by the ordinary dispersion line method is based on a certain moment of the pulse signal S_0 as the reference, without considering the width of the pulse itself. In this way, the time position of different frequency components of the microwave in the pulse time will be ignored, making it impossible to distinguish different frequency components in the pulse.

After using the circulating waveguide dispersion line, the waveforms of multiple pulse signals after 1, 2, …, and n dispersions can be measured, as shown in Fig. 8.5. At this time, the pulse parts of different frequencies in the circulating pulse after several times (generally only 2–3 times are sufficient) have been separated due to different delays, and the time intervals T_1 and T_2 of the corresponding f_1 and f_2 in each circulating output are fixed and determined by the length of the dispersion line. Therefore, we can calculate the time position of f_1 and f_2 in S_0 by measuring the size of T_1 and T_2 of adjacent circulating outputs and the number of cycles, thereby eliminating the measurement uncertainty caused by the pulse width and determining the time distribution of each frequency component in the pulse, and calculating the frequency size of each frequency component. However, it should be noted that what is measured here is different microwave frequencies at different times in the same pulse, not different frequencies of different harmonic components in the same pulse.

2. **Main performance and application of circulating waveguide dispersion line**
(1) **Main performance of circulating waveguide dispersion line**

The frequency range of circulating waveguide dispersion line measurement depends on the working bandwidth of the waveguide and the various components used in the device and the detector. The measurement uncertainty and resolution are also related to the uncertainty of the measurement time of the recording system. The circulating waveguide dispersion line developed by Chinese scholars has the following performance:

(a) WR90 waveguide dispersion line length L = 77.76 m, attenuation of circulating four times is $\leq$ 15 dB, frequency uncertainty is less than 2%.
(b) WR159 waveguide dispersion line length L = 46.0 m, attenuation of circulating six times is 3 dB, frequency uncertainty is less than 2%, and the uncertainty of frequency-time distribution measurement is less than 2 ns.

(2) **Measured waveform**

Figure 8.6 shows the waveform of the circulating waveguide dispersion line frequency measurement composed of WR159 waveguides. Figure 8.6a is the measurement waveform of a single-frequency microwave pulse, which shows the output results of four circulations. The first waveform is S_0, followed by S_1, S_2, S_3, and S_4 in turn. Figure 8.6b is the measurement result of a microwave pulse containing two frequency components. It can be clearly seen that after two circulations, the two frequency components have been separated.

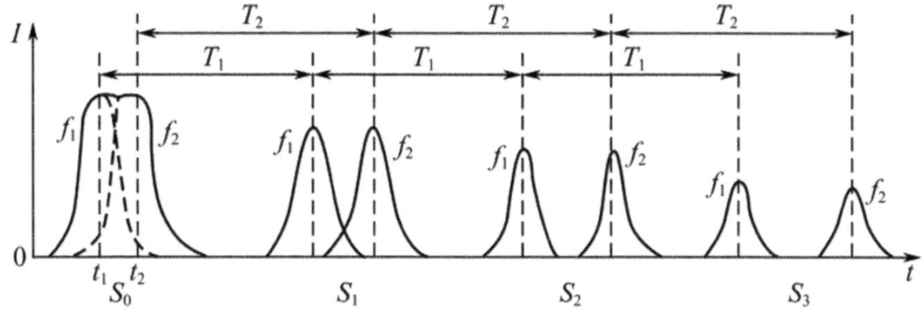

Fig. 8.5 Schematic diagram of the dispersion of dual-frequency microwave pulses transmitted in the circulating waveguide dispersion line

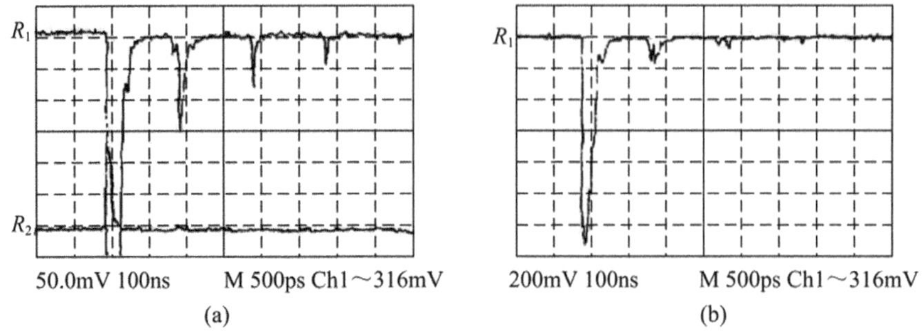

Fig. 8.6 WR159 circulating waveguide dispersion line frequency measurement results, **a** single-frequency microwave pulse, **b** dual-frequency pulse

8.2 Frequency Mixing Method (Heterodyne Method)

The dispersion line method is simple in equipment and convenient in measurement, and it is currently the most widely used method in high-power microwave frequency measurement. However, as we pointed out in Sect. 8.1, it also has some disadvantages and application limitations. To overcome the shortcomings of the dispersion line method for frequency measurement, the mixed frequency method can be used. It can perform real-time fast Fourier transform on the intermediate frequency signal obtained after mixing, so that the intermediate frequency can be directly read out to calculate the microwave frequency.

8.2.1 Measurement Principles of Frequency Mixing Method

The voltage-current characteristic of the mixer diode shows the characteristics of nonlinear resistance. Therefore, the microwave mixer is a nonlinear resistance frequency conversion circuit. The mixed frequency measurement method is a method of reducing the microwave signal frequency to intermediate frequency signal for frequency measurement, because the intermediate frequency signal processing, such as signal amplification, is more convenient and reliable than direct processing of high-frequency signals. The mixed frequency method can also be called the heterodyne method. The working principle of the mixed frequency method is to input a signal f_s of a microwave frequency to be measured and a local oscillator signal f_L of a known frequency into the mixer together. In the mixer, the two signals are mixed to produce a difference frequency $f_d = f_L - f_s$ (or $f_s - f_L$). The frequency f_d of the difference frequency signal obtained is much lower than the microwave frequency f_s, which is called the intermediate frequency. It can be detected directly by the oscilloscope, and the intermediate frequency can be obtained by Fourier transform, and the microwave frequency can be calculated from it.

(1) When the local oscillator signal is much larger than the signal to be measured

Assuming that the voltage added to the mixer diode is

$$V = V_s \cos \omega_s t + V_L \cos \omega_L t \tag{8.5}$$

where, $V_s\cos\omega_s t$ is the microwave signal of the frequency to be measured; $V_L\cos\omega_s t$ is the local oscillator signal. If the conductance of the mixer to the voltage V presented is $g(t)$, then the relationship between the current i flowing through the mixer and the voltage V can be expressed as

$$i = g(t)V = f(V) = f(V_s \cos \omega_s t + V_L \cos \omega_L t) \tag{8.6}$$

where,

$$g(t) = \frac{\mathrm{d}\, i}{\mathrm{d}V} \tag{8.7}$$

Since the voltage changes with time, so, the conductance $g(t)$ also changes with time, which is called the time-varying conductance. $g(t)$ can be expanded into a Fourier series. When $V_s \ll V_L$ (such as the front end of the radar receiver), the existence of $V_s\cos\omega_s t$ can be ignored, so $g(t)$ is approximately to

$$g(t) = g_0 + 2\sum_{n=1}^{\infty} g_n \cos \omega_L t \tag{8.8}$$

where, g_0 is the average mixing conductance of the diode; g_n is the mixing conductance of the n-th harmonic corresponding to the local oscillator signal $V_L\cos\omega_s t$.

Since $V_s \ll V_L$, it can be considered that the operating point of the diode changes with the local oscillator voltage, and the microwave signal to be measured is only a small voltage increment. Therefore, the loop current i can be expanded into a Taylor series at each operating point.

$$\begin{aligned} i &= f(V) = f(V_0 + V_s cos\omega_s t + V_L cos\omega_L t) \\ &= f(V_0 + V_L cos\omega_L t) + f'(V_0 + V_L cos\omega_L t)V_s cos\omega_s t \\ &+ \frac{1}{2!}f''(V_0 + V_L cos\omega_L t)(V_s cos\omega_s t)^2 + \cdots = i_g(t) + i_m(t) \end{aligned} \tag{8.9}$$

where the first term is the large signal current i_g flowing through the diode under the excitation of the local oscillator signal, which includes a DC part and a current part excited by the local oscillator fundamental wave signal; the other terms in the expansion are small signal currents i_m in the diode, when V_s is very small, the second-order and higher-order terms of V_s can be ignored in the i_m, and only its first-order term is taken, that is, the second term in the formula.

In this way, it can be seen from Eqs. (8.6)-(8.9) that the small signal component in the diode can be approximated as

$$\begin{aligned} i_m(t) &= f'(V_0 + V_L \cos \omega_L t) V_s \cos \omega_s t \\ &= (g_0 + 2g_1 \cos \omega_L t + 2g_2 \cos 2\omega_L t + \cdots) V_s \cos \omega_s t \\ &= g_0 V_s \cos \omega_s t + \sum_{n=1}^{\infty} g_n V_s \cos(n\omega_L \pm \omega_s)\, t \end{aligned} \tag{8.10}$$

where, the first term is the fundamental wave current of the microwave signal; the second term is the current of the intermediate frequency signal output ($n = 1$) and the high-order harmonic current. From Eq. (8.10), it can be seen that

(a) Countless combinations of frequency components are generated in the mixer. By using an intermediate frequency bandpass filter, the desired intermediate frequency component can be extracted while filtering out other combinations of frequencies.
(b) From Eq. (8.10), the amplitude of the intermediate frequency component ($n = 1$) can be obtained as

$$I_0 = g_1 V_s \tag{8.11}$$

The amplitude of the intermediate frequency current is proportional to the amplitude V_s of the measured microwave signal. That is, when the signal is small, there is a linear relationship between the amplitude of the microwave signal at the input of the mixer and the amplitude of the intermediate frequency signal at the output. Equation (8.10) also shows that the frequency of the intermediate frequency component should be ($\omega_L \pm \omega_s$).
(c) During the mixing process, when the local oscillator is a strong signal, countless harmonics will be generated, but the harmonic power changes approximately $1/n^2$ with the harmonic order n. Therefore, the intensity of the combined component of the mixing current decreases rapidly with the increase of the harmonic order n of the local oscillator signal. Usually, only a few main special frequency components will have a greater impact on the frequency conversion efficiency. These frequency components are: sum frequency $\omega_+ = \omega_L + \omega_s$, difference frequency $\omega_0 = \omega_s - \omega_L$ ($\omega_s > \omega_L$) or $\omega_0 = \omega_L - \omega_s$ ($\omega_L > \omega_s$), and the mirror frequency $\omega_i = 2\omega_L - \omega_s = \omega_L - \omega_0$ generated by ω_s and $2\omega_L$.

(2) **When the signal strength to be measured is quite large**

In the above small signal analysis, the initial phase difference between the local oscillator signal and the microwave signal was not considered, that is, it was assumed that their initial phase difference was 0, in fact, there will be a phase difference between the two. In the frequency measurement of high-power microwaves, obviously the condition $V_s \ll V_L$ will no longer be satisfied. When V_s is large enough, the higher-order terms V_s^2 and above can no longer be ignored. The harmonic components of the mixing current increase significantly, and there is still a phase difference between the microwave signal whose frequency to be measured and the local oscillator signal. In this case, the microwave signal and the local oscillator signal should be represented as

$$\begin{aligned} u_s &= V_s \cos(\omega_s t + \phi_s) \\ u_L &= V_L \cos(\omega_L t + \phi_L) \end{aligned} \tag{8.12}$$

where, φ_s and φ_L are the initial phase of the local oscillator signal and the microwave signal, respectively. We omit the tedious derivation process and directly write the final expression of the loop current $i(t)$ as follows:

$$\begin{aligned} i(t) &= \sum_{n=-\infty}^{\infty} \sum_{m=-\infty}^{\infty} \left|\dot{I}_{n,m}\right| \mathrm{e}^{\mathrm{j}(n\phi_L + m\phi_s)} \mathrm{e}^{\mathrm{j}(n\omega_L + m\omega_s)t} \\ &= \sum_{n=-\infty}^{\infty} \sum_{m=-\infty}^{\infty} \dot{I}_{n,m} \mathrm{e}^{\mathrm{j}(n\omega_L + m\omega_s)t} \end{aligned} \tag{8.13}$$

where, $\dot{I}_{n,m}$ is the complex amplitude of each $(n\omega_\mathrm{L} + m\omega_\mathrm{s})$ frequency component, and n and m are the harmonic orders of the local oscillator signal and the microwave signal, respectively.

From Eq. (8.13), it can be seen that the mixing current $i(t)$ contains all possible combinations of various harmonic number of times of the microwave signal ω_s and the local oscillator signal ω_L, which is much richer than the combined component of the microwave signal when it is a small signal. In addition to the intermediate frequency component we need, other harmonic frequency components should be suppressed as much as possible.

From Eq. (8.13), the real intermediate frequency current can be obtained as

$$i_0(t) = 2\left|\dot{I}_{-1,+1}\right| \cos[(\omega_s - \omega_L)t - \phi] \tag{8.14}$$

The frequency of the intermediate frequency signal can be (ω_s-ω_L) or (ω_L-ω_s), depending on the size of ω_s relative to ω_L.

If the frequency of the measured microwave signal is relatively high, in order to keep the frequency of the intermediate frequency signal relatively low and not to increase the frequency of the local oscillator signal too much, or to reduce the requirements for the frequency of the local oscillator signal when measuring higher microwave frequencies, the n-th harmonic of the local oscillator signal can be used for combination, that is, the difference frequency is made to be $f_\mathrm{d} = f_\mathrm{s}$-$nf_\mathrm{L}$ (or nf_L-f_s), where n is the harmonic number of times of the local oscillator signal.

8.2.2 *Experimentation of Frequency Mixing Method*

The mixing system must be calibrated by experiment before accurate measurement, and the influence of the power size of the input microwave signal on the IF signal also needs to be determined by experiment.

1. **Measurement system**

The mixed frequency method measurement system mainly consists of a mixer diode, a low-pass filter, a crystal local oscillator, a DC power supply, and an oscilloscope, as shown in Fig. 8.7. The high-power microwave (HPM) is input to the mixer through a receiving attenuator, and the local oscillator signal output by the crystal oscillator as the local oscillator source is also input to the mixer at the same time. If the relationship between the frequency of the local oscillator source and the voltage applied to the crystal is calibrated in advance, the frequency of the local oscillator source output signal can be directly known through the voltage value. A standard microwave signal source can also be used as the local oscillator signal, and its frequency can be directly read on the instrument [99, 102].

Since the harmonic nf_L of the local oscillator signal may also mix with the measured signal f_s to generate a high-frequency signal, the output signal of the mixer may not only contain intermediate frequency components, but also high-frequency components. In order to remove the high-frequency components, a low-pass filter is connected to the output end of the mixer to filter out the high-frequency signal, and the remaining intermediate frequency signal enters the oscilloscope, and the intermediate frequency signal frequency f_d can be directly measured on the oscilloscope.

Fine-tune the frequency of the local oscillator signal. If f_d decreases with the increase of f_L, it means that $f_d = f_s$-f_L, so the measured frequency $f_s = f_L + f_d$, conversely, if f_L increases, f_d increases accordingly, indicating that $f_d = f_L$-f_s, so $f_s = f_L$-f_d. In general, changing the frequency of the local oscillator signal several times can determine whether the signal frequency f_s to be measured should be calculated by adding or subtracting the intermediate frequency from the local oscillator's frequency.

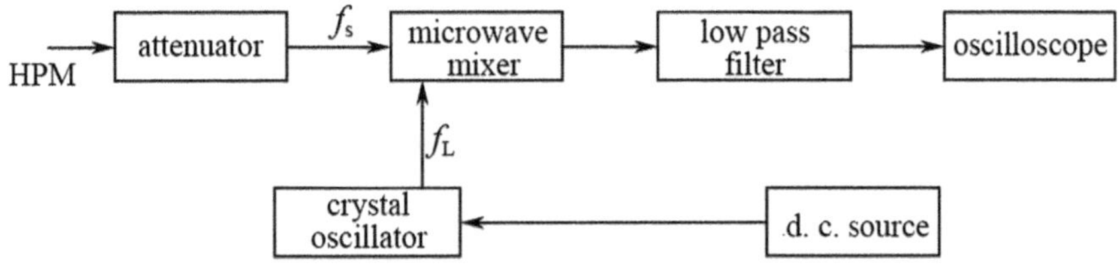

Fig. 8.7 Schematic diagram of the mixed frequency method measurement system

The oscilloscope for intermediate frequency indication should have a sufficient bandwidth, so that the intermediate frequency waveform can be directly displayed on the oscilloscope, and the frequency can be read from it.

When using the mixed frequency method for pulsed microwave frequency measurement, the frequency stability of the crystal oscillator or the standard signal source used as the local oscillator signal is required to be high, otherwise, the measured microwave frequency will drift.

2. **Power range of local oscillator signal and measured signal**

Since there are many types and models of mixers, the quality of the output signal will be directly affected by the choice of the mixer. First, the mixer should be selected according to the power level of the local oscillator signal. The basic condition for the mixer to work is that it needs to add a local oscillator signal with relatively high power, so as to minimize the distortion of the mixer; secondly, the working frequency range of the mixer that meets our requirements should be selected; the most important thing is to determine the dynamic range of the mixer. If it is used improperly and exceeds the dynamic range of the mixer, it may cause damage to the mixer or distortion of the intermediate frequency output signal.

When the power of the measured microwave signal input to the mixer is too small, the intermediate frequency signal may not be measured by the terminal display oscilloscope. Even if the intermediate frequency signal waveform can be displayed, distortion or serious interference signals may affect the accurate measurement of the intermediate frequency. Therefore, the power of the measured microwave signal should be adjusted to obtain a more complete waveform of the intermediate frequency output that is most convenient for measurement. Figure 8.8 shows the intermediate frequency output waveform when the input power of the microwave single pulse is too small or relatively appropriate. It can be clearly seen that when the power of the measured microwave pulse signal is only − 3 dBmW, the intermediate frequency waveform output has a smaller amplitude at the front end of the pulse and a larger amplitude at the tail, and the pulse waveform is obviously distorted; when the input power reaches 16 dBmW, the intermediate frequency waveform is relatively good, and the intermediate frequency can be conveniently measured on this waveform [102].

3. **Calibration of frequency measurement error**

The accuracy of frequency measurement is the most important technical parameter of the mixer system, which is commonly represented by the measurement error Δf. The calibration of frequency measurement error can be carried out by using two standard microwave signal sources, one of which simulates the measured signal and the other serves as the local oscillator signal. Assuming that the nominal frequency of the microwave output by the standard signal source simulating the measured signal is f_{s}, and the frequency of the signal output by the standard signal source serving as the local oscillator signal is f_{L}, then the microwave signal frequency obtained by measuring the intermediate frequency f_{d} through the mixer system is

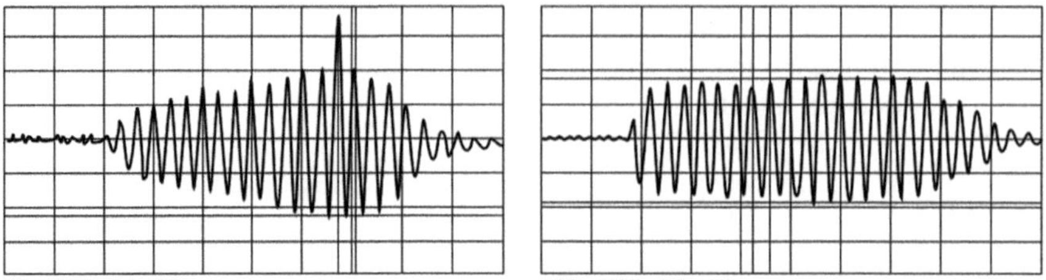

Fig. 8.8 Intermediate frequency waveform output when the input microwave power of the mixer is different, **a** input power – 3 dBmW, **b** input power 16 dBmW

$$f_s' = |f_L \pm f_d| \tag{8.15}$$

The relative error of the measurement can be expressed as

$$\Delta f = |f_s - f_s'|/f_s \tag{8.16}$$

The results of frequency measurement of X-band transit time tube oscillator by Chinese engineering and technical personnel show that when the intermediate frequency is below 1 GHz, the relative measurement error of the mixer system is less than 0.074%, and when the intermediate frequency is above 1 GHz, the relative measurement error is less than 0.015%.

4. **FFT transformation**

When studying virtual cathode oscillators at the Los Alamos National Laboratory in the United States, the microwave energy output from the radial coupling port of the frequency-locked resonant cavity of the virtual cathode oscillator is attenuated to a certain extent and then input into the mixer system. The oscilloscope displays the intermediate frequency signal output by the mixer as shown in Fig. 8.9a. The waveform is subjected to fast Fourier transform (FFT), and the result is shown in Fig. 8.9b. According to the intermediate frequency signal frequency given by the figure, the output pulse microwave frequency of the virtual cathode oscillator can be calculated to be 825 MHz, and the 3 dB bandwidth can be measured from the FFT waveform to be about 0.3%, which is about 2.4 MHz at the 825 MHz frequency [103].

8.3 Filtering Method

8.3.1 Filter Measurement Method

1. **Fixed bandpass filter**

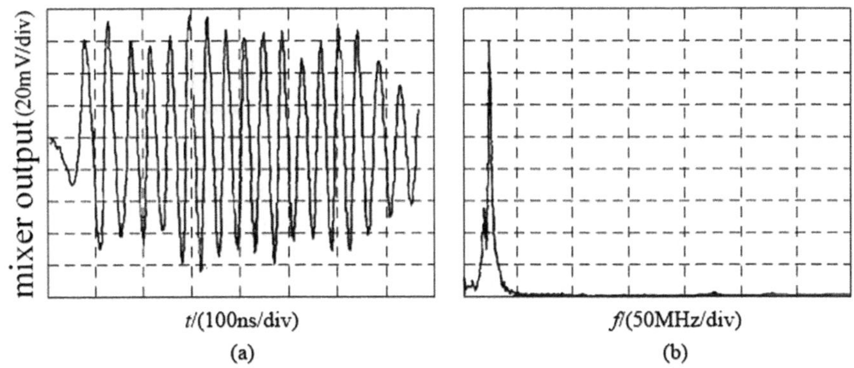

Fig. 8.9 Frequency measurement results of virtual cathode oscillator using mixing method, **a** Intermediate frequency signal time-domain waveform, **b** frequency-domain waveform after FFT transformation

The filtering method divides the frequency range where the microwave signal may be located into several frequency bands, each of which corresponds to a bandpass filter. When the microwave signal is input to these filters at the same time, which filter has a signal output indicates that the frequency of the microwave signal is in the passband frequency range corresponding to the filter. The bandpass filter method is the most direct way among filtering methods to measure frequency but a combination of cutoff waveguides can also be used because the waveguide itself is a high-pass filter. It can be seen that this method can only be used to determine the frequency range of the high-power microwave signal, and cannot directly and accurately determine the frequency size.

(1) **Band-pass filter method** [99]

The multi-channel bandpass filter detection method simulates the principle of the analog frequency spectrum analyzer, as shown in Fig. 8.10. The high-power microwave source of the frequency to be measured is attenuated and input into the measurement system. It is divided into *n* paths by a power divider, and enters the corresponding filter of each path. The passband range synthesized by *n* filters covers the entire frequency range of the signal to be measured. For example, if the output signal of the high-power microwave source is designed in the X-band of 12 GHz, the frequency range covered by the passband combination of *n* filters should not be less than $\Delta f = 12\text{-}8 = 4(\text{GHz})$, and the low-end frequency (f_L) of the first filter passband should start from 8 GHz, and the center frequency of the second, third...passband should increase successively until the high-end frequency (f_H) of the passband of the *n*th filter reaches 12 GHz. The passband bandwidth of *n* filters should be adjacent to each other but not overlap, as shown in Fig. 8.11. In ideal conditions, if the passband bandwidth of each filter is the same, then the passband bandwidth of each filter should be $\Delta f\ /n$. The passband bandwidth can be 0.5 dB bandwidth, 1 dB bandwidth, 3 dB bandwidth, or a bandwidth defined according to needs. Figure 8.11 shows the total frequency response curve of the measurement system composed of n filters when each filter has a 3 dB passband bandwidth.

(2) **Frequency measurement using band-pass filter method**

The high-power microwave signal is output through one of the n filters, and after being detected by the detector, it enters the oscilloscope for indication. According to the passband bandwidth of the filter that outputs the signal, the frequency of the microwave signal to be measured can be determined to fall within that frequency band. Obviously, under the same Δf, the narrower the passband bandwidth of each filter, the more filters are needed to cover the entire frequency range of Δf; therefore, the range of the measured frequency is also narrower.. In other words, the accuracy of frequency measurement is higher.

Obviously, in order to accurately measure the frequency using the bandpass filter detection method, it is necessary to require that the passband width of each filter in the

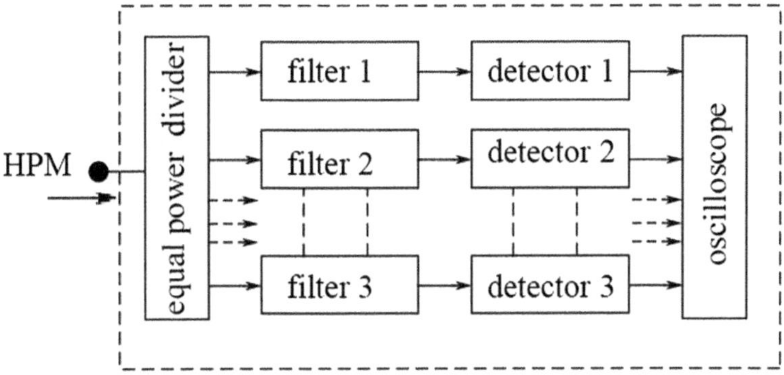

Fig. 8.10 The principle block diagram of the multi-channel bandpass filter method frequency diagnosis system

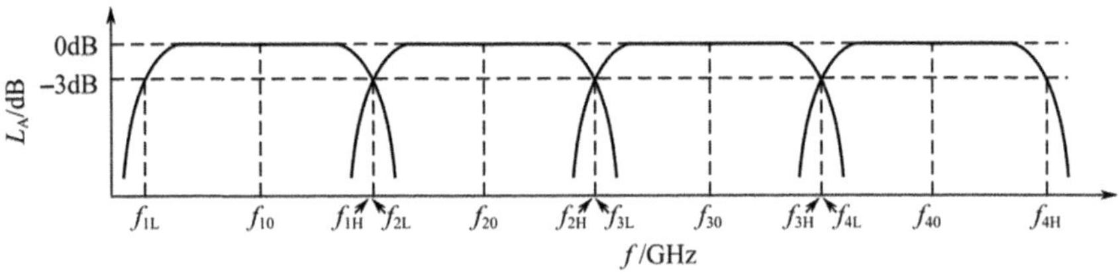

Fig. 8.11 The ideal relationship between the frequency responses of each filter in the multi-channel bandpass filter method frequency measurement system. (The subscript L represents the low end of the passband, the subscript H represents the high end of the passband, and the subscript 0 represents the center frequency of the passband)

measurement system be narrow enough to achieve the desired frequency measurement accuracy. For example, if a frequency measurement accuracy of 1 MHz is required, then the passband width of each filter must also be narrow to 1 MHz. If the passband width of each filter in the system is relatively wide, exceeding the frequency measurement accuracy we require, then the measurement obtained at this time will only be the range of the frequency under test, and the exact value of the frequency under test cannot be directly determined. In practical measurements, it can be graded to reduce the number of filters required. For example, when measuring the frequency of a microwave source with a frequency range of $\Delta f = 4$ GHz and a required measurement accuracy of 1 MHz, we can first use an array of 4 filters each with a bandwidth of 1 GHz to measure, so that the frequency being measured is determined within a certain 1 GHz bandwidth. Then we can use an array of ten filters each with a 100 MHz passband width to measure, narrowing the range of the measured frequency to within a 100 MHz bandwidth, and so on. By using filter arrays with bandwidths of 10 MHz and 1 MHz respectively for further measurements, the accuracy of the measured frequency can ultimately be improved to 1 MHz.

Since the frequency response curve of a filter cannot change instantaneously at the two ends of the passband, there will always be a slow change process, which means that at the junction of the frequency response curves of every two filters, there will inevitably be an overlap of the frequency response curves within a certain frequency range. This leads to the phenomenon where adjacent filters can work simultaneously, i.e., the microwave signal will be output simultaneously by these two filters, and the frequency under test will fall within the frequency range where the response curves of the two filters overlap. If the overlapped frequency range can be determined, this range is generally much narrower than the passband range of the filter itself, thereby improving the accuracy of frequency measurement. If the overlapped frequency range cannot be determined, in this case, it is generally advisable to take the passband of the filter with a relatively large output signal amplitude as the frequency range of the signal under test. Of course, the amplitude of the output signal should be calibrated in advance.

The improvement of resolution of frequency measurement using the bandpass filter method can be achieved through the design of the filter's passband width, the number of filters, and their frequency response curves.

2. **YIG tunable filter** [104, 105]

The above describes the measurement method where the passband of the filter is fixed. In order to adapt to finding the frequency under test within a wide frequency range, even with grading measurements, a large number of filters are still required to cover the entire frequency range where the frequency under test may be located. This makes the measurement system complex, large in volume, high in cost, and long in measurement time. By using a tunable filter with an adjustable center frequency, it is possible to quickly and broadly tune with a single filter, covering the entire frequency range where the frequency under test may be located, and thus measure the frequency under test. Obviously, using this method will be much simpler and more convenient, and the measurement system will be much simpler, smaller in volume, and lower in cost. This tunable filter is the YIG tunable filter.

(1) **Principle of YIG tunable filter**

YIG (Yttrium Iron Garnet) is a type of microwave ferrite, and YIG single crystal is a microwave ferrite obtained by natural crystallization after sintering high-purity Y_2O_3 and Fe_2O_3 in a certain proportion, adding a fluxing agent PbO.

All the spintronics in the microwave ferrite would produce a synthetic magnetic moment $\boldsymbol{M}$. Under the action of an external constant magnetic field $\boldsymbol{H}_0$, it will generate precession around $\boldsymbol{H}_0$, and the precession angular frequency ω_0 is linearly related to $\boldsymbol{H}_0$. Due to the presence of internal damping, the precession angular frequency of $\boldsymbol{M}$ will gradually decrease. When the ferrite is also subjected to a microwave electromagnetic field, when the precession angular frequency ω_0 equals the angular frequency ω of the external microwave electromagnetic field, $\boldsymbol{M}$ absorbs the energy of the microwave electromagnetic field to maintain the precession state. This phenomenon is called ferromagnetic resonance of the ferrite.

Microwave ferrite has been widely used in the microwave field, such as ferrite circulators, ferrite isolators, ferrite phase shifters, variant polarizers, modulators, etc. In Chap. 2, we also introduced a ferrite power meter used to measure power.

The principle of the YIG tunable bandpass filter is illustrated below using a single YIG ball. As shown in Fig. 8.12, two mutually orthogonal rings are placed around the periphery of the ferrite sphere. Ring A is placed in the y–z plane as the microwave input ring, while ring B is placed in the x–z plane as the microwave output ring. The two rings are orthogonal to minimize coupling and do not touch each other, but the two rings have the same coupling with the YIG ball. An external DC magnetic field H_0 is applied parallel to the planes of the two rings (in the axial direction). When the YIG sphere is subjected to H_0, its spin magnetic moment M will undergo precession around H_0 with a frequency ω_0. If a microwave signal is input into ring A, when the microwave electromagnetic field frequency ω does not match the precession angular frequency ω_0, the YIG ball will not undergo ferromagnetic resonance. Additionally, because the input and output rings are perpendicular to each other, the magnetic flux changes caused by the microwave magnetic field in ring A will not affect the magnetic flux in ring B, resulting in no microwave output in ring B. When the frequency ω of the input microwave electromagnetic field matches ω_0, the YIG single crystal ball undergoes ferromagnetic resonance, absorbing microwave energy strongly, and enhancing precession. The spin magnetic field of the YIG ball will change with the variation of the high-frequency current input into ring A, and its y component perpendicular to ring B will couple with ring B, inducing high-frequency current in ring B, resulting in microwave output in ring B.

When the external magnetic field H_0 is constant, the precession frequency ω_0 is also constant. Therefore, there is only one microwave frequency ω corresponding to ω_0 that can produce resonance. In other words, electromagnetic waves can only be output within a certain frequency range with $\omega = \omega_0$ at the center when ferromagnetic resonance is satisfied. This is the filtering principle of the YIG sphere filter. As the

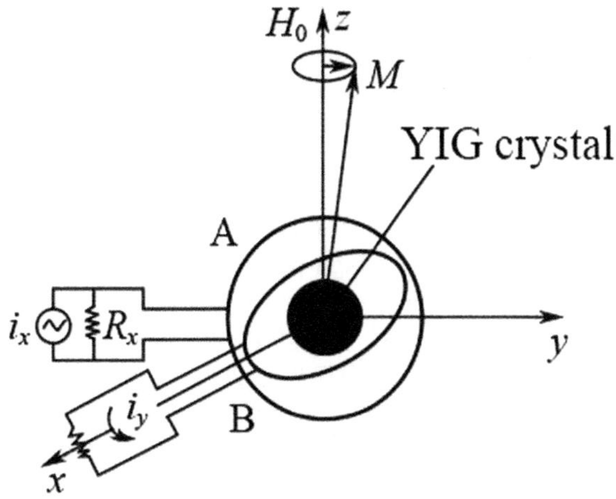

Fig. 8.12 Schematic diagram of the principle of the YIG tuning filter

external magnetic field H_0 changes, the precession frequency ω_0 of the YIG also changes, and the microwave frequency ω that can be output through the filter also changes. This means that the center frequency of the YIG filter changes as well. It can be seen that as long as the size of the external magnetic field is adjusted, the passband frequency of the YIG filter can be tuned, which is the tuning principle of the YIG filter. In practical use, the external magnetic field H_0 is generated using a coil, so by changing the current passing through the coil, the magnetic field can be adjusted in size. The tuning range of the YIG tunable filter is wide, such as 6–18 GHz, 18–26.5 GHz, 26.5–40 GHz, and even achieves bandwidths of multiple octaves, such as 1–12.4 GHz, 2–26.5 GHz, 1–18 GHz, etc.

For an isotropic ferrite sample of infinite size (i.e., the size of the YIG crystal is comparable to the wavelength), the relationship between the resonant frequency ω_0 and the external magnetic field H_0 is given by the formula:

$$f_0(\text{MHz}) = 2.8H_0(\text{Oe}) \tag{8.17}$$

where, Oe is the unit of magnetic field strength in Oersted. In practice, the size of the YIG ball used as a microwave YIG filter is much smaller than the microwave wavelength, so Eq. (8.17) is only an approximate formula.

(2) **Application of YIG tunable filters**

It is evident that when a microwave signal is input into a YIG tunable filter, and the current in the coil is adjusted to change the external magnetic field H_0, causing the precession frequency ω_0 of the YIG to change, the moment at which the YIG tunable filter has an output signal, the value of ω_0 at that moment is the microwave frequency ω. However, the practical application of the YIG tunable filter in high-power microwave frequency measurements is greatly limited; this is mainly limited by its tuning speed. Since most high-power microwaves operate in extremely short pulse states, with pulse widths often only a few tens of nanoseconds, to measure its frequency in such a short time, the tuning time of the YIG tuning filter must be shorter than the duration of the microwave pulse, otherwise, the YIG tunable filter may not have finished scanning the entire search range before the microwave pulse has ended, and it may not have had time to find the microwave frequency.

Currently, the scanning speed achievable by YIG tunable filters is only on the order of approximately 1 GHz for each 1 ms. To scan the entire frequency range of the output signal of a high-power microwave source, it would take several milliseconds or even tens of milliseconds, which is far wider than the pulse width on the nanosecond level of high-power microwaves. Clearly, this means that it still cannot meet the requirements for measuring high-power microwave frequencies. The reason we introduce this type of filter here is not only because it has the advantages of a simple measurement system, small size, low cost, and ease of use, but also because it can be applied to frequency measurements of certain long-pulse or high-repetition-frequency gyrotrons. Additionally, YIG components have wide applications in various aspects of the electronics field, and it's important for us to understand their frequency measurement capabilities.

8.3.2 Cutoff Waveguide Measurement Method

1. **Principle of cutoff waveguide method for measuring frequency**

Microwave technology tells us that a waveguide itself is a high-pass filter. Only when the frequency of the microwave mode is higher than the cutoff frequency of the mode ($f > f_c$), can the microwave be transmitted through the waveguide. If the microwave frequency is lower than the cutoff frequency ($f < f_c$), the waveguide will cut off this microwave, and as long as the longitudinal length of the waveguide is sufficient, it can be considered that the microwave of that frequency in the waveguide will be completely attenuated (reflected) and cannot be transmitted through the waveguide. The waveguide that can cut off microwaves below a certain frequency for a certain mode is called the cutoff waveguide of that mode.

It is based on this characteristic of the cutoff waveguide that we can measure microwave frequencies in a way similar to bandpass filter measurement. The determination of the cutoff frequency of the cutoff waveguide is well known.

For rectangular waveguides:

$$f_c = c\sqrt{\left(\frac{m}{a}\right)^2 + \left(\frac{n}{b}\right)^2} \Big/ 2 \tag{8.18}$$

where c is the speed of light in free space, m and n are the mode characteristic values, a and b are the dimensions of the wide and narrow sides of the rectangular waveguide.

For circular waveguides:

$$f_c = \frac{c\mu_{mn}}{2\pi R} \tag{8.19}$$

where,

c, m, and n are defined the same as in Eq. (8.18),

R is the radius of the circular waveguide,

For the TE mode, μ_{mn} represents the nth root of the derivative of the m order Bessel function $J_m'(k_c r)$, for the TM mode, it represents the nth root of the m-order Bessel function $J_m(k_c r)$, where $k_c = \mu_{mn}/R$ is the cutoff wave number;

r is the radial coordinate.

It can be seen from Eqs. (8.18) and (8.19) that the cutoff frequency of the waveguide is not only related to the microwave mode, but also to the waveguide dimensions. In the case where the microwave mode is already determined, as long as the waveguide dimensions are changed, the cutoff frequency can be altered. For example, for the TE_{10} mode, the cutoff wavelength of the rectangular waveguide is twice the width dimension a of the waveguide, i.e., $\lambda_c = 2a, f_c = c/2a$. Therefore, by manufacturing a series of rectangular waveguides with dimensions $a_1, a_2, a_3, \ldots, a_n$ where $a_1 > a_2 > a_3 > \ldots > a_n$, their respective cutoff frequencies are $f_{c1}, f_{c2}, f_{c3}, \ldots, f_{cn}$. Clearly, $f_{c1} < f_{c2} < f_{c3} < \ldots < f_{cn}$. These can be arranged to form a frequency spectrum, as is shown in Fig. 8.13. Each vertical line in the figure represents the width dimension of the waveguide, with the symbols $a_1, a_2, a_3, \ldots, a_n$ marked at the top of the line, and the corresponding cutoff frequencies $f_{c1}, f_{c2}, f_{c3}, \ldots, f_{cn}$ marked at the bottom. The rightward arrows at the top of each vertical line indicate the passband range of the rectangular waveguide for the TE_{10} mode. For example, if the width dimension of the rectangular waveguide is a_1, as long as the microwave frequency f for the TE_{10} mode falls within the range to the right of the arrow at the top of the leftmost vertical line, i.e., $f > f_{c1}$, it can be transmitted within that dimension of the waveguide; conversely, if it falls to the left of the vertical line, i.e., $f < f_{c1}$, it will be cutoff. If the width dimension of the rectangular waveguide is a_3, as long as the frequency f for the TE_{10} mode falls within the range to the right of the arrow at the top of the line corresponding to f_{c3}, i.e., $f > f_{c3}$, it can be transmitted, otherwise, if it falls to the left of the line corresponding to f_{c3}, i.e., $f < f_{c3}$, it will be cutoff, and so on.

2. **Measurement of cutoff waveguide method** [106]

When we use the cutoff waveguide method to determine the frequency range of microwave signals, two methods can be employed: one method is the same as the filter measurement method shown in Fig. 8.10, where the estimated range of the microwave frequency to be tested is divided into several segments, with the starting and ending points of each frequency range serving as cutoff frequencies. Corresponding cutoff waveguides are manufactured for each cutoff frequency, and then the parallel filters in Fig. 8.10 are replaced with these cutoff waveguides to conduct measurements in a similar manner. At this point, the microwave energy passed through and output is no longer from a single cutoff waveguide, but rather all cutoff waveguides with frequencies higher than the cutoff frequency of the waveguide will have output. Among all these cutoff waveguides with output, we can determine that the microwave

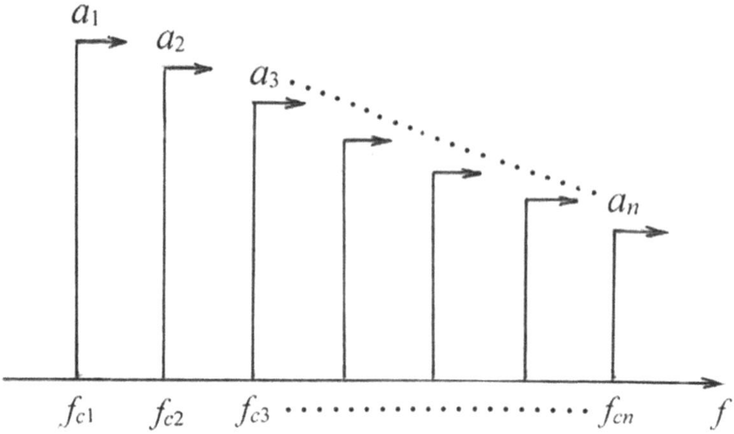

Fig. 8.13 Relationship between the cut-off frequency and the wide-side size of the TE_{10} mode of a rectangular waveguide

frequency is higher than the cutoff frequency corresponding to the smallest—sized cutoff waveguide and lower than the cutoff frequency of the adjacent cutoff waveguide with no microwave signal output. The other method is to individually connect these cutoff waveguides from largest to smallest to the output end of the microwave source, measure their detection output, until there is no microwave signal output. The microwave frequency then falls between the cutoff frequency of the last waveguide with signal output and the cutoff frequency of the first waveguide with no signal output. For example, still using the TE_{10} mode, it is easy to see from Fig. 8.13 that if the frequency f of the microwave to be tested falls between f_{c2} and f_{c3}, it can obviously pass through waveguides with width dimensions a_1 and a_2, but not through the waveguide with dimension a_3 (where $a_2 > a_3$), indicating that the waveguide with dimension a_3 is cutoff for it. Therefore, the frequency of the microwave to be tested is: $f_{c2} < f < f_{c3}$. Reference [106] used the cutoff waveguide method to determine the frequency range of a certain terahertz pulse, and the experimental system used is shown in Fig. 8.14.

The designed frequency of the terahertz radiation source is 0.143 THz. In the experiment, rectangular waveguides of two sizes, 1.2 mm × 0.6 mm and 1.0 mm × 0.5 mm, were used, with cutoff frequencies of 0.125 and 0.15 THz, and lengths of 50 mm for both. In the experiment, both waveguides were connected to the terahertz radiation source, and the detected pulse signals were displayed on the oscilloscope. The waveforms of the detection output in the two cases are shown in Fig. 8.15. Waveform 1 in the figure is the detected envelope waveform, and waveform 2 is the driving voltage of the terahertz source. A comparison between the detection output results in Fig. 8.15a and those in Fig. 8.15b clearly shows that when using a larger size and lower cutoff frequency waveguide, there is a significant amplitude of detection output, indicating that the measured terahertz signal frequency is greater than the cutoff frequency of the waveguide, 0.125 THz. Conversely, when using

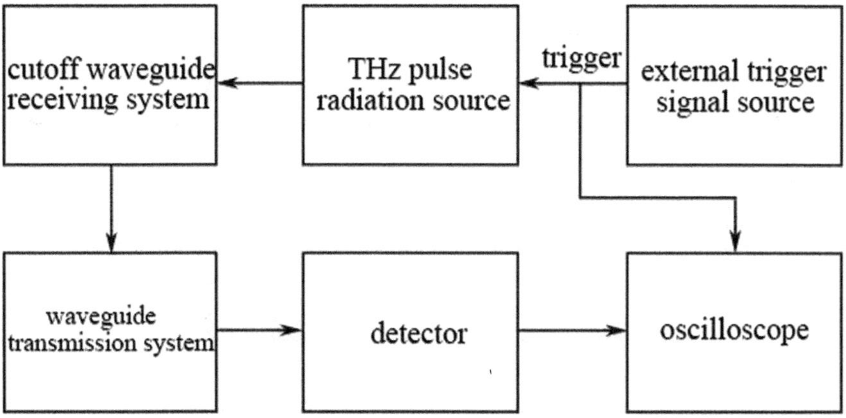

Fig. 8.14 Block diagram of an experimental system for measuring the microwave pulse frequency range by the cut-off waveguide method

a smaller size and higher cutoff frequency waveguide, the detector has no signal output, indicating that the frequency of the measured terahertz pulse is less than the cutoff frequency of the waveguide, 0.15 THz. Therefore, it can be inferred that the frequency of the measured pulse signal should fall within the range of 0.125 to 0.15 THz.

The cutoff waveguide method, like the bandpass filter method, can only determine the frequency range of the measured microwave signal and cannot directly measure the exact frequency unless the cutoff frequencies of the waveguides are divided according to the required frequency measurement accuracy, and cutoff waveguides are made based on these cutoff frequencies. Obviously, this not only greatly increases the number of waveguides to be manufactured but also imposes high requirements

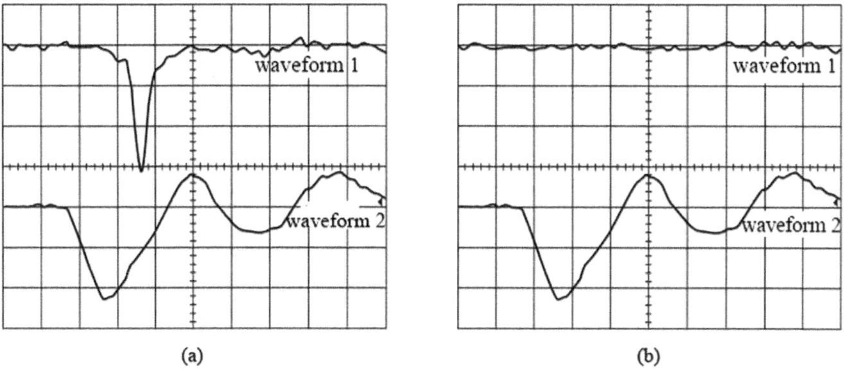

Fig. 8.15 Waveform of the detected output after passing through different cut-off waveguides, **a** when a waveguide with a cut-off frequency of 0.125 THz, **b** when a waveguide with a cut-off frequency of 0.15 THz

on the processing accuracy of the waveguide dimensions in order to achieve strict accuracy of the cutoff frequencies. Therefore, the cutoff waveguide method is generally only used to pre-estimate the frequency range of the signal before accurately measuring the frequency using other methods, providing guidance for the next precise measurement.

In fact, when the microwave working wavelength is greater than the cutoff wavelength of the waveguide, the microwave is in a cutoff state. However, in reality, the pulse signal does not immediately disappear in the cutoff waveguide; it only becomes an attenuated wave in the cutoff waveguide, with its amplitude exponentially decreasing with the increase in transmission distance within the cutoff waveguide. At this point, the cutoff waveguide exhibits inductive reactance, so if several parallel capacitive components (diaphragm or round-headed pins) are introduced into the waveguide, a resonant cavity can be formed, thereby constituting a bandpass filter. Such a filter is called evanescent mode filter. A bandpass filter composed of cutoff waveguides can be used for frequency measurement in the same way as the bandpass filter introduced in Sect. 8.3.1.

8.4 Other Methods of Frequency Measurement

There are many methods for microwave frequency measurement, but not all of them are suitable for measuring the frequency of high-power pulse microwave signals. There are the following several main methods that can be used for measuring extremely short pulse frequencies.

8.4.1 Direct Reading Measurement Using Oscilloscope

When the sampling rate of the oscilloscope is high enough, the carrier waveform of the microwave pulse can be directly displayed on the oscilloscope, that is, the waveform of the microwave. At the same time, if the microwave frequency is not very high, the number of complete microwave cycles contained in the envelope of the pulse and the time they occupy can be directly counted, thus the microwave frequency can be calculated. However, in general, it is difficult to directly and accurately count the number of microwave cycles on the oscilloscope, so the fast Fourier transform should be used to transform the time-domain waveform into a frequency-domain waveform, from which the microwave frequency can be directly read. Obviously, this kind of measurement is very simple and convenient, with the only requirement being that the oscilloscope must have a sufficiently wide analog bandwidth.

Chinese researchers used a DSA71254 oscilloscope (analog bandwidth of 12.5 GHz) to directly measure the frequency of the output pulses of a relativistic backward wave oscillator, as shown in Fig. 8.16. In the figure, Channel 1 is the microwave pulse signal waveform, and Channel M is the microwave frequency obtained by

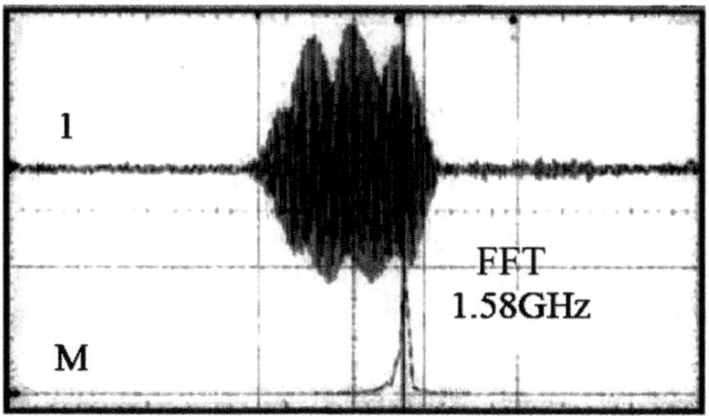

Fig. 8.16 The results of a direct measurement of the frequency of a relativistic backward wave oscillator using an oscilloscope DSA71254

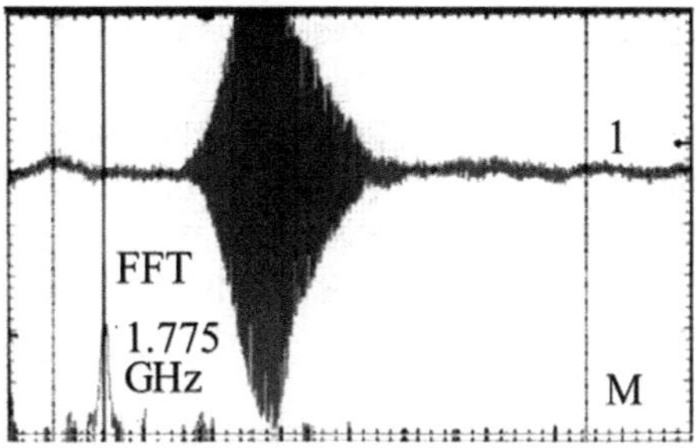

Fig. 8.17 Direct measurement of the frequency of a magnetically insulated line oscillator using an oscilloscope TDS7154B

performing a fast Fourier analysis on this signal waveform, with a center frequency of 1.58 GHz and a 3 dB bandwidth of less than 30 MHz. Figure 8.17 shows the results of directly measuring the frequency of the signal from a magnetically insulated line oscillator using a TDS7154B oscilloscope. Similarly, Channel 1 is the microwave signal, and Channel M is the frequency obtained by performing FFT analysis on this signal, with a measured microwave center frequency of 1.775 GHz [107, 108].

8.4.2 Measurement Method with Evanescent Mode Attenuator

The cutoff waveguide allows mode waves with frequencies higher than the cutoff frequency of the waveguide to pass through. While electromagnetic waves with frequencies lower than the cutoff frequency are either attenuated or reflected. After a certain length, the amplitude of the wave will approach zero. In other words, mode waves with frequencies lower than the cutoff frequency in the waveguide will disappear. Thus, the waveguide in this situation is often referred to as an evanescent mode waveguide. A filter made using the cutoff waveguide is called an evanescent mode filter or evanescent mode waveguide filter. An attenuator made using the evanescent mode waveguide is called an evanescent mode attenuator or evanescent mode waveguide attenuator.

1. **Working principle of evanescent mode attenuator**

Based on the characteristics of the evanescent mode waveguide attenuator, scholars at Tomsk Polytechnic University in Russia proposed a method for measuring high-power microwave frequency and power using the evanescent mode waveguide attenuator. This method does not require a high sampling rate and wide analog bandwidth oscilloscope, nor does it require advanced computer calculations, and it can measure the frequency of nanosecond-level single or low-repetition frequency pulses. They conducted actual frequency measurements of a virtual cathode oscillator (vircator) in the 2.7–3.7 GHz frequency range.

When the microwave frequency is lower than the waveguide cutoff frequency, or when the microwave wavelength is greater than the waveguide cutoff wavelength, the microwave will change from a transmission state to an attenuation state in the waveguide (from a traveling wave to an attenuated wave). Whether it is a rectangular waveguide or a circular waveguide, the attenuation constant α can be expressed as:

$$\alpha = \frac{\omega}{c}\sqrt{\left(\frac{f_c}{f}\right)^2 - 1} = \frac{2\pi}{\lambda}\sqrt{\left(\frac{\lambda}{\lambda_c}\right)^2 - 1} \tag{8.20}$$

At this time, the expression for the electromagnetic wave in the $+\,z$ direction becomes:

$$\boldsymbol{E} = A\boldsymbol{E}(x, y)\mathrm{e}^{\mathrm{j}\omega t}\mathrm{e}^{-\alpha z} \tag{8.21}$$

It can be seen that the electromagnetic wave becomes a wave that decays according to the e^{-az} law along the $+\,z$ direction. Obviously, the amount of decay of the wave increases with the distance z, that is, the length of the evanescent mode waveguide. It is precisely because of this characteristic of the cutoff waveguide that we can use it to make an attenuator, namely a cutoff waveguide attenuator or an evanescent mode waveguide attenuator.

The attenuation amount L of the evanescent mode attenuator with a length l can be calculated according to the following formula:

$$L = -20\lg e^{-\alpha l} = 8.686\alpha\, l\,(\mathrm{dB}) \tag{8.22}$$

The evanescent mode attenuator has the following characteristics:

(a) The attenuation amount L is proportional to the length l, and the attenuation amount L can be accurately calculated based on the waveguide size, transmission mode, and frequency. Therefore, there is no need to calibrate the attenuation amount directly, making it an absolute attenuator that can be used as a standard for attenuation.

(b) Since α is not only related to the mode, but also to the microwave frequency f, the attenuation amount L is a function of frequency, making the microwave amplitude after passing through the evanescent mode attenuator also a function of frequency.
(c) When $\lambda \gg \lambda_c$ $(f \ll f_c)$, the attenuation constant α can be large, so a large attenuation amount can be obtained in a short length.
(d) Because the evanescent mode attenuator is a reflective attenuator, there is a significant reflection at both the input and output ends, leading to severe mismatch.

It is precisely based on characteristics (a) and (b) that the evanescent mode attenuator utilizes the strict and accurate relationship between its output microwave amplitude and frequency for frequency measurements.

2. **Measurement method of evanescent mode attenuator** [109]
(1) **Measurement diagram**

The system diagram for measuring the frequency of high-power microwave pulses using an evanescent mode attenuator is shown in Fig. 8.18. The system includes a coaxial-waveguide transition joint 1, a matching coaxial branch 2, two wideband coaxial fixed attenuators 3 and 4, a circular waveguide evanescent mode attenuator 5, and two wideband coaxial microwave diode detectors 6 and 7. The both ends of the evanescent mode attenuator 5 are coaxial lines, and the middle section is a cutoff circular waveguide. The inner conductor of the coaxial line serves as the excitation and coupling probe of the circular waveguide, and the distance that the inner conductor of the coaxial line extends into the circular waveguide is adjustable. Adjusting the extension distance effectively changes the length of the cutoff circular waveguide, thereby changing the attenuation amount.

High-power microwaves emitted by the high-power microwave source are received by a receiving horn placed in the far field, and become low-power TE_{10} mode microwaves entering the measurement system. They are then transmitted through a coaxial-waveguide transition joint to the coaxial line. The matching coaxial branch acts as a power divider to divide the microwave power into two equal-power paths, entering two measurement channels: one is called the wideband channel, denoted as

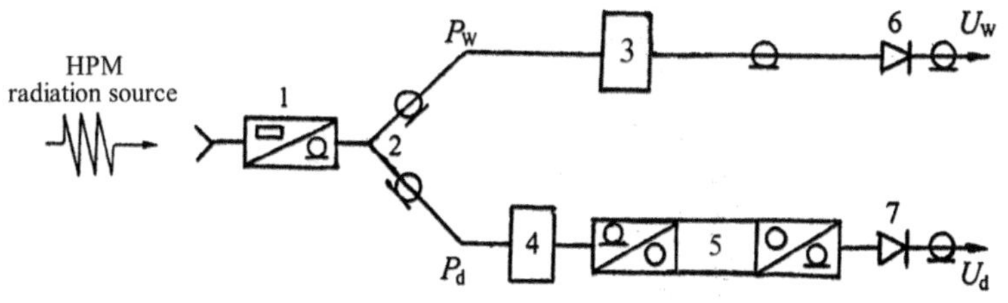

Fig. 8.18 Block diagram of the evanescent mode attenuator measurement frequency system, 1-coaxial-waveguide transition joint, 2-coaxial branch, 3, 4-fixed attenuator, 5-circular waveguide evanescent mode attenuator, 6.7-detector

W, which includes a fixed attenuator and a detector; the other is called the dispersion channel, denoted as d, which includes a fixed attenuator, evanescent mode attenuator, and detector. Since the attenuation amount of the evanescent mode attenuator is frequency-dependent, it exhibits dispersion characteristics, hence the name dispersion channel. The outputs of the two channels are input to the two display channels of the dual-channel oscilloscope.

(2) **Measurement method**

The measurement is carried out using a virtual cathode oscillator. The distance from the output horn of the virtual cathode oscillator to the receiving horn is 3 m, and the size of the receiving horn is 10 cm × 10 cm. The oscilloscope has a sampling rate of 2 ns (bandwidth 500 MHz).

(a) Before the measurement, the relationship between the input power P_w, P_d and the detection output voltages U_w, U_d of the two channels should be calibrated. That is, by using a standard signal source, we change the input power at different frequencies, i.e., changing P_w, P_d, and measuring the detection output voltage U_w, U_d, thus plotting the curves of $P_w = F\ (U_w, f)$ and $P_d = F(U_d, f)$. The calibration curves obtained in the experiments by Russian scholars are shown in Fig. 8.19. During the calibration process, the fixed attenuator in the wideband channel is set to an appropriate attenuation level, also determined according to the power range allowed by the detector. In the dispersion channel, the attenuation of the evanescent mode attenuator is adjusted to an appropriate level, determined also based on the power range allowed by the detector. When the attenuation of the evanescent mode attenuator is not sufficient, the attenuation of the fixed attenuator of that channel is used to supplement it.
(b) Measure the envelope waveform of high-power microwave pulse signals. The high-power microwave source is connected to the system. The microwave signals received by the antenna are input to two channels. The fixed attenuator

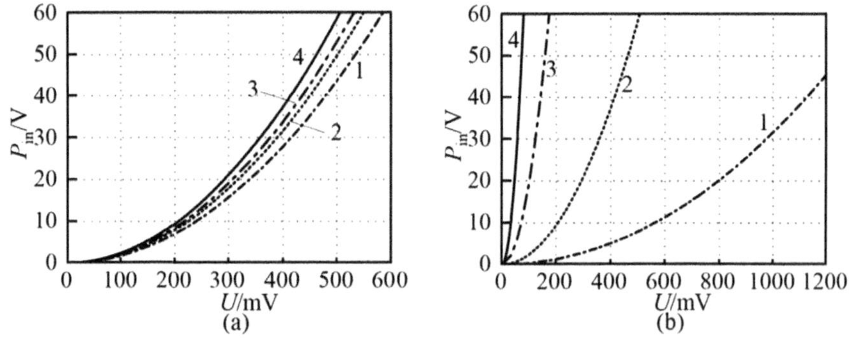

Fig. 8.19 Calibration curves of the relationship between input power and detector output voltage, **a** calibration curves for broadband channels, **b** calibration curves for dispersion channels. Parameter f: curve 1: 3.7 GHz, curve 2: 3.4 GHz, curve 3: 3.1 GHz, curve 4: 2.7 GHz

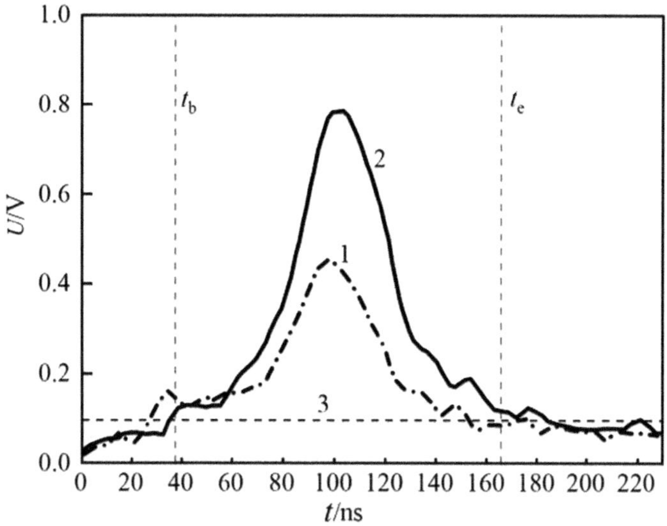

Fig. 8.20 The envelope waveform of the detection voltage U_w and U_d of the broadband channel and the dispersion channel, Curve 1: U_w: Curve 2: U_d, Curve 3: Noise voltage

and the evanescent mode attenuator in both channels maintain the same attenuation level as in step 1. After the signals from the two channels are detected, the envelope waveforms of the detection voltages U_w and U_d for the broadband channel and dispersion channel are displayed on the dual channel oscilloscope as shown in Fig. 8.20.

(c) To determine the frequency and power of the microwave signal at a certain moment t_i in the pulse width, we read the values of U_w and U_d corresponding to time t_i on the display of Fig. 8.20. Then, on the calibration curve given in Fig. 8.19, we find the corresponding sizes of U_w and U_d at that moment and read the P_w and P_d values at different frequencies under the same U_w and U_d. A curve is then plotted to show the relationship between P_w, P_d and frequency f. Since P_w and P_d are equally divided by the coaxial power divider, they are of equal magnitude. Therefore, the intersection of the two curves gives the frequency and power at that moment, as shown in Fig. 8.21.

(d) If the instantaneous frequency throughout the microwave pulse width needs to be measured, the above measurement process should be performed for each sampled instant in Fig. 8.20. The pulse width is determined by the condition that the detected voltage of both channels exceeds the noise voltage, i.e., the difference $\tau = t_e - t_b$ at times t_e and t_b in Fig. 8.20 is used to determine the pulse width.

(e) Measurements show that the total power of the virtual cathode oscillator is 600 MW, with a pulse width of 20 ns and a working frequency range of 2.7 to 3.7 GHz. When compared with the reference frequency meter (accuracy $\delta = \pm 0.05\%$), the frequency measurement error is less than 1%.

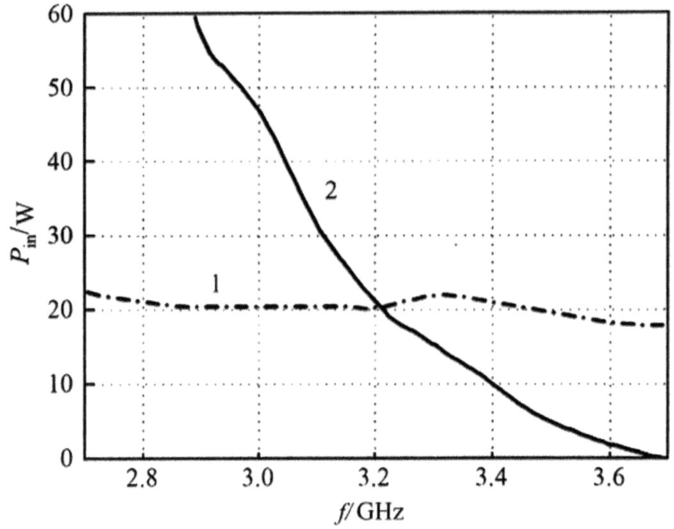

Fig. 8.21 Relationship between input power and frequency of broadband and dispersive channels, Curve 1: P_W, Curve 2: P_d

References

1. James Benford, John A. Swegle. High-Power Microwave, (Second Edition). Trans. Jiang Weihua, Zhang Chi. Beijing: The National Defense Industry Press,2009. (Chinese).
2. Robert J. Barker, Edl Schamiloglu. High-Power Microwave Sources and Technologics. Trans. 《High-Power Microwave Sources and Technologics Translation group》. Beijing: Tsinghua University Press,2005. (Chinese).
3. A. S. Gilmour, Jr. . Klystrons,Traveling Wave Tubes,Magnetrons,Crossed-Field Amplifiers and Gyrotrons. Trans. Ding Yaogen, Zhang Zhaochuan, etc. . Beijing: The National Defense Industry Press,2012. (Chinese).
4. Granatstein V. L, Alexeff I. High-Power Microwave Sources. Boston, London:Artech House, 1987.
5. Wang Wenxiang. Microwave Engineering Technology, (Second Edition). Beijing:The National Defense Industry Press,2014. (Chinese).
6. Li Zhenyuan, Feng JinJun, Liang Youhuan. Microwave Measurement Technologies for Travelling Wave Tubes. Beijing: The National Defense Industry Press,2013. (Chinese).
7. Electronic Industry Standards of the People's Republic of China, SJ 20769-1999, Measurement methods for peak pulse power. (Chinese).
8. Electronic Industry Standards of the People's Republic of China, SJ/T 10182-91, Power Measurement for Waveguide and Coaxial Components. (Chinese).
9. Electronic Industry Standards of the People's Republic of China, SJ/T 10182-91, Power Measurement for Waveguide and Coaxial Components. (Chinese).
10. Wang Zhitian. Radio-electronic measurement, Beijing: Atomic Energy Press, 2002. (Chinese).
11. Zhang Jiucai. Study on Power Frequency Meter of High-Power Pulse Microwave. Master's Thesis, Chengdu: University of Electronic Science and Technology of China, 2006. (Chinese).
12. Gu Xiaoming. Power Meter and The Measurement of Microwave Power. Electronic Technology, 2004(1):14–16. (Chinese).
13. Ji Liang. Microwave Power Measurement Method. Beijing: The National Defense Industry Press,1973. (Chinese).
14. М,И.Билько,А.К.Томащевский, Измерение Мощности На СВЧ. Trans. Zhang Lun. Beijing: Posts and Telecom Press,1979. (Chinese).
15. Zhang Lun. Review on Microwave Power Measurement. The Journal of Electronic Tube Technology, 1976(3): 1–13. (Chinese).
16. Lu Yihui. Microwave Power Standards and a Milliwatt-Level Micro-Calorimetric Power Meter. Acta Electronica Sinica, 1963(4):114–118. (Chinese).
17. Fang Yan,Wu Youzhang. Automatic Regulation System for Micro-Calorimetric Power Meter, Acta Metrologica Sinica, 1988,9(3):227–232. (Chinese).
18. Zhang Keqian,Wang Dun, Wang Zhenyu, Measurement of Single-Pulse Microwave Power and Energy, Vaccum Electrnics, 2003(1): 8–11. (Chinese).

W. Wang et al., *High-Power Microwave Measurement*,
https://doi.org/10.1007/978-981-96-7763-4

19. Valeri Vountesmeri, K. Gura K and Alexander Chenakin. A X-band Magnetoresistive Sensor for Active Pulse Power Measurements. 27th European Microwave Conference,1997: 838–841.
20. Khanna R. K, Sharma P. K. Novel Technique for Investigating High-Power Microwave Pulses Using Kerr Effect. Electronics Letters,1989, 25(5):311–312.
21. Hossian A., Rashid M. H. Pyroelectric Detectors and Their Applications. IEEE trans. on Industry Applications, 1991,.27(5):824–829.
22. Wang Chengyang, Chen Zhibin, Zhuo Jiajing, et al. Comprehensive Test Technology of Pulsed Laser Peak Power in Common Wavebands. Journal of Applied Optics,2007,28(1):72–76. (Chinese).
23. Fang Weihai, Wen Xin, Zhang Lu, et al. Review on Developments of High-Power Microwave Measurement and Development. Journal of Astronautic Metrology and Measurement, 2013,33(3): 1–6. (Chinese).
24. John D Kraus,Ronald J. Marhefka, Antennas: For All Applications(Third Edition), Trans. Zhang Wenxun. Beijing: Publishing House of Electronics Industry 2011. (Chinese).
25. Chen Yu . Research on Several Problems in High Power Microwave Power Measurement by Far Field Method. Master's Thesis, Changsha:National University of Defense Technology, 2005. (Chinese).
26. Shu Ting, Wang Yong, Li Jijian, et al. Measurement of High Power Microwave in The Far-Field Zones. High Power Laser and Particle Beams, 2003,15(5):485–488. (Chinese).
27. Yang Jianhua. Study of Cherenkov oscillator-conical amplifier tube of low magnetic field resonance. Doctoral Dissertation, Changsha. National University of Defense Technology, 2002. (Chinese).
28. Yan Junkai, Liu Xiaolong, Ye Hu, et al. X-Band HPM Feed Total Radiation Power Measurement Using Array Method. High Power Laser and Particle Beams, 2011,23(11):3149–3153. (Chinese).
29. Qu Jin, Liu Qingxiang, Hu Jinguang,et al.Measurement System on Power Density of High Power Microwave Radiation. High Power Laser and Particle Beams, 2004,16(1):77–80. (Chinese).
30. Chen Daibing.Theoretical and Experimental Researches on Bifrequency Magnetically Insulated Transmission Line Oscillator (BFMILO). Doctoral Dissertation, Mianyang: China Academy of Engineering Physics, 2008. (Chinese).
31. Shao Hao. Characterization of Resonances in Coaxial Vircator. Doctoral Dissertation, Xi'an: Northwest Institute of Nuclear Technology, 2006. (Chinese).
32. André Angot, Complements de Mathématiques Al'usage des Ingénieurs de L'élégtroteghnique ét des Télécommunigations.Trans. Lu Zhigang, Xie Xianglin, Zeng Deji. Beijing: Posts and Telecom Press,1979. (Chinese).
33. 《Math manual》, Math manual. Beijing: People's Education Press,1979. (Chinese).
34. Xie Wenkai, Qian Guangdi, Meng Lin. Experimental study on 3cm vircator. Proceedings of the High-Power Microwave Symposium (HPM-91), Chengdu,1991:1–9. (Chinese).
35. Yuan Chengwei, Peng Shengren, Zhang Qiang. Radiation Power Measurement of Conical Horn Antennas Excited by TE_{1n}/TM_{1n} Modes. High Power Laser and Particle Beams, 2016,28(8):64–68. (Chinese).
36. Chen Yu, Li Zhiqiang, Ma Yuzhong, et al.Measurment of Output Power of Microwave Sources Using Cutoff Characteristic of Waveguide. High Power Laser and Particle Beams, 2007,19(6):901–904. (Chinese).
37. Hu Yongmei, Researchs of High Power Microwave Power Measurement Technology. Doctoral Dissertation, Xi'an,Northwest Institute of Nuclear Technology,2000. (Chinese).
38. Gu Maozhang, Zhang Keqian. Microwave Technique. Beijing: Tsinghua University Press.1990. (Chinese).
39. Simon Ramo, John R. Whinnery. Field and Wave in Modern Radio. Trans. Zhang Shilin, Xiao Duchi. Beijing: Posts and Telecom Press,1958. (Chinese).
40. Liu Guozhi, Liu Jingyue, Huang Wenhua. A Simple Method of Measurement of HPM Mode and Power. High Power Laser and Particle Beams, 1998,10(4): 606–610. (Chinese).

41. Hu Yongmei, Liu Guozhi, Chen Changhua, et al. Preliminary study on the methods to enhances power capacity in an online coupled probe measurement system. Proceedings of the fifth National High Power Microwave Symposium, Zhuhai,2002:560–565. (Chinese).
42. Li Yongzhong, Zhang Yazhou, He Chongjun. Investigation of Electromagnetic Probe for Measurement of HPM. Journal of National University of Defense Technology, 1999, 21(5):72–74. (Chinese).
43. Liu Jinliang, Zhong Huihung, Tan Qimei. Coaxial E-Field Probe Couping for High Power Microwave Measurement. Journal of Microwaves, 1997, 13(1):83–87. (Chinese).
44. Bethe H. A. Theory of diffraction by small holes. The Physical Review,1944, Second Series, 66(7 and 8): 163–182.
45. Huang Hongjia. Microwave Principles. Beijing: Science Press,1984. (Chinese).
46. F. Sporleder, H.-G.Unger. Waveguide tapers transitions and couplers. Trans. Qin Yaokun. Beijing: Science Press,1984. (Chinese).
47. Wang Wenxiang, Xu Meisheng, Yu Guofen. The Design of a New Type of Directional Coupler, Journal of University of Electronic Science and Technology of China,1991, 20(5):497–502. (Chinese).
48. W. X. Wang, M. S. Xu, G. F. Yu, et al. The Design of a Waveguide–Coaxial Line DirectionalCoupler. Int. J. Electronics, 1993, 74 (1): 111–120.
49. W. X. Wang, Y. B. Gong,J. H. Sun, et al. Analysis and Design of a Waveguide–Coaxial Lline Single–Hole Directional Coupler. Int. J. Electronics, 1996, 81(3):311-319.
50. McDermott D. B, Song H. H, Hirata Y., et al. Design of a W–Band TE_{01} Mode Gyrotron Traveling–Wave Amplifier with High Power and Broade–Band Capabilities. TEEE Trans.on PS,2002,30(3): 894–902.
51. Ching-Fang Yu, Tsun-Hsu Chang. High Performance Circular TE_{01} Mode Converter. IEEE Trans. on MTT, 2005, 53(12): 3794–3798.
52. Marie G. R. P. Mode Transforming Waveguide Transition.U.S.Patent, 2 859 412, 1958.
53. Tang Jingxian, Xu Runmin, Wang Youmin, et al. Design and Development on Cruciform Transducer from H_{10}–H_{01}. Journal of Sichuan University,1980(4):115–124. (Chinese).
54. 野田健一. 円形導波管中の各種モード励振器. 研究実用化報告,1960, 9(9): 1065–1075. (Japanese).
55. M. K. Thumm, W. Kasparek. Passive High-Power Microwave Components, IEEE Trans. on PS, 2002, 30(3): 755–786.
56. Hartmut Stickel,Bernd Jödicke. Generation of TE_{0n}^{O} Mode by Diameter Steps in Circular Waveguides. Conference Digest. 14th Int. Conf. On Infrared and Millimeter Waves, PISA, 1986: 602–604.
57. Huang Minzh, Zhao Guoqing, Wang Wenxiang, et al. The Design and Testing of TE_{01}-TE_{0n} Mode Converter of Step Waveguide.Vaccum Electrnics, 2004(5): 12–15. (Chinese).
58. Mclachlan N. W,Bessel Function for Engineer (Second Editions). Oxford University Press, 1955.
59. Zeng Hongjin, Wu Xiaolong, Liu Xiaolong, et al. Design of a Vacuum Diode Based High-Power Microwave Detector. High Power Laser and Particle Beams, 2014,26(9): 214–217. (Chinese).
60. Yu Huilong, Hu Haiying. High-Power Microwave Pulse Power Measurement. Annual Science and Technology Report of the China Academy of Engineering Physics: Electronics and Photonics, 1998: 12. (Chinese).
61. Chen Daibing, Fan Zhikai, Meng Fanbao, et al. A High Power Microwave Detector Besed on Thermionic Diode. High Power Laser and Particle Beams, 2006,18(5): 871–875. (Chinese).
62. Mindaugas Dagys, Zilvinas Kancleris, et al. High-Power Microwave Pulse Measurement Using Resistive Sensors. IEEE Trans. on Instrumentation and Measurement, 1997, 46(2): 499–502.
63. Z. Kancleris,R. Simniskis,M. Dagys,et al. High Power Millimetre Wave Pulse Sensor for W-Band. IET Microw. Antennas Propag., 2007 (3): 757–762.
64. M. Dagys,Z. Kancleris,et al. Resistive Sensor for High Power Microwave Pulse Measurement. 2011 Int. Conf. on Electronmagnetic in Advanced Applications, 2011: 43–46.

65. Z. Kancleris, G. Slekas, V. Tamosiunas, et al. Polarisation Insensitive in-situ Resistive Sensor for High-Power Microwave Pulse Measurement in TE_{11} Mode Circular Waveguide. IET Microw. Antennas Propag.,2012:6(1): 67–71.
66. Z. Kancleris, G. Slekas, V. Tamosiunas, et al. Modelling of Resistive Sensor for High-Power Microwave Pulsr Measurement of TE_{01} Wave in Circular Waveguide. IET Microw. Antennas Propag., 2010, 4(6): 771–777.
67. Z. Kancleris,P. Ragulis,R. Simniskis,et al. Resistive Sensor for High Power Microwave Pulse Measurement in Double-Ridged Waveguide. Ultra-Wideband Short-Pulse Electromagnetics 10, Springer
68. Liu Xiaolong, Liu Guozhi, Qiu Shi, et al. A Novel High Power Microwave Detector. High Power Laser and Particle Beams, 1997, 9(4): 585–590. (Chinese).
69. Wang X F, Wang J G, et al. High-Power Terahertz Pulse Sensor with Overmoded Structure. Chin. Phys. B,2014, 23(5): 634–638. (Chinese).
70. Huang Wenhua, Liu Jingyue, Fan Juping, et al. New Type of High Power Microwave Detector. High Power Laser and Particle Beams, 2002, 14(3): 449–452. (Chinese).
71. Chen Ming, Fan Dongyuan, Li Suilao. Acoustic surface wave detector. Xian: Northwestern Polytechnical University Press Co.Ltd. 1997. (Chinese).
72. Zailsev B D, Kuznctsova I E, Joshi S G, Acoustic wave devices for measuring characteristics of single high-power microwave pulses . Ultrasonics, 1998, 36(1): 397–401
73. Zailsev B D, Ermolenko A V, Fedorenko V A, et al. An electroacoustic method for measuring the intensity of high-power microwave pulse. IEEE Trans.on Ultrasonics,Ferroelectrics and Frequency control, 1996, 43(1): 30–35.
74. W. X. Wang, L. N. Yue, G. Q. Zhao et al. Discrimination and analysis of microwave modes in high power system. Int.J.of Infrared and Millimeter Waves, 2005, 26(2): 147–161.
75. W. X. Wang, M. S. Xu, G. F. Yu. Mode analysise and distinction of waveguide system. Vaccum Electrnics, 1993,(1):1–6. (Chinese).
76. Pereyaslavets M,Idehara T,Ogawa I,et al. Quick estimation of mode content in a submillimeter-wave gyrotron output. Int.J.of Infrared and Millimeter Waves, 1999, 20(6): 1195–1205.
77. Thumm M. High-power millimeter-wave mode converters in overmoded circular waveguides using periodic wall perturbations. Int.J.Electronics, 1984, 57(6): 1225–1246.
78. Bratman V L,Denisov G G,Kol'chugin B D,et al. Powerful millimeter-wave generators based on the stimulated cerenkov radiation of relativistic electron beams. Int. J. of Infrared and Millimeter Waves, 1984, 5 (9): 1311–1332.
79. Z. X. Huang, J. M. Yao, G. W. Dou. The mode diagnosis of high-power microwave. Internal-Combustion Engine and Powerplant, 2009, No. B06: 31–33. (Chinese).
80. Kiwamoto Y,Kira F,Saito T,et al. Analysis of multimode microwaves in oversize waveguide using an infrared camera image of radiated power distribution. Rev. Sci. Instrum, 1991, 62(12): 3075–3081.
81. Z. X. Zhang. A method for identifying gyrotron output mode compositions. Acta Electronica Sinica,1985, 13(5): 45–50. (Chinese).
82. Z. X. Zhang, G. Janzen, G. Műller, et al. Mode analysis of gyrotron radiation by far field measurements. Int.Conf.on Infrared and Millimeter Waves, Miami U.S.A: 1983: T4.3.
83. Miyake Shoji,Wada Osami,et al. Focusing of high-power millimeter-wave radiation by a guasi-optical antenna system. Int. J. Electronics, 1991, 70(5): 979–988.
84. Idehara T,Ogawa I,S.Maede S,et al. Observation of mode patterns for high purity mode operation in the submillimeter wave gyrotron FUVA. Int. J. of Infrared and Millimeter Waves, 2002, 23(9): 1287–1295.
85. Stone D S. Mode analysis in multimode waveguides using voltage traveling wave ratios. IEEE Trans. MTT, 1981, 29(2): 91–95.
86. Kiwamoto Y, Kira F, Saito T, et al. Microwave-mode analyzer with dispersive reflector antenna. Rev. Sci. Instrument, 1992, 63(5): 3167–3173.
87. Janzen G,H.Stickel H.Mode selective directional couplers in overmoded circular waveguides. TEEE Int. Conf. on Infrared and Millimeter Waves, Miami, U.S.A: 1983: TH4.6.

88. Janzen G,H.Stickel H.Mode selective directional couplers for overmoded wavguide systems. Int. J. of Infrared and Millimeter Waves, 1984, 5(7): 887–917.
89. Janzen G,H.Stickel H.Improved directional couplers for overmoded waveguide systems.Int. J. of Infrared and Millimeter Waves, 1984, 5(10): 1405–1417.
90. W X Wang, Y B Gong, G F Yu, et al. Mode discriminator based on mode-selective coupling.IEEE Trans.on MTT, 2003, 51(1): 55–63.
91. W X Wang, L N Yue, P Y Jiang, et al. Design of wide-band mode discriminator based on mode-selective coupling.Int. J of Electronics, 2008, 95(2): 99–110.
92. W. X. Wang, M. S. Xu, G. F. Yu,et al.The design of broad band mode selective directional coupler.Journal of University of Electronic Science and Technology of China, 1993, 22(Supp.): 125–133. (Chinese).
93. W X Wang, Lawson W and Granatstein V L. The design of a mode selective directional coupler for a high power gyroklystron.Int. J. Electronics, 1988, 64(3): 705–716.
94. W X Wang.Improved design of a high power mode selective directional coupler.Int. J. Electronics, 1994, 76(1): 131–142.
95. L N Yue, W X Wang, et al.The Calibration of the Coupling Coefficients of the Mode-selective Coupler without the need of Single Mode Exciter.Progress in Electromagnetics Research Letters, 2019, 87:115–121.
96. Kasparek W,Müller G A. The Wavenumber Spectrometer—an Alternative to the Directional Coupler for Multimode Analysis in Oversized Waveguides.Int. J. Electronics, 1988, 64(1): 5–20.
97. Barkley H J, Kasparek W, Kumric H, et al. Mode purity measurements on gyrotrons for plasma heating at the stellarator W VII-AS. Int. J. Electronics, 1988, 64(1): 21–28.
98. G. Q. Zhao, W. X. Wang. Study on radiation characteristics of small hole array on circular waveguide wall. The Proceedings of the 7th National Youth Academic Exchange Conference on Laser Science and Technology. Harbin: 2003: 478–481. (Chinese).
99. Wang Hongjun, Huang Wenhua, Liu Guozhi.Several diagnostic methods for high-power microwave frequencie frequencies. Journal of Microwaves, 2001, 17(1): 36–41. (Chinese).
100. Zhang Xiaoping. Study on a novel magnetic insulated line oscillator. Doctoral Dissertation, Changsha: National University of Defense Technology, 2004. (Chinese).
101. Liu Guozhi, Huang Wenhua, Wang Hongjun. A ring of waveguide dispersive line for single short pulse microwave frequency analysis. High Power Laser and Particle Beams, 1999, 11(2): 234–238. (Chinese).
102. Liu Jinliang, Zhang Yazhou, Chen Guoqiang, et al.An expernental research of using heterodyne technique formeasuring frequencies of single microvave pulse.High Power Laser and Particle Beams, 1997, 9(4): 631–635. (Chinese).
103. Fazio M V,Freeman B L.et al.A microsecond-pulse-length,frequency-stabilized virtual-cathode oscillator using a resonant cavity[J].IEEE Trans.on Plasma Science, 1992, 20(3): 305–311.
104. Yan Zhigang.Engineering design of numerical control high-selectivity low-group-delay YIG bandpass filter component.Master's Thesis, Chengdu: University of Electronic Science and Technology of China, 2013. (Chinese).
105. Feng Huiyu.Research on numerical control high-speed YIG filter technology.Master's Thesis, Chengdu: University of Electronic Science and Technology of China, 2011. (Chinese).
106. Wang Guangqiang, Wang Jianguo, Li Xiaoze, et al.Frequency measurement of 0.14THz high-power terehertz pulse.Acta Physica Sinica, 2010, 59(12): 8459–8464. (Chinese).
107. Ge Xingjun. Study of the L-band tunable coaxial relativistic backward wave oscillator. Doctoral Dissertation, Changsha:National University of Defense Technology, 2010. (Chinese).
108. Ge Xingjun, Gao Liang, Cao Yibing, et al. Primary experiments of L-band relativistic backward wave oscillator. High Power Laser and Particle Beams, 2010, 22(3): 609–611. (Chinese).
109. Babichev D A, Shiyan V P, Mel'nikov G V. An instrument for measuring the frequency content of high-power nanosecond microwave pulse.Instrument and experimental techniques, 2003, 46(3): 369–372.

GPSR Compliance

The European Union's (EU) General Product Safety Regulation (GPSR) is a set of rules that requires consumer products to be safe and our obligations to ensure this.

If you have any concerns about our products, you can contact us on ProductSafety@springernature.com

In case Publisher is established outside the EU, the EU authorized representative is:

Springer Nature Customer Service Center GmbH
Europaplatz 3
69115 Heidelberg, Germany

Batch number: 10704307

Printed by Printforce, the Netherlands